RaumFragen: Stadt – Region – Landschaft

Reihe herausgegeben von
O. Kühne, Tübingen, Deutschland
S. Kinder, Tübingen, Deutschland
O. Schnur, Berlin, Deutschland

Im Zuge des „spatial turns“ der Sozial- und Geisteswissenschaften hat sich die Zahl der wissenschaftlichen Forschungen in diesem Bereich deutlich erhöht. Mit der Reihe „RaumFragen: Stadt – Region – Landschaft“ wird Wissenschaftlerinnen und Wissenschaftlern ein Forum angeboten, innovative Ansätze der Anthropogeographie und sozialwissenschaftlichen Raumforschung zu präsentieren. Die Reihe orientiert sich an grundsätzlichen Fragen des gesellschaftlichen Raumverständnisses. Dabei ist es das Ziel, unterschiedliche Theorieansätze der anthropogeographischen und sozialwissenschaftlichen Stadt- und Regionalforschung zu integrieren. Räumliche Bezüge sollen dabei insbesondere auf mikro- und mesoskaliger Ebene liegen. Die Reihe umfasst theoretische sowie theoriegeleitete empirische Arbeiten. Dazu gehören Monographien und Sammelbände, aber auch Einführungen in Teilaspekte der stadt- und regionalbezogenen geographischen und sozialwissenschaftlichen Forschung. Ergänzend werden auch Tagungsbände und Qualifikationsarbeiten (Dissertationen, Habilitationsschriften) publiziert.

Weitere Bände in der Reihe http://www.springer.com/series/10584

Jessica Matloch

The Assessment of German Cultural Landscapes

Evidence from Three Regions Located in the Metropolitan Area of Hamburg

Springer VS

Jessica Matloch
Ernst-Moritz-Arndt-University of Greifswald
Germany

zgl. Dissertation, Gottfried Wilhelm Leibniz University of Hannover, 2017
Gefördert vom Bundesministerium für Bildung und Forschung (BMBF) im Rahmen der Fördermaßnahme „Innovationsgruppen für ein Nachhaltiges Landmanagement", Innovationsgruppe „Regiobranding" (FKZ 033L121C).

RaumFragen: Stadt – Region – Landschaft
ISBN 978-3-658-21415-9 ISBN 978-3-658-21416-6 (eBook)
https://doi.org/10.1007/978-3-658-21416-6

Library of Congress Control Number: 2018936524

Springer VS

Printed on acid-free paper

This Springer VS imprint is published by the registered company Springer Fachmedien Wiesbaden GmbH part of Springer Nature
The registered company address is: Abraham-Lincoln-Str. 46, 65189 Wiesbaden, Germany

Danksagung

An dieser Stelle möchte ich mich für die Unterstützung bedanken, ohne die die Dissertation bestimmt eine andere wäre. Ich möchte Prof. Dr. Daniel Schiller für seine stets konstruktiven Anmerkungen und seine weiteren Ideen danken, die immer wieder neue Perspektiven in meine Arbeit gebracht haben. Darüber hinaus möchte ich ihm für sein Vertrauen bedanken, welches er mir stets entgegen brachte. Ohne dieses Vertrauen wären die Zusammenarbeit und damit auch die Ausarbeitung der Themen sicher weniger spannend gewesen. Auch Prof. Dr. Stephan Thomsen möchte ich danken, von dem ich lernen durfte, wie publizierte Artikel aufgebaut werden und worauf ich in der wissenschaftlichen Bearbeitung achten sollte. Das hat mich in der Bearbeitung der einzelnen Aspekte der Arbeit immer begleitet.

Zusätzlich möchte ich mich bei dem gesamten Projektteam von Regiobranding bedanken. Dort war immer ein konstruktiver Austausch gegeben, mit dem auch neue Ideen in die Arbeit eingearbeitet werden konnten. Besonders hervorheben möchte ich Linda Lange, von der ich immer wieder zu einzelnen Aspekten auch kritisches Feedback erhalten habe. Auch der persönliche Austausch mit ihr war sehr wichtig, um Projektinhalte richtig interpretieren zu können. An dieser Stelle möchte ich auch Prof. Dr. Jörg Schwiebert danken, der mir bei Fragen rund um ökonometrische Modelle und Stata immer mit Rat und Tat zur Seite stand und die statistischen Zusammenhänge immer verständlich erklärte. Auch bei meinen ehemaligen Kollegen aus dem Niedersächsischen Institut für Wirtschaftsforschung (NIW) möchte ich mich bedanken, die mir im Vorfeld wichtige Hinweise zum Projektinhalt gegeben haben und mir, soweit möglich, bis zur Abgabe zur Seite standen. Nicht zu vergessen sind auch die damaligen studentischen Hilfskräfte, ohne die eine Digitalisierung und Aufbereitung der eingegangenen Fragebögen in dieser kurzen Zeit nicht möglich gewesen wäre.

Persönlich möchte ich mich für die vor allem mentale Unterstützung meines Ehemannes Andreas und sowohl meiner als auch seiner Eltern bedanken, aber auch bei meiner guten Freundin Rabea, die mich immer unterstützt haben, auch wenn es mal Schwierigkeiten gab. Ohne diese Personen hätte ich manchmal vielleicht das große Ganze aus den Augen verloren.

Contents

Index of Abbreviations

BPI:	Buying Power Index
CER:	German Civic Engagement Report
CVM:	Contingent Valuation Method
IIA:	Interdependence of Irrelevant Alternatives
L:	Region of Lübeck
LD:	Region of Lüchow-Dannenberg
LNWM:	Region Lübeck/ Nordwestmecklenburg
LP:	Region of Ludwigslust-Parchim
LPLD:	Region Ludwigslust-Parchim/ Lüchow-Dannenberg
MR HH:	Metropolitan area of Hamburg (Metropolregion Hamburg)
NWM:	Region of Nordwestmecklenburg
OECD:	Organization for Economic Co-operation and Development
SOEP:	German Socio-Economic Panel Study
STB:	Region Steinburger Elbmarschen
WTA:	Willingness to Accept
WTP:	Willingness to Pay
WTV:	Willingness to Volunteer

List of Figures

List of Tables

Summary

This doctoral thesis examines the meaning of regional cultural landscape for their residents. Especially regarding the demographic change and the declining population in more rural areas, it is important to understand how the residents perceive their cultural surrounding. The aim is not to do a comparison between the perception of the residents and experts rather than characterize the region by the perception of its residents. This is important, because economic values like prices become less important than social values like accessibility to public services or environmental conditions.

The analysis uses the method of willingness to pay (WTP), an approach stemming from the research area of environmental economics, and explains different results by using theories derived from psychology and environmental science research. Furthermore, because of the object of interest is the landscape, environmental planning is also included. This is new to the research because WTP analysis usually focused only on the environmental science than integrating other disciplines. For instance, the theoretical background is often missing in the existing analyses.

To get a better understanding of the perception of landscape, three different areas of the Hamburg metropolitan region were chosen. The areas are Lübeck/ Nordwestmecklenburg (LNWM), Steinburger Elbmarschen (STB) and Ludwigslust-Parchim/ Lüchow-Dannenberg (LPLD). They differ in their amount of residents as well as in economic factors, such as the intensity of farming and the mix of different landscape types. This is shown in the second part of Chapter 2. As a result of these differences, it is possible to compare regions and their landscapes to one another and identify several similarities and differences between them.

The analytical chapters 6 to 9 examine the meaning of the cultural landscapes. The first of these chapters examines the perception of the landscape in the regions and does a comparison of the results. This is followed by a deeper look into the differences of landscape perception by the groups of newcomers, returnees and locals. The theoretical framework is then expanded to the approach of the willingness to volunteer (WTV) and analyze the different impact for a decision of landscape support of the WTP or WTV. The last analytical chapter performs a cluster analysis to get to know different groups in these regions. These determined clusters are then analyzed deeper with respect to the WTP, WTV, life satisfaction and use of the region.

Some results of these chapters are that for each region the water landscapes is preferred by almost each resident, no matter if the resident moved into the region or lived there since birth. This is a common finding in other analyses and based on theoretical aspects. Furthermore, the landscape perception is often influences by education as well as the relationship the resident has with nature. In some cases it could be assumed that the educational level and the relationship to nature are interdependent. It could be shown that the impact of the relationship to the region differs in its impact between regions and resident groups. There might be some indirect influences that are responsible for the impact of the regional identity. Regarding the involvement in or for the landscape, evidence suggests that specific groups of residents are more willing to invest volunteer work in and for regional landscapes than others. For instance, residents who moved back into the region with a bond to the corresponding region are most willing to volunteer in the field of the regional history, while highly educated young women are more willing to work voluntarily for animal welfare. The use of the region is highest for relaxing and lowest for cultural purposes.

These results suggest that regional management could promote their regions by using their landscapes and could influence perception by strengthening the relationship to nature or addressing special groups for volunteer work in or for the landscape. Further research could examine the relationship between WTP and life satisfaction. As in the WTV, it might be the case that people who are more satisfied are more willing to pay for something than people with less satisfaction in life. Another psychological approach would be the analyses of the motives behind WTP and WTV.

Keywords: Cultural landscape, landscape perception, Contingent Valuation, willingness-to-pay

Zusammenfassung

Diese Doktorarbeit beschäftigt sich mit der Bedeutung der regionalen Kulturlandschaft für ihre Bewohnenden. Gerade im demographischen Wandel und einer abnehmenden Zahl der Bevölkerung in ländlicheren Gebieten ist es wichtig zu verstehen, wie die Bewohnenden ihr kulturelles Umfeld wahrnehmen. Das Ziel der Arbeit ist nicht einen Vergleich zwischen der subjektiven Wahrnehmung der Bewohnenden und Experten anzustellen, sondern die Charakterisierung der Region durch die Wahrnehmung der Bewohnenden. Das ist vor allem wichtig, weil die ökonomischen Werte wie Preise unwichtiger werden und die sozialen Werte wie die Erreichbarkeit von Institutionen der Daseinsvorsorge oder die landschaftlichen Gegebenheiten zunehmen.

Die Analysen dieser Arbeit werden mit Hilfe des willingness to pay (WTP) Ansatzes bestimmt, einem Ansatz aus der Umweltökonomik. Die unterschiedlichen Ergebnisse werden mit Hilfe von Theorien aus den Bereichen der Psychologie oder Umweltwissenschaft erklärt. Da das Objekt der Untersuchung die Landschaft ist, wird auch die Umweltplanung in die Analysen einbezogen. Dies stellt eine Forschungslücke dar, da die meisten Analysen, die WTP nutzen, nur die Umweltwissenschaften einbezieht und nicht noch weitere Disziplinen. Häufig wird beispielsweise der theoretische Hintergrund außen vor gelassen.

Um ein besseres Verständnis der Landschaftswahrnehmung zu bekommen, wurden drei Regionen aus der Metropolregion Hamburg ausgewählt. Diese Regionen sind Lübeck/ Nordwestmecklenburg (LNWM), Steinburger Elbmarschen (STB) und Ludwigslust-Parchim/ Lüchow-Dannenberg (LPLD). Diese Regionen unterscheiden sich in ihrer Anzahl von Bewohnenden, in ökonomischen Faktoren wie der Intensität der Landwirtschaft, aber auch in ihren unterschiedlichen Flächenanteilen der verschiedenen Landschaftstypen. Diese Regionen werden detaillierter im zweiten Teil von Kapitel 2 vorgestellt. Durch die Unterschiede in den Regionen ist es möglich, die Regionen und ihre Landschaften miteinander zu vergleichen und Gleichheiten und Unterschiede zu identifizieren.

Die analytischen Kapitel 6 bis 9 betrachten die Bedeutung der Kulturlandschaften. Das erste dieser Kapitel untersucht die Wahrnehmung der Landschaften in den Regionen und vergleicht die Ergebnisse miteinander. Das folgende Kapitel analysiert Unterschiede in der Landschaftswahrnehmung von verschiedenen Gruppen (Einheimische, Rückkehrende, Zugezogene). Das theoretische

Konstrukt der WTP wird im 8. Kapitel erweitert durch den Ansatz von willingness to volunteer (WTV) und betrachtet die unterschiedlichen Einflussfaktoren für die Entscheidung für ein Engagement in oder für die Landschaft in Form von WTP oder WTV. Das letzte analytische Kapitel nutzt eine Clusteranalyse, um unterschiedliche Gruppen in den Regionen zu identifizieren. Diese Cluster werden darauffolgend tiefer hinsichtlich der WTP, WTV, Lebenszufriedenheit und der Nutzung der Region analysiert.

Einige der Ergebnisse dieser Kapitel sind, dass in jeder Region die Wasserlandschaft über den anderen Landschaften präferiert wird. Hierbei ist es unwichtig, ob jemand in die Region gezogen ist oder dort seit Geburt lebt. Das ist auch ein Ergebnis, welches in anderen Analysen zu finden ist, aber auch auf der Theorie begründet werden kann. Die Landschaftswahrnehmung ist häufig durch das Bildungsniveau und die Beziehung zur Natur beeinflusst. In manchen Fällen kann angenommen werden, dass das Bildungsniveau und die Beziehung zur Natur interagieren. Es kann gezeigt werden, dass die Bindung an die Region sich innerhalb der Regionen und Gruppen der Bewohnenden (Einheimische, Rückkehrende, Zugezogene) unterscheidet. Dort könnten indirekte Einflüsse für die Unterschiede verantwortlich sein. Hinsichtlich des Ehrenamtes in oder für die Landschaft suggerieren die Ergebnisse, dass spezifische Gruppen bereiter sind als andere, sich zu engagieren. Zum Beispiel sind Rückkehrende mit einer Bindung an die Region eher bereit, sich für den Bereich der Heimatkunde zu engagieren, während vor allem junge, hoch gebildete Frauen sich ehrenamtlich eher im Tierschutz einbringen. Über alle Cluster hinweg konnte eine häufigere Nutzung der Region für Entspannungszwecke als für kulturelle Zwecke identifiziert werden.

Diese Ergebnisse deuten an, dass das Regionalmangement die Region durchaus fördern könnte, indem es die Landschaften nutzt. Darüber hinaus kann es die Wahrnehmung der Landschaft stärken, indem es die Beziehung zur Natur stärkt oder für ehrenamtliche Tätigkeiten gezielt Gruppen anspricht. Weitere Forschung wird in dem Zusammenspiel von WTP und Lebenszufriedenheit gesehen. Wie im Bereich WTV könnte es sein, dass zufriedene Menschen auch eher bereit sind, sich finanziell für die Landschaft engagieren, als weniger zufriedene. Eine mehr psychologische Ausrichtung zukünftiger Forschung bietet die Analyse von Motiven für WTP oder WTV.

Schlagwörter: Kulturlandschaft, Landschaftswahrnehmung, Kontingente Bewertungsmethode, Zahlungsbereitschaft

1 Introduction to the Assessment of Cultural Landscapes

This thesis assesses the meaning of cultural landscapes for the residents of three regions. The term 'landscape' could be used for every landscape that exists in the world (Schein, 1997). In general, a landscape is the result of practices, relationships and interaction of nature with humans (Butler, 2016; Council of Europe, 2000; Swanwick, Hanly & Termansen, 2007). Therefore, a landscape might be a valuable source towards understanding the past of a culture or region (Kuechler, 1993). In literature, the landscape was no longer present until the European Landscape Convention published the revised landscape convention that integrated all types of landscapes (Council of Europe, 2000). Some authors differentiate between natural and cultural ones (e.g., Fritz-Vietta, De la Vega-Leinert & Stoll-Kleemann, 2015; Swanwick et al., 2007), others see all landscapes as cultural landscapes, given that these are influenced by interactions between humans and nature (Groth, 1997; Job, 2003; Roschewitz, 1999).

Even because a cultural landscape is formed by the past land uses in the region, different people could interpret these landscapes differently due to ones individual's subjective perception (Butler, 2016; Meinig, 1979). This subjective perception relies first on the own experiences in the past (Butler, 2016) and second is formed by the usage and memories of each individual (Stephenson, 2008). These two aspects are therefore important for landscape valuation and individual perception (Carson, Flores & Meade, 2001; Kim, Kim & Doh, 2015; Morrison & Dowell, 2013). Consequently, the landscape cannot be perceived and valued by an individual without any cultural context (Stephenson, 2008). Because of the importance of cultural context, it could be that individuals living in the same cultural context might perceive the landscape in similar ways (Paasi, 2002).

These landscapes or cultural landscapes are – in most cases – public property that one person can use the same time than another does (Einig & Dosch, 2008; Hartje, Degenhart & Dehnhardt, 2003; Job, 2003; Marangon & Visintin, 2007). Landscape changes might not be realized directly by the people living in the landscape – since it is a process; but if the change is recognized it is often seen as a negative development (Lankia, Neuvonen, Pouta & Sievänen, 2014). This might be because the people are happy with the current landscape when asked about it, and they might be afraid that life within the landscape might influence their environment in a negative way. This implies that changes in the

J. Matloch, *The Assessment of German Cultural Landscapes*, RaumFragen: Stadt – Region – Landschaft, https://doi.org/10.1007/978-3-658-21416-6_1

landscape might also change the ways its residents perceive the landscape (Tveit, 2009). Especially landscapes with high biodiversity or characteristic elements of the regional history are perceived as more important to protect (Bohnet & Konold, 2015) than landscape of the everyday life of the residents.

Because of the demographical development, rural region have often a declining population growth, and thus von Reichert, Cromartie and Arthun (2014) argue that all aspects to gain newcomers in these regions are really important. One solution for the declining population might be that the regions get more attractiveness by their natural resources as the amenities (McGranahan 1999, Rasker & Hansen, 2000). Additionally, Deller, Tsai, Marcouiller and English (2001) conclude that the occurrence of these amenities could have a positive impact on the economic growth for these regions. This link between the regional development and amenities are in previous research used to explain why people move to a region (Knapp & Graves, 1989). The preferences for regions with special amenities are lower crime rates and higher quality schools. Additionally, these regions are more attractive for labor (Knapp & Graves, 1989). But there might also some weak spots by using the natural resources for the development of regions. For instance, Green (2001) found that in rural areas natural resources and economic development in these regions might be difficult because on the one hand, these regions want to attract tourists that might not be valued positively by residents and on the other the economic development might destroy some region's natural resources.

These developments in the landscape perception, the previous findings in the research of landscapes and its importance for especially rural areas are one of the most crucial factors to consider when analyzing the perception of residents and their regional landscapes. To get the focus more on their natural resources, it might be the case that these allow the regions to experience growth. But before regional management could do more marketing with these natural resources or amenities, they need to know how important the landscape is to its residents. This doctoral thesis should be a basis – especially for the selected regions – for obtaining information about how residents perceive their landscape.

The thesis is comprised of ten chapters. A short introduction into the topic and the importance of the assessment of cultural landscapes is given in the first chapter. The second chapter describes the setting of the assessment. It starts with the conceptual setting in Section 2.1. In this part, the term cultural landscape is defined. Furthermore, the different values in the perspectives of culture and economics are presented. In the next chapter, the theoretical background for a trans-disciplinary approach is described because the analysis looks at a cross-section of different research areas and does not focus on the usual micro or macro level. The conceptual setting closes with a description about the research objectives and gaps in more detail. Following is Chapter 2.2, which describes the setting of the project and the chosen regions. The first three chapters characterize each region with their features and the last section of this part closes this description by comparing the regions with each other using some operating numbers in comparison with Germany.

Chapter 3 explains the survey data that is used in the following analyses. Section 3.1 describes the implementation of the questionnaire. It presents the issue of the survey, the sample drawing, print and shipping and the digitization and the post-coding of the questionnaires. The response rates are shown in Chapter 3.2, and a characterization can be found in Chapter 3.3. First, are presented the representative status and afterwards a comparison of some operating numbers of the survey and Germany are described. The difficulties that are in associated with the survey could be found in Chapter 3.4.

To get deeper into the complex construct of the landscape valuation, Chapter 4 explains the methodology used for evaluating landscape perception. The first part (Chapter 4.1) shows different approaches that are suitable to determine a value for a landscape and justifies why the contingent valuation (CVM) and the willingness to pay (WTP) methods were chosen for the survey. It also explains why the approach is used as an indicator for the perception of the landscape instead of a monetary value[1]. Properties and limitations of the chosen method of the WTP are presented in Chapter 4.2. Section 4.3 discusses some critique about the usage of the WTP for the determination of a value for such public property as a landscape. These critics could all be found in the literature, and an attempt was made to use these to improve the use of the data in the following analyses. The chapter closes with Section 4.4, which describes the practical application of the method.

The variables of the survey that are used more often are presented in detail in the following part. Section 5.1 describes the answers regarding the WTP questions. All prompted landscapes are presented in each category that was examined. The differences between the variables newcomers, returnees, and locals are shown in Chapter 5.2. Since the creation of the variable for volunteering as contribution uses more than one question, these are presented in Chapter 5.3. Chapter 5.4 then discusses how respondents answered the life satisfaction question. The section closes in Chapter 5.5 with a description of the manifestations for use of the region.

The first analytical chapter is Chapter 6. It analyzes the WTP for three regional landscapes that could be found in all of them. The aim is to identify different influences of the perception of landscape and differences between the

1 For the following, valuation, WTP and perception are used synonymously.

regions. To get an idea about the importance of the valuation of landscapes, Chapter 6.1 introduces the topic by explaining the research gaps. The theories important towards understanding why individuals value landscapes differently are presented in Chapter 6.2. In this section, previous research and the hypotheses that should be analyzed by this part can also be found. Chapter 6.3 describes the variables that are used and the statistical methodology. The empirical results are shown in Chapter 6.4. This chapter starts with the results of each landscape in the regions and ends with a comparison of the results of the landscapes per region. The results show that having a bond to nature is the most important factor in landscape perception – no matter the landscape type or region. Additionally, the impact of the neighborhood varies for different landscape types, while the impact of having a regional identity differs by regions. Socioeconomic variables and the use of the landscape do not seem to be important for the perception of the landscapes. The results are discussed in Chapter 6.5. This part also discusses further variables, the pictures of the landscape, and the generalizability. A summary of this part of the thesis can be found in Chapter 6.6.

The next chapter analyzes the differences between newcomers, returnees and locals regarding their landscape perception with respect to two types of landscapes (called everyday and symbolic ones) in the Steinburger Elbmarschen (STB) region of Germany. It starts with an introduction of the importance of migration biographies for the landscape perception in Chapter 7.1. The theoretical background of migration theories and the landscape perception are presented in Chapter 7.2. This part describes some selected theories regarding migration and discusses some previous findings. Additionally, the analyzed landscape types of the symbolic and the everyday landscape are described. Based on these backgrounds, hypotheses are created. Chapter 7.3 describes the landscapes of the STB in more detail to get an idea how these look. The followed Chapter 7.4 describes the used variables and the methodology in more detail. The results are displayed in Chapter 7.5. These show the results for a general perception as well as for each landscape and compare it in the end to answer the hypotheses. In general, newcomers and returnees have a higher perception of the regional landscapes than locals do. It could be identified that newcomers and returnees perceive the symbolic landscapes (special landscape types of the examined regions) differently, depending on the type of landscape, but in both cases the locals gave the second highest valuation. Additionally, the results show that the educational level is important for the perception of the landscapes; no matter if it is a symbolic or an everyday one. The discussion in Chapter 7.6 discusses the results, especially for the background of the small sample used for the analyses. Chapter 7.7 provides a summary of the chapter.

The analyses in Chapter 8 integrate the WTP and willingness to volunteer (WTV) for the regional landscapes. The aim of this paper is to identify the individual characteristics that are important for the decision in which form (financially or by volunteering) an individual supports the protection of landscapes. Chapter 8.1 explains why a combination of these two constructs seems to be important. Chapter 8.2 follows with some theories about giving behavior. (The theories regarding the WTP approach are already described in Chapter 6.2.1.) In this part some findings of previous literature and the formulation of the hypotheses analyzed in this chapter could also be found. The sample regarding the variables used and the methodology are described in Chapter 8.3 and the empirical results are shown in Chapter 8.4. First the results are presented and second these are used to answer the hypotheses. One's gender, age, educational level and bond with nature determine the decision to contribute something towards regional landscape protection. Generally, older people and women are less willing to contribute to landscape protection. A conclusion summarizes the results in Chapter 8.6. In addition to the analyses done in this chapter, there are two additional analyses written. The first excursion describes the monetization of volunteer work for the used sample to get to know about the potential of volunteering for the region. The second one compares the data of the sample regarding volunteering to the German Civil Engagement Report (CER) to get an idea about the structure of the sample in regard to the German average.

The last chapter aims to identify different clusters of individual characteristics to better understand the decision for WTP or WTV, life satisfaction of the people, and use of the region by these clusters. Therefore, Chapter 9.1 introduces the importance of the regional features assessment. The theoretical constructs regarding life satisfaction and the use of the region are described in Chapter 9.2. (The other constructs are already defined in Chapter 6.2.1 for the WTP and in Chapter 8.2.1 regarding the WTV). Chapter 9.3 describes the variables used for the cluster analysis and the methodology. The results of the cluster analysis are shown in Chapter 9.4. Seven clusters were identifies that are differing especially in age, educational level and their bonds to the region and/or to the nature. Following is Chapter 9.5 with a description of the variables analyzed within the determined clusters. Also some hypotheses are formulated in this part. Chapter 9.6 shows the results of the deeper analyses of the clusters. The analyses show that especially a higher level of education leads to more WTP, WTV, higher satisfaction rates and a higher rate of use of the region. The clusters with the retired people seem to be the clusters with the less support of the regional landscapes in terms of WTP or WTV. A discussion about the findings can be found in Chapter 9.7, as well as a part of the further work. The part ends with a summary in Chapter 9.8. Additionally to the cluster analysis and its analyses, there is done an excursion with a statistical model to get an idea about the impact of WTP and WTV for different clusters.

The last chapter summarizes all findings. Section 10.1 shows the key results of each of the text chapters and shows the contributions made to previous literature. It can be seen that the results of these chapters are quite similar in some cases – especially with respect how educational level and the relationship to nature are important drivers in of landscape perception. Most of the contribution to the literature is the type of analyses and the object of analyses itself. Chapter 10.2 suggests some political recommendations, especially for the analyzed regions regarding future interactions with the landscape and its residents. Chapter 10.3 concludes by posing some further research questions.

2 Setting of the Assessment

The following sections describe first the conceptual setting of the analyses. This includes not only the overall research questions but the aims of this research and the classification of the research field as well. The second part of this chapter explains the background of the project and the analyzed regions in more detail.

2.1 Conceptual Setting

This section of Chapter 2 describes the theoretical aspects of the further analyses. The first part starts with some definitions about the construct of a cultural landscape. It ends with some remarks about the valuation of a landscape as aspect of the research. Following is the description of the term values in a trans-disciplinary perspective that is important for the research done here. It explains some interfaces to research disciplines that are combined in this work. Based on the theoretical concepts and values in a trans-disciplinary perspective, the research gap is addressed with the subsequent research questions.

2.1.1 Cultural Landscapes

The term *landscape* is generally denoted as "…land at the interface of the earth´s surface and atmosphere" (Unwin, 1975, p. 130). More precise, the landscape is generated by people's perceptions about land as well as its use of. It covers the interaction and the bond between people and places (Swanwick et al., 2007). With the publication of the European Landscape Convention, the landscape came back into focus of the people, because this convention includes all kinds of landscapes (Council of Europe, 2000). Essentially, landscapes can be differentiated between natural and cultural landscapes (Fritz-Vietta et al., 2015; Roschewitz, 1999; Swanwick et al., 2007). While natural landscapes are characterized as those with no – or just a few changes – stemming from beyond natural developments, a cultural landscape was changed by humans (Groth, 1997; Job, 2003; Roschewitz, 1999). Job and Knies (2001) stated that there is no natural landscape anymore and all landscapes are cultural landscapes as a consequence of human influences.

The term *cultural* can be defined on different stages, such as regional or national. Culture connects different persons (Throsby, 2007) and is expressed in

J. Matloch, *The Assessment of German Cultural Landscapes*, RaumFragen: Stadt – Region – Landschaft, https://doi.org/10.1007/978-3-658-21416-6_2

shared practices (Cosgrove, 1989). It can also express aesthetics, symbols or a spirit of a group (Klamer, 2003).

Almost all landscapes reflect traditional usage by the population and are related to cultural aspects (Antrop, 2005; Cosgrove, 1989; Schein, 1997). Because a *cultural landscape*[2,3] covers the human influences on a landscape (Schein, 1997), Job and Knies (2001) argue that there is no natural landscape anymore, but cultural landscapes everywhere. Therefore, there is no uniform definition what exactly a cultural landscape is (Antrop, 2005; Fritz-Vietta et al., 2015; Gailing, Kühn & Vetter, 2008; Schein, 1997). One of the most common aspects is that cultural landscapes are public goods that could be used by more than one person at the same time (Einig & Dosch, 2008; Hartje et al., 2003; Job, 2003; Marangon & Visintin, 2007). Especially landscapes with high bio or cultural diversity and characteristic elements of history seem more worthy of protection than landscapes without these elements (Bohnet & Konold, 2015).

In the understanding of these analyses a cultural landscape is understood as a regional landscape that is designed by cultural and social influences but also of the world around the landscape. It is comprised of a variety of elements that involves natural components, particular landscape forms, land use and structural conditions. Beyond these aspects, a cultural landscape is also characterized by regional culture that includes language, products, dishes, crafts, history and common experience and perceptions (referring to Gailing & Röhring, 2008; Ray, 1998; World Cultural and Natural Heritage, 2015)[4].

Because of the interaction of the cultural landscape and the humans living in it, landscapes change over time (Antrop, 2005). Each new application of the landscape results in a decreasing degree of older, more traditional influences (Miles, Sullivan & Kuo, 1998; Palang et al., 2006), but most of these are creeping and not recognized by the residents. As a consequence of the decreasing visibility of the traditional usages, the identity of the resident population that could be building out of the history could get lost (Antrop, 2005). This might be one reason why residents – if they recognize these changes – often perceive it as negative (Lankia et al., 2014). The residents might be used to the actual status of their environment and fears a negative development for their living environmental surrounding (Dale, Ling & Newman, 2008; Schultz & Zelezny, 1999).

2 In the following, cultural landscape and landscape are used synonymously.
3 An historical overview about the term cultural landscape is given by Gailing (2014).
4 A similar definition, but for landscape quality, is given by Daniel (2001).

2.1.2 *Values in a Cultural and Economical Perspective*

By protecting, preserving and designing the landscape, the government has costs for the regional landscapes. In some cases, it wants to know the benefit of these for the landscapes. These values are economic values, expressed by concrete numbers (Klamer, 2003).

As stated by Klamer (2003) cultural goods have more than an economic value, as they are special because of the treatment of the residents and their intrinsic value to society-at-large. The values of a cultural good might be more implicit and might be expressed by the residents through stories or emotional statements (Klamer, 2003). Such values are called social values or social capital because they integrate the interpersonal relationships like the identity or emotions to the good (Klamer, 2003). These values are usually depending on the context (Klamer, 2003) and therefore different individuals can have different values for a cultural good (Peacock, 1992).

It seems to be adequate to use the perspectives of culture and economics at the same time because these constructs can co-exist (Throsby, 2001) and in some cases depend on one another. Throsby (2001) explains this by using values and norms as expression of the culture and the currency and prices for goods in regard to economics in the same system. However, Klamer (2003) describes that the culturalists are mostly people form the art field using the cultural good in their work, while the economists highlight the economic output of the good for society. This confirms the coexistence of these two perspectives.

Because of the widespread use of the perspectives of the economists in the measurement of values, investments in the cultural good are usually justified by an increasing economic output for the region and hence regional growth (Klamer, 2003). A cultural good might be more than just a scarce resource as it might be seen by the economists, because pricing in the perspective of economists and the subjective value by culturalists are not the same, and the interaction of these two disciplines is quite rare (Klamer, 2003).

But sometimes it is important to have a measurement that is able to display the value of a public good like the landscapes (Job & Knies 2001). But because of the public character, it is methodologically challenging in economics to measure the value, so it needs be done carefully (Cornes & Sandler, 1986; Haab & McConnell, 2002; Karkow & Gronemann, 2005; Spash, 2008). Hence, Peacock (1992) speaks for mechanisms in which individuals can express their preferences for cultural goods instead of methods or measurements in terms of the social values of a good. But measurements of the social values are rare – if a measurement would be possible at all (Klamer, 2003). One possibility is the measurement of life satisfaction using different aspects in the region as an alternative measurement for wealth (Kahnemann & Krueger, 2006; Veenhoven,

2011). The economic methods of measurements[5] usually use a monetary value because money might have an equal value for everyone (Hossack & An, 2014). Furthermore, it might be easier to compare the costs for preservation, protection, and design of the landscapes by the government with the benefits the residents see and use this for their economic output.

The used method in this thesis is the economic approach of the WTP, which is a typical measure for economic values (Klamer, 2003). Economically, the cultural good is a scarce resource, but the analyzed landscapes in this thesis are these that are mostly not a scarce resource because they could be found in many regions. This would be more the perspective of a culturalist, because a landscape can be something special for the residents if it is spoken about as something special. To use the WTP in a broader context of integrating the cultural perspective, the monetary value is used as an indicator for the valuation of the good. Hence, the thesis uses in the following WTP valuation preferences and perceptions as synonymous. Even if, the aim of such WTP analyses is to get a monetary value for an improved good and calculate the consumer surplus, in this thesis the current appearance of the landscape is under examination.

2.1.3 Theoretical Background for the Trans-Disciplinary Approach

The varieties of capitalism, a theory of the political economy, describe the different appearances of capitalism and its theoretical and political implications regarding the economic relationships (Peck & Theodore, 2007). The political economy analyzes the rationality of decision-making processes of political actors by using them as groups of different interests (Schefold, 1994). The aim is to support the understanding of the political decisions that can change constantly (Schefold, 1994). The varieties of capitalism theory can be understood as a macro-economic perspective (Hall & Soskice, 2001) and based on one firm, and the capital that is available on the market for this firm (Hall & Soskice, 2001; Jessop, 2011). Mostly, the comparative advantage of the institution is the center of attention (Hall & Soskice, 2001; Jessop, 2011). In classical economic theories, working counter to rationally acting individuals are the varieties of capitalism –assumed to be embedded in bigger systems in the variegated capitalism perspective (Peck & Theodore, 2007). The focus here is on the strategic behavior of actors in their institutional surrounding in regard to production costs like wages (Jessop, 2011; Peck & Theodore, 2007). Actually, since economic geographers share similar views about globalization as the economists, they are not included in the discussion about the varieties or variegated capitalism (Peck & Theodore, 2007).

5 A description of some selected measurements is presented in Chapter 4.1.

The geographical political economy, as described by Sheppard (2011), extends these theories to the assumptions that capitalism itself is only one way to characterize the economic outcomes of a society, but at the same time, the geography is endogenous to the economy in some ways and processes of the biophysical, cultural and social norms and values are important as well. The importance of informal rules for the society was already integrated in the varieties of capitalism approach (Hall & Soskice, 2001). Additionally, on a regional level, the geographical political economy uses mathematical models that were not common before some researchers found the theory of Marx to be helpful to explain aspects on this level (Sheppard, 2011). Using mathematical methods, the approach focused on representative cases rather than on getting a lot of information (Jessop, 2011).

These theories can be transferred to this thesis because the production described in them is bounded to biophysical, social, cultural and political processes (Smith, 1984) as also an individual could be (Hall & Soskice, 2001; Jessop, 2011). The costs for the individual in the case of this thesis would then be losing the landscapes. In this case the individual as starting point values the landscape with his or her emotions and feelings (micro level). These can also change like the decisions made in the political economy described by Schefold (1994). The results of the valuation have impact on the region the individual lives in (meso level). In the thinking and development of further improvements of the region, this could also have an impact on society or the understanding of culture (macro level). Most of the analyses of the WTP are on the macro or micro level, depending on the purpose of the examination. The aim of the research is to combine these two levels by answering the meso-analytical questions regarding the meaning of the landscape for the region in which the individual lives in. Also, it does not seem to be common to use statistical methods for analyzing on a regional level, especially not in geography as described by Sheppard (2011).

Additionally to these levels that are combined in these analyses, there are also some interactions between different disciplines. First, the WTP is used as a measurement method. This method originated in the discipline of environmental economics and due to the monetary value in the end, it could also be allocated to economics. The analyses have a strong relationship to the landscape because the definition and other important constructs were developed by a discussion with mostly people in the field of environmental planning. The theoretical background for the constructs of the landscape perception and willing behavior are mostly located in the area of psychology. These interactions of the variety of disciplines could not be identified in other, previous analyses.

2.1.4 Research Aim and Gap

The construct of the valuation of the cultural landscape came again into the focus of research with the revised landscape convention in 2000 (Council of Europe, 2000). It seems like a change in the perception of the landscape regarding its importance in the lives of its inhabitants can be found. It is not only the house or rental prices that are important for the choice of residency, but also some soft factors. These soft factors are usually related to the quality of life and include public services, the infrastructure as well as environmental conditions. The landscape can be understood as a part of the environmental conditions and is therefore an interesting aspect to delve into deeper.

Because of this development, this doctoral thesis analyzes the meaning of cultural landscapes for residents. The focus of the research is based on the subjective perceptions of residents of three different regions in Germany. To obtain this, the cultural and the economic value of the landscape are included in the analyses (Chapter 2.1.2). This implies a self-selection because only residents that are still living in the regions are integrated. The perceptions are not ascertained in order to compare them with the perceptions of experts. The aim is, in fact, to get an idea about the perception of regional landscape to use for the characterization of the landscapes in these regions. Therefore it is necessary to analyze the perception of the regional landscapes on the meso-level so that the region can use the results (Chapter 2.1.3). Perception in this case can be understood as synonymous to valuation, because it is indeed used as an environmental economics valuation method to get an idea about the perception of these regional landscapes. It is necessary to integrate not only the perspective of the environmental economics, but also the environmental planning and psychology to understand the results (Chapter 2.1.3). Most of the analyses using WTP are not using any kind of theory to get to know which results could be expected for the WTP. Such an analysis using all disciplines and values to answer the meso-analytical question about the meaning of regional landscapes to their residents was not done before. The more detailed research gabs are describes in the following.

Previous analyses examined special landscapes like vineyards or protected areas (e.g., Job & Knies, 2001; Karkow & Gronemann, 2005; Molina, Silva & Herrera, 2016) as well as more common landscapes like agricultural land or old buildings (e.g., Bamert, Ströbele & Buchecker, 2016; Kämmerer, Schmitz & Wiegand, 1996; Vanderheyden, Horst, Rompaey & Schmitz, 2013). But most of these cultural landscape analyses used future developments of the landscape to get to know which development of the landscape was preferred among respondents. But this process does not determine the current value of the regional landscape in or for the region. Additionally, these examinations compared either landscapes or regions with each other (Campbell, 2007; Kämmerer et al., 1996; Marangon & Visintin, 2007), but none looked deeper into comparing the perception of different landscapes and regions at the same time. The comparison might help to understand the importance of regional conditions for the perception of landscape. Nonetheless, none of the analyses could proceed this way because the landscapes were too different to each other. The data of these analyses allows this comparison because the questions were in each region the same and the photographs of each landscape were also similar, only taken in each region.

The research regarding the literature to locals and newcomers are often focused on the characteristics why an individual moves into another or out of a region or about the motives for moving (Barcus & Brunn, 2010; Brennan & Cooper, 2008; Kondo, Rivera & Rullman, 2012; Ní Laoire, 2007), but there could not identified any analyses of the differences in landscape perception by these groups. The theories regarding the landscape perception and the migration biographies might also show some indications for different perceptions between these groups because of their different cultural backgrounds (Fielding, 1992; Halfacree & Rivera, 2011; Ní Laoire, 2000; Tveit, 2009). To analyze the differences of these groups might be interesting – especially for people working in regional management, as they could build some target groups for improving the perception of the regional landscapes. These groups might have different needs with respect to their surroundings.

The combination of WTP and WTV is usually used in developing countries, due to the higher income constraints than found in more developed countries (Echessah, Swallow, Kamara & Curry, 1997; Stone, Bhat, Bhatta & Mathews, 2008; Vásques, 2014). For developed countries, only one analysis by Lankia et al. (2014) was found. However, this analysis looked only at a privately owned landscape and not public landscapes that could be found in every region. It might be very interesting to get to know what kind of people are more interested in volunteering or paying for the landscapes. The regional management could use the results to more directly address concerns about the regional landscape to potential volunteers or financial contributors. This could improve the involvement of people for their region and would help the landscapes to preserve for further generations.

The development of clusters for the intensity of the perception of the region itself could not be identified in the previous literature. There are analyses about specific characteristics of the individual (e.g., *Bundesministerium für Umwelt, Naturschutz, Bau und Reaktorsicherheit* (BMUB), 2014 for the relationship to nature; Soini, Vaarala & Pouta, 2012 for the regional identity) and the clustering of these, but for the region as a whole, it might be show a more holistic perception of the region itself. Even because there are little evidence about the perception of the region, its impacts of the regions attractiveness is not analyzed yet as well.

In addition to the added value of the research in regard of the topics above, the management of a region might benefit from the analyses as well. For instance, the regional management might be able to use the results to design an improved marketing platform for tourist, regional advertisement for newcomers or returnees, or improve the image of firms or the value chain in the region. The aims might be different for different fields of work in the region, but it could be a basis for deeper analyses regarding the characterization of the region itself.

2.2 Setting of the Project and the Chosen Regions

Every cultural landscape[6] has individual qualities. It tells a story, using the defining elements of the landscape and the usage of it. Therefore, a cultural landscape is not only adding some value to the regional quality of life, but also influence the attraction of a region. The potential of landscapes with respect to the forming of regional identity has not been exhausted yet (Kempa, 2014; Kempa & Herrmann, 2014).

The research project REGIOBRANDING analyzes how quality and identity-establishing regional landscapes could be used to strengthen identity in these regions (Kempa, 2014). In the focus of this project are special resources like cultural landscapes of rural areas that could be better used for regional development (Projektgebiet, n. d.). There are three phases of the project to match the project goals: first, the inquiry of knowledge inclusive production of knowledge, second, the reprocessing of the knowledge and finally, the development of concepts, their implementation and evaluation (Kempa, 2014, Ziele, n. d.).

Based on the results, scientific and regional players work on concepts for highlighting cultural landscapes characteristics in three selected regions[7]. These concepts should illustrate how these features could be used for systematic and reliable development in these regions. Using specifics get more and more important for being competitive with other regions and therefore the project Regiobranding also involve economic, social and cultural aspects instead of taking

6 In this case, a cultural landscape is understood as a regional landscape that is designed by cultural and social influences. It is comprised of a variety of elements that involves natural components, particular landscape forms, land use and structural conditions. Beyond these aspects, a cultural landscape is also characterized by regional culture that includes language, products, dishes, crafts, history and common experience and perceptions (referring to Gailing & Röhring, 2008; Ray, 1998; World Cultural and Natural Heritage, 2015).

7 The regions are described in more detail in Chapter 2.1.

only physical aspects into account. This holistic perspective should help by building, communicating and stabilizing a new regional Identity using the cultural landscapes qualities (Kempa, 2014; Kempa & Herrmann, 2014).

As a basis for the branding process, the scientific project partners first have to ascertain their comprehensive knowledge about the selected regions (Ziele, n. d.) by examining the sensibility of the residents regarding the regional landscapes. There should be a representative sample survey about the positive and 'needs improvement' aspects of the region. This procedure should expose possible starting points for the development of the region (Kempa, 2014).

The regions of this analysis are located in the metropolitan area ("Metropolregion") of Hamburg (MR HH), Germany. These regions embrace 17 rural districts and two urban districts in four federal states. Challenges like the demographic change or the infrastructure should be treated using the combination of urban and rural areas (Projektgebiet, n. d.).

Altogether, three regions in the metropolitan area have been chosen as models. While the first region, called STB, is located in the northwest, the region Lübeck/ Nordwestmecklenburg (LNWM) is in the northeast and the third region Ludwigslust-Parchim/Lüchow-Dannenberg (LPLD) is located in the eastern sector of the Hamburg metropolis (Fig. 2-1).

These regions were selected because of their location in the hinterlands of the metropolis Hamburg. Often, inhabitants of the hinterlands of a metropolis identify themselves with the landscapes of even this city – in the case of Hamburg, the Hamburg Harbor, for instance. But actually these people are living in other mostly rural regions and have their "own" local landscapes.

Fig. 2-1: Metropolitan area of Hamburg and the three chosen regions

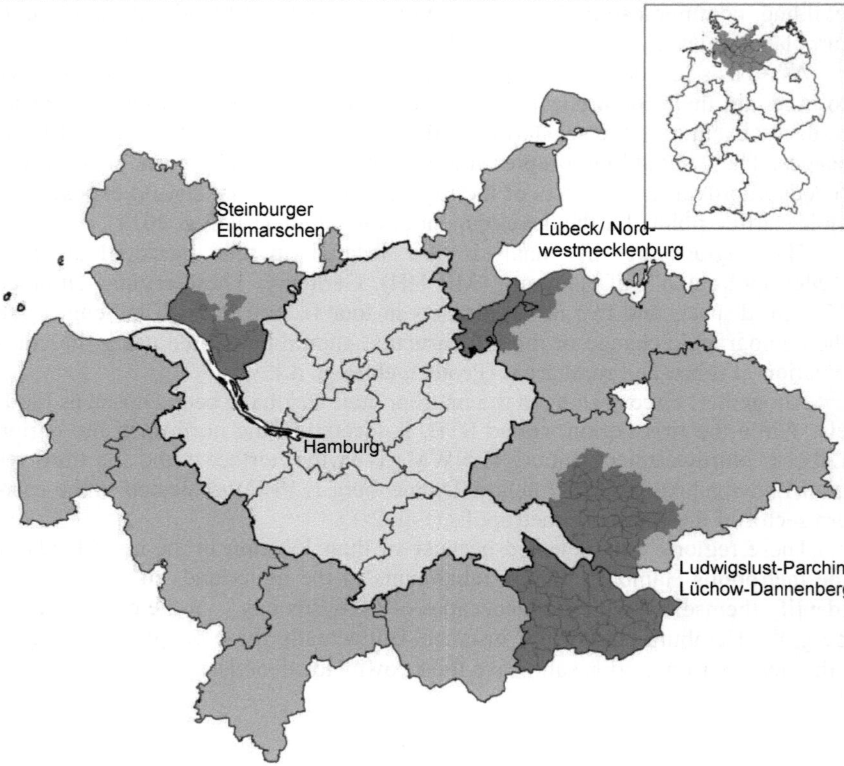

Note: The map in the right of the picture displays the map of Germany as a whole. The marked area is the MR HH. The big map shows the MR HH and its districts. The brighter areas are these districts that are in contact with the metropolis Hamburg, the darker areas are not. The darkest areas show the three regions examined.
Source: Own illustration.

2.2.1 Area of Target: Lübeck/ Nordwestmecklenburg

The region of LNWM is characterized by the Ice Age and located in the valley of eastern Holstein and Nordwestmecklenburg (NWM) along the Trave river at the Bay of Lübeck (L). It compromises of the Hanseatic city of L, Schönberger

Land district (municipals of Selmsdorf, Lüdersdorf and Dassow) and the municipality of Kalkhorst[8].

While NWM is relatively sparsely inhabited, L is one of the regional centers of Schleswig-Holstein and in the past was one of the most important commercial centers in the Baltic Sea. The area of this region is about 423 km² and 226,682 people lived in the region in 2013. 12.3 % of the residents are under 15 years of age, 65 % between 15 and 65, and 22.6 % are older than 65. The people living in the region are slightly older than the average German population (Regional Database Germany).

The structure of the city is scattered by the Trave and Wakenitz rivers, the Elbe-Lübeck-Canal as well as a lot of green spaces. The medieval city center is a UNESCO World Cultural Heritage site. The advantage of this city structure is the production of fruits and vegetables, but the disadvantage is the conflict between the industrial and green areas.

As typical elements of this region are semi open and dry habitats, the cliff coast with the stone reef in the front and the salty wetlands. Heaths, bogs and open farmland with hedgerows are crossed by water. Every spring and fall a lot of migrating birds are hatching and nesting, especially in natural reserves of the region. Because of this, this region is important for international bird migration (Fokusregion Lübeck und Nordwestmecklenburg, n. d.).

2.2.2 *Area of Target: Steinburger Elbmarschen*

The region STB is located in the Holsteinian Marshland of the Elbe in the district of Steinburg. It consists of the cities Itzehoe, Wilster and Glückstadt and all municipalities of Horst-Herzhorn, Krempermarsch and Wilstermarsch.

In 2013, 78,410 people are living in the region that comprises 488 km^2. 12.9 % of the residents are under 15 years of, 64.7 % are between 15 and 65, and 22.4 % are over 65. The people living in this region are also slightly older than the federal average (Regional Database Germany). A small area analysis of the population development showed a significant decrease in this region (-6.5 %)[9].

During the 12th century, the cultivation of the marshland of the Elbe started with cooperation of Dutch colonists. Today the region is characterized by narrow and long farmland with surface drains and terpenes. The so-called "Barghus" shows the typical building culture of this time. With the development of windmills that are able to scoop water, the area of Steinburg could be cultivated

8 The municipality Kalkhorst did not participate on the survey and could therefore not be analyzed with the data.

9 The data originates to the national account system of the federal states and the Federal Employment Agency.

for population and agriculture. The only windmill during this time was found in Honigfleth, close to Wilster. Before the founding of Glückstadt in 1617, the cities of Krempe and Wilster were the main cities of this region. Glückstadt is famous for its historical city center as well as the "Glückstädter Matjes" and the "Glückstädter vegetables". Today the railroad embankment, bridges out of steel framework and lockage are important for regional identification.

Since the 1970, the visual landscape changed due to restructuring in the agriculture sector and as a consequence thereof, the decrease in agricultural, expansion of the infrastructure, and trenches that are surrounded by pipes and modern instead of historical farmhouses. The expansion of regenerative energies also changes the cultural landscape in this region and causes conflicts with the residents regarding their living quality. Furthermore, the newcomers into the region were given the opportunity to use the older dwellings and consequently preserve the historical farmhouses (Fokusregion Steinburg, n. d.).

2.2.3 Area of Target: Ludwigslust-Parchim/ Lüchow-Dannenberg

The third region consists of the district Lüchow-Dannenberg (LD) as a whole in the southeast and a part of the district Ludwigslust-Parchim (LP) in the northeastern section of the Hamburg metropolis. The region is split by the Elbe (river). The part of LP is comprised of Dömitz-Malliß, 13 municipalities of Hagenow-Land, five of Ludwigslust-Land, three in Grabow and the cities Ludwigslust, Hagenow and Lübtheen. This region is called Ludwigslust-Parchim/Lüchow-Dannenberg (LPLD).

In comparison to the others the region is relatively large with 2,077 km^2. In 2013, 100,726 people lived in this region. The percentage of people under 15 years of age are 12.1 %, while 63.6 % are aged between 15 and 65 and 24.3 % are over 65. The population development shows a negative trend given that the development coefficient from 2008 to 2012 was -27.9 (STB -14.5, LNWM 3.6)[10].

Forests and bogs are formative habitats of this region. The use of agriculture and the backfilling of drawdowns changed the visual landscape. The close to nature areas in the South and the North of the Elbe are protected as UNESCO biosphere reserve. Across the region, there was the inner German border and the areas around the border are even more unimpaired natural areas. But the past location of the inner German border accounts for the poorer infrastructure and could be therefore be a negative influence on the emigration – especially of younger people.

10 The data originates of the Regional Database Germany, the State office for statistics of Niedersachsen, Hamburg, Schleswig Holstein and Mecklenburg-Vorpommern.

The settlement development and the cultivation of the landscape of this region are documented in the chronicle. Some archeological sites show settlements since the Bronze Age, and in particular the circular villages in the part of LD are UNESCO World Culture Heritage sites. In this part of LP there are sundry monuments that document previous settlement.

Especially the sandy grounds in the part of the region of LP have ideal conditions for the growth of sea buckthorn. Additionally, in the south part of the city Ludwigslust there is bog iron ore mining. For this part of LD, the large amount of artists is indicative because it is popular due to the recurrent cultural event, the so-called "Kulturelle Landpartie" (Fokusregion Griese Gegend-Elbe-Wendland, n. d.).

2.2.4 *Some Operating Numbers of the Regions*

Regarding the socioeconomic data, there are some differences between the regions that show it is meaningful to compare them. The regions differ in terms of the residents per area and therefore in their urbanity: In LNWM, there are living 536 people in one km² (Tab. 2-1). This figure is lower in the other regions (STB 161 and LPLD 48). The number of people younger than 15 or older than 65 are almost similar in each region, but the demographic development per 1,000 residents differs. While the index is positive for LNWM, for STB and LPLD it is negative. That indicates that both regions are shrinking – LPLD even more than STB. The Buying Power Index (BPI) per resident differs between the regions: The BPI of LPLD is about 10 units lower than that of STB, while LNWM is in between. The job density is a number that shows further differences between the regions: its figure is about 62 % for LNWM and 45 % for LPLD. STB is closer to LNWM, with a job density of 55 %. As a consequence of the different rate of urbanity and the job density, the economic structure differs: L has the highest specialization in the service sector, while LP, LD, NWM and STB still have a relatively high rate of jobs in the agriculture sector.

The description of the regions could also be characterized from the covered areas in each. The region of LNWM is the most urbanized area, with also 37 % for the district of L and 9 % for NWM of residential estate and traffic, followed by the district of Steinburg. While the percentage of the land coverage for the region of LNWM is slightly higher than for Germany, it is lower for the district of Steinburg. The districts of LP and LD show the lowest percentage of areas covered with these land uses.

Tab. 2-1: Overview of some of selected operating numbers of the selected regions

	LNWM		STB	LPLD		Germany
Residents per square kilometer (2013)[1]	536		161	48		226
Residents younger than 15 years (2013)[1]	12,3%		12,9%	12,1%		13,1%
Residents older than 65 years (2013)[1]	22,6%		22,4%	24,3%		20,8%
Demographic development per 1,000 residents (2008-2012)[1,2]	3,6		-14,5	-27,9		-3,1
Buying Power Index per resident (2012)[3]	92,1		96,6	86,0		100,0
Job density (2013)[1,4]	62,3%		55,3%	45,7%		55,5%
	L	NWM	Steinburg	LP	LD	Germany
Working Population Primary Production (2012)*[5]	0,2%	4,1%	5,5%	5,8%	8,4%	1,6%
Working Population Secondary Production (2012)*[5]	17,4%	29,9%	24,7%	32,4%	23,8%	24,8%
Working Population Tertiary Production (2012)*[5]	82,3%	66,1%	69,8%	61,7%	67,9%	73,6%
Share of areas covered with areas of residential estate and traffic (2015)*[1]	36,9%	8,8%	11,1%	7,4%	8,2%	13,7%
Share of areas covered with recreational areas (2015)*[1]	5,5%	2,0%	0,7%	1,3%	0,3%	1,2%
Share of areas covered with agricultural areas (2015)*[1]	32,5%	71,4%	72,5%	59,5%	51,9%	51,6%
Share of areas covered with forests (2015)*[1]	14,4%	13,5%	9,4%	28,2%	37,1%	30,6%

Note: * These operating numbers are only available for the whole district.
Sources: 1 Regional Database Germany, 2 statistical administrative bodies of Lower Saxony, Hamburg and Schleswig Holstein as well as Mecklenburg-Hither Pomerania (*Landesamt für Statistik Niedersachsen, Statistisches Amt Hamburg und Schleswig Holstein, Statistisches Amt Mecklenburg Vorpommern*), 3 GfK (Buyer Power Index D=100), 4 Federal Employment Agency, 5 Federal Office of Statistics (*VGR der Länder*).

With respect to recreational areas, the region of LNWM has the highest percentage of covered land with these areas (L 5.5 %, NWM 2 %). This might be accounted for with the areas covered by water and the neighboring areas. For LP alone, the land coverage of this areas is higher (same as for Germany) than for the other two districts, but because this district is considered along with LD, the percentage might be somewhere around the same percentage as that of Steinburg district.

The highest cover of agricultural areas can be found in the district of Steinburg. This is not surprising, given that the cultural landscape in the south is mostly covered with meadows of marshland in the south. The second highest percentage of the cover of the agricultural areas is shown by the district of NWM, but considered with L as a whole, the region gets the lowest percentage.

The region of LPLD is in between the other two regions when the districts are seen as one region. These percentages are in line with the German average.

With the exception of the district of LD, the land coverage of forests is lower than the average for Germany. It might also be slightly higher for the considered region of LPLD because it consists of the whole district of LD and only a small part of the region of LP. Forest could also be found in the region of LNWM (L 14 %, NWM 14 %), but the percentage is only half of what is standard for Germany. The district of Steinburg shows the lowest rate in the region.

3 Survey Methodology

To capture the perception and the use of the residents, a survey is conducted as paper pencil survey or as an online alternative.

3.1 Implementation of the Questionnaire

First, every group working in the project and was interested in the survey collected ideas about the question that would help them answering the question. These questions were discussed in several meetings and everyone agreed with the draft of the questionnaire. Afterwards the questionnaire was under review by the project partners working in the chosen regions described above. Following, there was a discussion with them about the questions and a further version was developed and again discussed. At the same time the layout was done and the online version rendered. In the end, there was a pre-test done with persons living in the region and experts regarding doing a survey (altogether 30 questionnaires). The results of the pre-test were worked into the latest draft.

There were comprised five topics in the questionnaire: use of the region, perception of the region, civic engagement, financial engagement and socioeconomic data of the respondent. Most of the questions could be answered using a checkmark, but a lot of questions also provided the possibility to write something that was not mentioned. On the last sheet survey collection/ drop-off points in the region were listed. This offered another opportunity (besides the self-addressed stamped envelope) to give the filled questionnaire back to us. To motivate to participate in the used survey, almost every region conducted a lottery where the respondents could win some regional prizes. To participate in the lottery, the respondent needed to provide an email address or telephone number. The question about further conservation closed the survey.

Originally, the plan was to mail 5,000 questionnaires per region. The project partner working in the chosen regions were able to connect with the person in charge in the registration office. If not possible, the identification happened by the web and by phone. During this phone call the project and the desire was described.

The first sampling was done for the region LNWM. In this case the first written request of surrender of data was done. It included a cover letter, the reason and conditions for the sampling and the description of the project. The con-

J. Matloch, *The Assessment of German Cultural Landscapes*, RaumFragen: Stadt – Region – Landschaft, https://doi.org/10.1007/978-3-658-21416-6_3

ditions for the sampling were a birthdate between the first of January 1926 and the 31st of December 1996 as well as the region having been the first place of residency. This age limitation could be explained by using the arguments of Tuan (1990), who stated that young children have a growing perception of specific objects in the environmental surrounding. The age limit of 18 was also used from Verbič and Slabe-Erker (2009) and Howley, Hynes and Donoghue (2012). Additionally, there was a description of the required traits and the amount of the sample regarding districts or municipalities. This request needed to be permitted by the Ministry of Internal Affairs. The transfer of the data was done differently depending on the region. The data of each sample were not saved digitally to ensure the privacy of the data.

The data from the registration offices were first edited to use the data for the cover letters (e.g., supplement of the salutatory address, change of text formatting, addition of online-access). The aim was to mail 3,000 questionnaires to a random sample in each region. To achieve this, a random number was generated in Excel and the sample was sorted again. The first 3,000 people of this table were used for the creation of the cover letters. For the print the addresses were sorted by zip code because only with this order could the questionnaire be mailed as an "info-brief" (a cheaper method to mail a high amount of letters to one region). The cover letters were printed in the Niedersächsischen Institut für Wirtschaftsforschung (NIW), and the questionnaires from a printing company. Previous surveys suggested that respondents are willing to answer more often if the survey is in their mailbox on a Friday (SurveyMonkey Germany, 2014). To achieve the highest response rate possible, the shipping of the questionnaire was done in the beginning of the week so that it would arrive at people's homes in time for the weekend.

The period for answering the questionnaire was limited to six weeks. Before the summer holidays the questionnaires for LNWM and one part of LPLD were mailed. For the part L of region LNWM, the mailing was done at the 8th of June in 2015 (Tab. 3-1). The rest of the region, consisting of 700 questionnaires, was sent the survey on the 18th [11]. On the same date, the 2,000 questionnaires for LPLD were shipped. In the beginning of September of the same year, another 2,000 questionnaires for the other sector of the region LPLD were mailed. The questionnaires for the last region STB were shipped on the 24th of September.

11 More about the divergent number can be found in Chapter 3.4.

Tab. 3-1: Data about the questionnaire and its response rates

Region	Shipped	Packed	End of process time	Response absolut	Response-rate	Additional questionaires
LNWM	2.700			244	9.0 %	none
L	2.000	8-Jun-2015	24-Jul-2015	193	9.7 %	none
NWM	700	18-Jun-2015	7-Aug-2015	51	7.3 %	none
STB	3.000	24-Sep-2015	6-Nov-2015	427	14.2 %	56
LPLD	4.000			599	15.0 %	29
LP	2.000	18-Jun-2015	7-Aug-2015	193	9.7 %	8
LD	2.000	4-Sep-2015	23-Oct-2015	406	20.3 %	21

Source: Resident survey.

For the digitization of the data of the questionnaire, a codebook was developed. It describes the variables and their manifestations. The data was digitized using MS Excel to offer the opportunity to import the data in almost every statistics program. The plausibility of the codebook was proven from student assistants at the NIW, who checked the logic of the book and answered the questions immediately. Moreover, based on the previous experiences with other projects, a cost and time projection for the digitization of the questionnaires was done by the student assistants.

In the beginning, there were more meetings regarding the clarification of question of the data digitization. These clarifications of some specifics of a question were than written down into the codebook. The most often source of error was scribbled writing or more crosses than demanded.

Once the data was digitized, it was reviewed for plausibility using Stata (e.g., mistakes). Additionally, the other answers were reviewed (assignment to existing variables, possibility of categorization of answers, etc.). All changes made during the post-coding were entered into the attachment of the codebook. This offers the opportunity to track the changes in the future.

The more regions were post-coded, the more complicated and elaborate was the decoding of the other regions, given the alignment of the already post-coded data and the consistency of the decoding in all regions.

3.2 Responses

All questionnaires arrived via business replay envelope got a receipt stamp. The envelopes collected in the regions had mostly no receipt stamp. The response rates in the regions are between 9 and 15 % (Tab. 3-1).

More people than the random sample had interest in participating in the survey, so that the questionnaire offered the possibility for them to be part of the survey. This questionnaires were marked to identify these people later on because these questionnaires could be distorted the sample.

Before the collection of the questionnaire in boxes for the digitization, the sections for the lottery and their willingness to accept future contact were separated to warrant the anonymity of the survey. The contact details were entered in another table than the data of the questionnaire. With the end of the survey in a region, the box with the sections of the lottery was mailed to the partners in the region to pick a winner.

Connelly, Brown and Decker (2003) found that response rates of surveys with the subject of natural resources dropped over time. They concluded that increasing junk mail and market survey might be the reason for this negative development of the response rates. Furthermore, the response rate might also been influenced by the surveyed population (here: the general public), the length of the questionnaire (here: 10 pages), month of first mailing (Tab. 3-1), height of print type and length of hypothetical questions (Connelly et al., 2003; Green, 1996; Green, Boser & Hutchinson, 1997; Heberlein & Baumgartner, 1978).

Surveys that analyze a specific topic are usually getting higher response rates than those that contain broad subject matter (Connelly et al., 2003; Green, 1996; Green et al., 1997; Heberlein & Baumgartner, 1978). This might also be the reason why the response rates are relatively low. But Heberlein and Baumgartner (1978) and Green et al. (1997) found that government-sponsored research got better response rates that could positively affect the response rate of the survey. The months of the mailing might also have an influence, as stated by Connelly et al. (2003). The first part of the questionnaires was mailed in June, defined by Connelly et al. (2003) as a 'low response rate' month. The second part of the survey was mailed from September through November. Compared with the first part, the response rates were higher. This result confirms the categorization of Connelly et al. (2003). Higher rates could be reached if the questionnaires could be mailed from January through March. The font of print type and the complexity also could have affected the response rate – but complexity would not have a high influence on the response rate (Connelly et al., 2003). The questionnaire was printed in print type 12, but the complexity might be a reason for the response rate. During the pretest the partner working in the regions noted that some of the question might be difficult to answer, but because of the research interest of the scientists these questions required asking. Heberlein and Baumgartner (1978) found no effect from length or complexity.

3.3 Characterization of the Sample

The following section characterized the sample by mostly socioeconomic data by using some data of the Regional Database Germany (Regionaldatenbank) and the German Socio-Economic Panel Study (SOEP) for comparison. Hence it is necessary that the survey data of is used on almost the same scales as the German sample. Only in this case, this comparison allows showing if the sample will be similar when distributed to the German population. The representative status could be described only for the gender and age of the samples, because this data can be found in greater detail than other data. The representative status will be the first part of this chapter, followed by the second one, describing the sample in more detail.

3.3.1 Representative Status of Gender and Age

A test of the homogeneity of the survey data was implemented to say something about the representative status of the statistical data. The significance level of 10 % was chosen because with this limit, the regions could be clearly distinguished between representative and non-representative ones. With a value lower than 10 % the sample of the corresponding region is differently distributed than the statistic says. If this is the case the statements about the whole population in this region could be only interpreted with caution.

The statistical values regarding the gender are equally distributed in all regions than it is in the used sample (Fig. 3-1). This indicates that the data of the survey can be interpreted as representative.

Fig. 3-1: Representative status of the gender in each region (in percent)

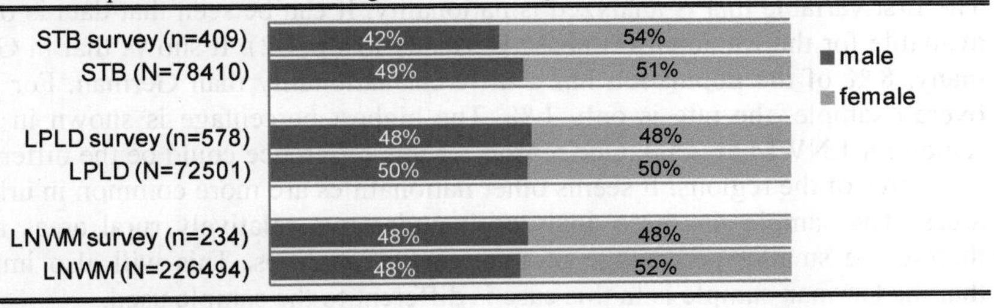

Source: Resident survey, Regional Database Germany (Effective 2013-12-31), own illustration.

The picture differs with age. A lot of differences between the age classes of the survey data and the statistical values could be identified (Fig. 3-2). These are more obvious in those classes in the middle. None of the regions, except LPLD, is representative on the significance level of 10 %. In reality, there are more people living in the region with one of the lower age classes than those who answered the questionnaire. The interpretation of the variable age and others should be done very carefully.

Fig. 3-2: Representative status of the age in each region (in percent)

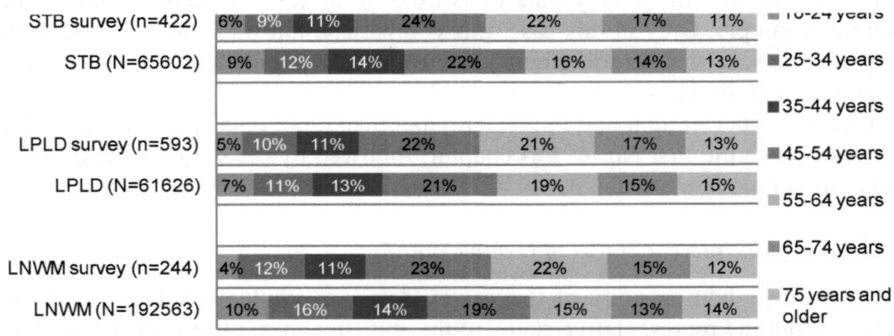

Source: Resident survey, Regional Database Germany (Effective 2013-12-31), own illustration.

More variables could not be analyzed to get information about the representative status of the survey data because most of the variables are only available on the level of the whole district. Because the regions consist of smaller units of the districts, the data could not used and compared to the data.

3.3.2 Characterization by Using Some Other Variables

The first variable that is analyzed is nationality. It can be seen that data is only available for the Regional Database Germany (Tab. 3-2). It shows that in Germany, 8 % of the population has a different nationality than German. For the overall sample, the rate is only 1 %. The highest percentage is shown in the region for LNWM (1.2 %). One reason for this difference could be the different structures of the regions. It seems other nationalities are more common in urban areas. The sample for these analyses includes two relatively rural areas and therefore a smaller percentage of different nationalities. This will also imply that the German sample is in this case is different to the sample used.

The living conditions show some over and underrepresentation. While in the survey sample, the individuals living alone are about 15 % in each sample, for Germany it displays values of about 40 % of the SOEP and 37 % for the Regional Database Germany. This indicates an underrepresentation of people living alone. An overrepresentation is shown for the households with two persons. The sample of Germany shows in almost both sources a rate of about 35 %. The survey sample displays values of about 54 %. This implies that there is an overrepresentation in the case of people living in households with two persons. These two phenomena might be explained by the over and underrepresentation of age. Usually, younger people more often live alone while older people usually with their spouse. The condition to live with more than three or more persons is almost in line with the German statistics (Fig. 3-2). In most cases, these are middle-age families. This class seems neither over nor underrepresented in the survey sample.

Tab. 3-2: Numbers for the characterization of the sample

	Sample	LPLD	STB	LNWM	Regional Database Germany (2011)	SOEP (2013)
Other nationality	0,9%	0,8%	0,7%	1,2%	8%	(.)
Lives alone	16,1%	17,7%	14,1%	15,6%	37%	40%
Lives with two persons	52,0%	52,0%	49,9%	55,3%	33%	34%
Lives with three or more people	30,4%	27,7%	35,3%	28,7%	30%	26%
Lives in own property	73,8%	73,7%	77,9%	67,1%	46%	46%
No graduation	0,7%	1,0%	0,7%	0,0%	5%	3%
Mainschool	19,2%	17,8%	23,1%	15,5%	36%	32%
Middleschool	36,8%	37,4%	36,2%	36,6%	29%	24%
High school	16,3%	15,6%	18,3%	14,3%	20%	27%
College	27,0%	28,2%	21,7%	33,6%	8%	(.)
Fulltime work	45,9%	45,9%	45,3%	47,1%	31%	37.7%* 39.1%$^{+}$
Part-time work	13,0%	10,8%	16,1%	12,8%	10%	11.6%* 11.1%$^{+}$

Note: (.) indicates data that was not available in the database, *only West Germany, $^{+}$ only East Germany.
Source: Resident survey, Regional Database Germany (Zensus 2011), German Socio-Economic Panel Study (SOEP) Group (2015a, 2015b).

The rate of living in the own property is relatively high for the survey sample compared with the data out of the Regional Database Germany and the SOEP. While both rates for Germany show rates about 46 %, it is about 73 % in the survey sample. The values indicate that in more rural settings people live in their own property while in more urban areas most people rent. Because the selected regions are almost rural except the part of L out of LNWM, the rate of the property owners could be explained.

The survey sample display some over and underrepresentation in the case of the educational level. Especially, the lower educational levels of no graduation or high school are underrepresented while the higher education like graduation from college is overrepresented. The rate for Germany shows about 4 % of people without any high school degree; the survey sample has the highest rate for the region of LPLD and the lowest one for LNWM. Regarding the main school, the result is similar. The German rate is about 34 % but in the survey sample it is underrepresented by about 20 %. Regarding middle school, the rates are the opposite. The survey samples show rates of about 36 %, while the German rates are lower with about 30 % for the Regional Database Germany and of about 24 % of the SOEP. This implies an overrepresentation of the middle school graduation in the survey sample. In comparison with the Regional Database Germany (20 %), the rates of high school graduation is relative low (about 17 %), while the college degree is definitely overrepresented with about 30 % in the survey sample and 8 % for Germany as the whole. In comparison to the SOEP where high school and college are added, the rates of the survey sample are even higher (about 43 % while 27 % for the SOEP and 28 % for the Regional Database Germany, respectively. This implies that not only the middle school is overrepresented, but also people with higher degrees were as well. It might be the case that people with higher education are more interested in their region and therefore answered the questionnaire more often than people with lower levels of education.

The last variables that could be compared are regarding the full and part-time work. Regarding the full-time work, the rates of the survey sample are higher than these for Germany (31 % Regional Database Germany). The data of the SOEP were only available separately for East and West Germany, but with almost the same rate (39 % East respectively 38 % West). Regarding part-time work, the rates of the survey sample are slightly higher than for Germany (10 % Regional Database Germany, 11 % SOEP East, 12 % SOEP West).

Overall, the sample seems to be in more cases to be different from the German population than to be similar. It might be the case that there are other aspects that are not considered, but with the data analyzed this cannot be confirmed.

3.4 Difficulties Regarding the Survey

There were some difficulties regarding the implementation of the questionnaire. One the one hand it was not possible to get any response from one of the districts of the region LNWM that we would like to integrate. Therefore the full amount of 3,000 questionnaires in this region could not be realized.

When the envelopes were packed with the cover letter, the questionnaire and the business reply envelope for the region of LNWM and LP, the postal service went on strike. As a result, some of the questionnaires arrived late in the region and in some cases the allotted time for answering the questionnaire expired. This might be one explanation why the response rate of LNWM is relatively low in comparison to the other regions. The region of LPLD had a relative high response rate for, so that the response rate is still relatively high.

Because there were already mailed 2,000 questionnaires to the residents in region LP, the same amounts were mailed to the other part of the region. Therefore, in this region 4,000 questionnaires were mailed to the residents instead of the 3,000 originally planned.

Moreover, the partners in STB and LPLD wanted people who did not get a questionnaire to be involved via random sampling. These questionnaires were labeled and mailed out to these interested people. Because of the sign these questionnaires could be picked out. Otherwise the random sample could be biased by these answers, since usually these people might be more interested in the topic or at least in the region than the people in the random sample. But these questionnaires might be relevant for some of the project partners to gather more information.

The last difficulty regarding the questionnaire itself was the form of representation for the regions and their residents, because there was the desire to want to display results that are relevant for one or more doctoral thesis. There was only done a descriptive analysis to make sure that every project partner will be the first to use the results for their doctoral thesis.

4 Measurement of the Valuation of Cultural Landscapes

This part of the doctoral thesis is about the method chosen for the valuation of the cultural landscapes in the selected regions. The first section describes some approaches that are used in the literature to determine a value in the case of monetary value. Additionally, the choice of the method is described. Properties and limitations are shown in the second part of this chapter. Following these, critics that can be found in the literature about using the approach of the WTP are presented. These critics were considered in the creation of the questionnaire and the aim was to avoid the mistakes that could be followed by the critics. The chapter ends with a short discussion of its suitability in practice.

4.1 Approaches, Methods and Techniques of the Valuation of Cultural Landscapes

Because of the public character of the landscape, it is tough for the participants to convert their landscape valuation into a concrete number (Hackl & Pruckner, 1999; Ryan & Spash, 2011,) without having an idea of a concrete value as comparison. There are different options to determine the value of a public good like a cultural landscape: indirect and direct approaches to measure the valuation. While indirect approaches assume that the consumption of the public good needs a purchase of another private good, the price of this private good is assumed to be the value of the public one. The direct approach asks the individuals directly about a value for a public good (Job & Knies, 2001; Karkow & Gronemann, 2005). For each approach a short explanation of two methods is presented: hedonic pricing and choice experiments for indirect measurements and travel costs and CVM as direct approaches.

Hedonic pricing is a method that deals with market outcomes because the sold good (e.g., a house) has qualities that are special to the market. The individual values the good by using different components of it (Pruckner, 1995). As example for this, Haab and Connell (2002) named the effects of air pollution (non-market value) on house prices (at least market-induced). The respondents are asked for different prices for an identical house with only a difference in air pollution, for instance. The price variation between the house with less or higher air pollution is interpreted as the value of this difference in the prices (Haab & Connell, 2002; Pruckner, 1995).

J. Matloch, *The Assessment of German Cultural Landscapes*, RaumFragen: Stadt – Region – Landschaft, https://doi.org/10.1007/978-3-658-21416-6_4

Choice experiments trace back to conjoin analyses used mathematical psychologies. By using choice experiments, the utility is confronted with the costs of a specific good. In this method the valuation is a derivative of different decisions of the participants regarding different scenarios of qualities of the analyzed good. For instance, when a study analyzes a landscape, there are different scenarios for this landscape regarding a higher degree of buildings in the landscape, a high degree of maintenance and care, a low degree of maintenance and care, with new roads crossing the landscape and the landscape if nothing would happen. The participant does not need to give a concrete value for the good; he or she just stated the preferences between two scenarios that are changing. Afterwards the researcher is able to determine a ranking of the different scenarios. One big advantage of this method is the possibility to consider the multidimensionality of environmental goods (Schmitz, Schmitz & Wronka, 2003).

Travel costs method is based on a specific good and the question to the sample is how far the sample would drive for a visit of this good. The value of this good is displayed by the distance and the driving costs to reach it (Burt & Brewer, 1977; Mitchell & Carson, 1989; Pruckner, 1995). This method could be seen as critical for the valuation of residents because the values get higher if the person lives far away, implying that the values for the residents is lower than for tourists (Bishop & Heberlein, 1979; Haab & Connell, 2002). This method also does not include changes in quality of the good over time (Bishop & Heberlein, 1979).

CVM[12] was especially developed to compare costs and benefits of environmental goods (Throsby, 2007; Venkatachalam, 2004) and to determine the value of protecting cultural landscapes (Job, 2003). This method traces back to the theory of economic value of a good (Hicks, 1939) and the utility model described by Mäler (1974). The utility model describes individual´s preferences for consumption of private goods with a given budget and has the aim to determine welfare for the population[13]. The risk of a failure through the price mechanism is still given as in all utility models (Marangon & Visintin, 2007). The determination of a consumer surplus (calculated by using the hypothetical value of the sample and the actual costs of the government for the public good) usually is the objective of a CVM analysis (Heyne, Meannig & Süßmuth, 2009; Job & Knies, 2001). Using this method, there are hypothetical markets created where the public good is tradable like a private good because a public good is not tradable in an actual market (Heyne et al., 2009; Job, 2003; Karkow & Gronemann, 2005; Mitchell & Carson, 1989; Pruckner, 1995). This supports

12 The history of the development of the method is described in detail by Mitchell and Carson (1989).

13 The formal derivation of this method is described by Hanemann (2001).

the determination of a value for the public good. The participants are asked about a maximum payment for an improvement of the good (WTP) or the minimum compensation of the good that remains constant or gets worse (willingness to accept, WTA) (Carson et al., 2001; Karkow & Gronemann, 2005; Pruckner, 1995; Venkatachalam, 2004). Both answers reflect the personal valuation, individual preferences and the indifference of a person regarding a change in the good in form of money (Haab & Connell, 2002; Mitchell & Carson, 1989; Pruckner, 1995; Ryan & Spash, 2011).

The CVM assumes that each individual is able to judge landscapes as valuable for their welfare (Throsby, 2003), but it could also used for different or similar purposes (Carson et al., 2001). For instance, there are other studies, using the values of events like the World Cup (e.g., Heyne et al., 2009) or for building a new sports stadium (e.g., Johnson & Whitehead, 2000). The most advantage of this method is the flexibility and simplicity (Howley et al., 2012).

The analyses use CVM because it is suitable to ascertain the individual's valuation of residents for different landscapes. It offers the possibility to illustrate a monetary value of the individual valuation, preferences and indifference concerning changes of the good (Haab & Connell, 2002; Mitchell & Carson, 1989; Pruckner, 1995; Ryan & Spash, 2011). This method makes it possible to examine the landscape perception as a whole while other methods examine the perceptions of specific landscape elements (Swanwick et al., 2007). Additionally, a comparison between different regions or landscapes is only feasible with a holistic view on the landscape. Additionally, the aim of this thesis is to determine a value of current landscapes instead of improvements. This might help the respondents to determine the value of the landscape, because they are able to go there right away. Furthermore, its use is seen as an appropriate approach because money has an almost equal value for everyone (Hossack & An, 2014).

Travel Costs are not adequate because usually residents are not traveling to the landscape like tourists do. Hedonic prices are not reflecting the assessment of the landscape in subjectivity of the residents and further, the prices for the houses are changing over time. The method of the choice experiments was not possible because of the possibility to access the landscape and the analyses of the perception of real landscapes in each region. CVM is also the most frequently used method to estimate a value of an environmental good in the literature (e.g., Karkow & Gronemann, 2005; Kim et al., 2015; Molina et al., 2016; Santos, 2001).

There are different CVM techniques to determine the WTP. One technique is the *bidding game*, based on an auction, where the sample needs to answer the question if he or she would buy a good for stated prices (Mitchell & Carson, 1989). The stated prices increase or decrease depending on the answer of the respondent. The game stops when the individual would be willing to pay the stated amount of money and this amount is interpreted as the valuation of the good (Haab & Connell, 2002). The *discrete choice* approach just asks yes or no questions with selected amounts of money (Haab & Connell, 2002). If the respondent is able to personally express an opinion on the price for the public good, the direct method is conducted (Haab & Connell, 2002). This is also known as the *open-ended* approach. In the *closed-ended* approach the respondent get different prices presenting by a scale (Völkner, 2006). A *payment card* uses a direct question approach by using a visual aid (Mitchell & Carson, 1989). The respondent has the possibility to choose stated amounts of money (stated values or between the stated values) included the "don't know" and "don´t answer" (Hackl & Pruckner, 1999; Howley et al., 2012; Mitchell & Carson, 1989). Usually, its answers are lower than these of closed-ended approaches (Hackl & Pruckner, 1999). Another technique asking for a monetary valuation is the *referendum* format. The sample is asked if they would pay a specific amount of money for an improvement for instance by using an entry fee (Boyle, 2003; Morrison & Dowell, 2013).

The doctoral thesis uses the WTP approach as an indicator for the perception of the current landscape. The data show that this treatment is more applicable than using the monetary value in the following the analyses. Because of the use as an indicator, during the analyses, WTP or valuation can be understood as synonymous with perception of the landscape.

4.2 Properties and Limitations of Willingness to Pay

Questions related to CVM can be asked in a personal or a written survey. Most frequently, the questions of interests are asked during a *personal interview* (Mitchell & Carson, 1989). As stated by Carson et al. (2001) personal interviews carry the risk that the interviewee answers in a way that is social desirable (Mitchell & Carson, 1989) or that they do not have enough time to come to a value for the landscape that could be adequate for them (Swanwick et al., 2007). This could result in an overestimation of the real WTP. Moreover, in personal interviews complex questions are often explained from the interviewer in their own words, which could distort the results. Therefore, other studies use a *mail survey* (e.g., Job, 2003; Kämmerer et al., 1996; Marangon & Visintin, 2007). Kaltenborn and Bjerke (2002) mixed both survey methods to get more answers. They just mailed the questionnaire to these people who were interested during the personal interview to fill a questionnaire. Carson et al. (2001) stated that respondents of mail surveys tend to be more interested in the good than the non-respondents (called sample nonresponse bias). On the other hand, all information needed has to be provided in the mail survey because the sample has no possibility to ask question about the analyzed good (Mitchell & Carson, 1989). Therefore, both methods of data gathering have advantages and disadvantages. Within the project, the partners decided to gather data by using a mail survey, because the aim was to avoid the bias of socially desirable answers.

The hypothetical *market* should be described as *plausible* as possible (Mitchell & Carson, 1989) – otherwise an information bias can occur (Ventakachalam, 2004). The interpretation of the results of a survey with incomplete information should be carefully because only if all information is given, the results could be interpreted in the right way (Carson et al., 2001). Therefore, it is necessary to give enough information about the public good even if the information is not complete (Throsby, 2003). For the question of the WTP, the questionnaire shows photographs of the landscapes in the examined regions that already exist and the sample were pleased to answer the question how much they were willing to pay for preservation, protection and design of the imaged landscapes, additional to the taxes they already paid (Fig. A 1). Hence, there was not a future development created or an improvement shown, the questionnaire tries to sustain the status quo of the good and used therefore current landscapes. This procedure could minimize the risk of a too hypothetical situation and therefore decrease the risk of a low external validity as described by Völkner (2006).

The questions in the questionnaire used a scale from 0 euros to more than 100 euros for determining the valuation of the landscapes in the regions (Fig. A 1). This scale should support the sample to find an answer of this question, but especially the risks of non-bidders might be relatively high (Spash, 2008). For this reason, the possibility of the value 0 euros was offered. But this approach (*closed-ended*) of the WTP leads to the problem, that the sample might give only vague answers or do not think about the real value of the landscape because the prices are already given (Bateman & Turner, 1992; Mitchell & Carson, 1989; Völkner, 2006). In contrast, the *open-ended* approach can have a high amount of zero WTP and some values above all others and could be positively skewed (Carson et al., 2001; Ryan & Spash, 2011; Völkner, 2006). But in the case of the open-ended approach there is no starting bias (the start of the scale can influence the answer of the sample) (Ventakachalam, 2004). Both approaches are criticized in the literature because of its validity, but they are still appropriate to estimate the value of a good because they are easy to implement and have lower costs as other approaches (Völkner, 2006). As stated by Hackl and Pruckner (1999) and Arrow et al. (2001), the format of the answers combined to some extent the closed-ended and the referendum format.

Hypothetical bias occurs due to the technique itself and is seen as one of the most problems of this method (Bateman & Turner, 1992; Mitchell & Carson, 1989). An attempt was made to minimize this bias by using real landscape in each region and asking what the residents of the region were willing to pay in additional taxes for the preservation, protection and design of their own landscapes. They did not need to assume a change in the landscape.

Aggregating WTPs for different landscapes in one single measure leads to an *overestimation* of the WTP, because theoretically each new good decrease the individual´s available income (Carson et al., 2001). The questionnaire stated explicitly that the amount of money would be additionally for taxes and that each cultural landscape should be rated independently (Fig. A 1). For the analyses, a dummy variable that was created showing if a person was willing to pay for at least one landscape – but this is not an aggregate like Carson et al. (2001) had talked about.

4.3 Critics of the Measurement

The question of the WTP is challenging for the respondents because they usually do not need to know how much they would pay for a specific public good that they have never bought before (because they do not have to). Therefore the statement of the amount of payment might not be *stable*. This can be tested through asking the question at different times and comparing the data (Bateman & Turner, 1992; Völkner, 2006). The questionnaire was designed to ascertain the recent situation in the regions where the participated people are living. Because of the relatively low response rate in some cases, it does not seem to be adequate to ask the same sample again later on.

Additionally, the respondents have *no incentive* to answering their *true* WTP (Völkner, 2006), even because the true WTP is typically an unobserved value of the good (Verbič & Slabe-Erker, 2009). By comparing the results of WTP and the true WTP by Verbič and Slabe-Erker (2009), the same variables displayed as significant. Therefore the following analyses do not make a difference between WTP and true WTP. But the true WTP has to deal with another problem: Usually, the respondent has stated in advance that the actual cost is lower than the WTP (Bishop & Heberlein, 1979; Ryan & Spash, 2011). This was proved by different examinations (Cummings & Harrison, 1994; Neill, Cummings, Ganderton, Harrison & McGuckin, 1994) and could also be the case in these analyses. It would lead to the conclusion that the WTP estimates might be a poor predictor for financial transactions (Campbell, 2007). But the aim was to identify the drivers of the valuation of the different landscapes instead of a comparison of the utility and costs, and therefore this problem might not be important to the following analyses.

The order of the goods (*order bias*) plays an important role for setting the WTP; the first goods on a list will be valued higher than a good in last place (Kahnemann & Knetsch, 1992). Therefore, Verbič and Slabe-Erker (2009) reminded the respondents of their limited income and that they had to distribute goods differently, as described in the utility model. The order bias could lead to lower amounts of money for the landscapes showing a second or third ranking (Mitchell & Carson, 1989) and might explain why the amount of money is higher for water (first position) than the other two landscapes (second-last and last position), and why the landscapes at the bottom of the list might get lower valuations.

Embedding effect is present if the value of the WTP is widespread because some respondents see the good in itself while others see the good as part of the whole (Kahnemann & Knetsch, 1992; Pruckner, 1995; Venkatachalam, 2004). Kämmerer et al. (1996) proved this result by finding a higher WTP for major lands. Given that the presence of the embedding effect affects the validity, it seems to be necessary to consider the internal validity (Arrow et al., 2001). The internal validity for gender (each region) and age (LPLD) was given previously in Chapter 3.3.1. The content validity was verified for bond to nature and region and the different landscapes. Convergent validity could not be tested, therefore it could be stated that at least a medium validity is given. It can be concluded that the answers are at least representative for gender in the analyzed regions as claimed by Mitchell and Carson (1989).

The statement of a WTP seems to be *demand-based*. If the demand of a person is higher, they might be willing to pay more than a person with lower demand (Völkner, 2006). The questions about the bond to nature and to the region are able to control for this, but if these variables are adequate to find out the demand of the sample is not clear.

It might be the case that the sample has answered the question of WTP different from the own perception because they think the public good will get a price in the future and the sample tries to avoid this by saying he or she is willing to pay nothing (McFadden, 1998, Völkner, 2006) or answered it in a social desirable way (Kim et al., 2015). This worry already was discussed in the pretest of the survey. Another alternative could be the perception of unfairness regarding the allocation of the money by the government for the protection of the environment and the landscapes (Jorgensen, Syme, Bishop & Nancarrow, 1999). As payment method named at least 2 % of the whole sample actively that the government should pay for the landscapes instead of the population. Hartje et al. (2003) and Job and Knies (2001) also found a high proportion of respondents who thought that the government is responsible for the preservation and protection of landscapes. This could be the case in these analyses as well because the questionnaire reminded the respondents of German government taxes. Verbič and Slabe-Erker (2009) reminded them of the challenges in protecting landscapes and Howley et al. (2012) spoke of the importance of agriculture landscapes on nutrition. The question in the implemented survey was if a person would be willing to pay extra money for the landscape. Carson et al. (2001) described it as desirable when the individual would think about an actual situation. But *strategic behavior* could lead to a lower WTP and could also explain the high amount of 0 Euros and protest bids[14]. Tests of strategic behavior in a CVM survey have been developed, but could not yet be proven (Cummings & Harrison, 1994; Pruckner, 1995). This bias could lead to the argument stated by Schmitz et al. (2003) that the population does not aspire to the best quality landscapes.

According to Sagoff (1988, cited by Bateman & Turner, 1992), a high proportion of "protest voters"[15] exposes weaknesses in the research. *Protest bids* were almost 20 % in each region or landscape and these bids were left out of the analyses, although these people might value the landscape in the same way than some of the zero bidders do (Jorgensen et al., 1999), or without giving any value (Hartje et al., 2003). It could not be controlled for the exact meaning behind a protest or a zero bid could not be ascertained.

Another argument could be *free-riding,* meaning an individual might be selfish and therefore assume that others will give money to the public good and hence it does not matter they give any money (Bateman & Turner, 1992; Carson et al., 2001; Samuelson, 1954). Carson et al. (2001) argued because the sample just includes the value of the landscape for their own generation instead of seeing the landscape as a resource also for future generations it might be a lower WTP.

The aspects of strategic behavior and free-riding would more obvious if the focus is on the *payment methods* questions in the survey. Most of the respondents (almost 63 %) answered they would like to pay the actual amount of WTP as donations. Almost every fourth respondent of the survey of Kämmerer et al. (1996), Hartje et al. (2003) or Job and Knies (2001) also voted for donations. This method does not seem to be adequate for protection, preservation and design of cultural landscapes because there is no constant monetary value for the landscapes and only a few individuals would actually pay. More interesting are the second common answers: entrance or parking fees (both 31 %). Entry fees

14 Spash (2008) classified protest bidders as bidders of zero Euros but for other reasons than the factual value of the environment. The amounts of respondents who protested selecting any of the given options do not have the same motivation as respondents who voted for zero WTP.

15 These are people who are stating nothing instead of using zero.

were also the second most given answer in the analyses of Hartje et al. (2003). This could be a possibility for continual financial support of the landscapes.

The *comprehensibility* is one of the most problems in a CVM examination. All respondents have their own life experiences and educational background and therefore understand questions in different ways (Mitchell & Carson, 1989; Spash, 2008) and therefore the valuation of a landscape could only be a subjective valuation of each individual (Campbell, 2007; Job & Knies, 2001; Meinig, 1979; Schein, 1997). This subjectivity could be observed in the discussion with the project partners: while one person understood the question, another one asked about the reason of this question or did not understand the context of this question.

Throsby (2007) stated that every CVM has an information failure because all information necessary for the valuation could not give (e.g., preferences) and not every aspect can be expressed by a monetary value. This could be happened in the questionnaire as well, because people value a landscape for different reasons (Kaltenborn & Bjerke, 2002). For instance, motives or attitudes of each individual seem to have an influence on the WTP (Davis, 1963; Ryan & Spash, 2011). As Morrison and Dowell (2013) and Kim et al. (2015) found, the visits of the landscapes could also be important for the valuation, because only an experienced good could be valued (Carson et al., 2001; Hartje et al., 2003; Karkow & Gronemann, 2005).

4.4 Practical Application of the Method

In the first place, there was done a literature review to understand how a valuation of landscape could be done. After deciding for the WTP approach, the first questions were developed. There was to place two questions into the questionnaire: one about the WTP and one about the WTA. These questions were first presented in a monthly meeting with the other scientific partners involved in the project. First comments about the applicability of the WTA approach were done from the scientific partners in other disciplines. The comments were about the understanding of the question for ordinary people.

After these comments, both types of question were changed before it was presented to the partners that are working for the regions. They had the possibility to comment on each question. In the first place, these partners crossed out both types of question. These partners argued that they were inapplicable to ordinary people. Another important argument was that the partners were afraid that their residents could think that the public administration would first asked about some values for the landscapes and afterwards implement some additional costs for the preservation, protection and design of the landscapes. This argument was not discussed with the scientific partner, but it was understandable for each of the partners. The question with respect to the WTA was how to come up with answers without any knowledge of the topic.

Therefore, the question about the WTA was removed from the questionnaire. It was again revised since one of the aims of the researchers still was to determine the value of the landscapes in the region. The question was further discussed with each region separately to understand what the regional conditions were and why the partners had doubts about asking the about the WTP.

The most discussed issue with the partners of the regions was the topic of taxes. They thought that people were not willing to give any additional money for their landscapes because of the taxes they already were paying. The taxes seemed to be a sensible topic in the regional contexts. It was hard to make them understand that the question is only a hypothetical one and that it is not the goal to implement additional taxes for improving the regional landscape. Furthermore, the partners feared that people working together with them could be put off by this question. This would lead to bad press for the project, which was only one year in progress when the questionnaire was deployed.

After some discussions, the partners in the regions agreed that the question could be asked. But they attached some restrictions regarding the wording of the question: The text needed to be included that the stated amount of money for the preservation, protection and design of the landscape is additional to the taxes the residents are already paying (Fig. A 1). However, there needed to be a disclaimer before the question was asked that at the moment these were being funded by the government. Additionally, this text described some examples for what the government uses the money for in cases of preservation, protection and design of the regional landscapes. The last sentence of this description verbalized that an additional amount of money from the population could be done with further investments for preservation, protection and design of the landscapes.

This process shows that the practical application might be more difficult than a researcher would think. In some cases, research might be too theoretical to understand the hardships that are present in practice. The work with partners in the area of public administration reveals these hardships. Research must take into account that such partners are politicians who need to be careful with what policies they support. In this case, research might benefit from a discussion with partners working in the actual settings/locales, because for the respondents the questions might then be easier to understand.

5 Variables of Interests

The most frequently used questions and their manifestations are presented here in more descriptive detail than in the analytical chapters. These evaluations are just to get a feeling about the questions and the possible answers.

5.1 Willingness to Pay

For the perception of the landscapes the *Question 4.2* of the survey (Fig. A 1) was used. The question asked what amount of money a person would be willing to pay privately and monthly (in addition to the taxes they already pay) for the preservation, protection and design of the landscapes that are presented in the pictures below. The landscapes are typical landscapes that were photographed in each region and so the landscapes differ by region. The questionnaire showed different photographs for the region of LPLD because the region is bigger than the others and the people might have a relationship with the landscape they know. Given that the regional landscapes differ, the results are presented by landscape type. The possible answers were identical in each region: Respondents had the possibility of answering with one of the following: "0 euro", "1–5 euros", "6–10" euros, "11–20" euros, "21–50' euros, "51–100 euros", "more than 100 euros".

The WTP for the water landscape was asked in each region. The differences between the regions are minimal (Fig. 5-1). In the LNWM region and STB 54 % of the sample are not willing to pay any money for the preservation, protection and design of the landscapes; the amount is a bit higher for LPLD. Between 29 % and 23 % of the respondents, depending on the region, are willing to give at least 1 to 5 euros for the water landscape. This indicates that in the region LNWM 17 % and in the region of LPLD 18 % of the sample is willing to give more than 5 euros monthly. This amount of people willing to give more than 5 euros monthly for the landscapes is with 20 % even higher for the region of STB.

J. Matloch, *The Assessment of German Cultural Landscapes*, RaumFragen: Stadt – Region – Landschaft, https://doi.org/10.1007/978-3-658-21416-6_5

Fig. 5-1: Valuation of the water landscape (in percent)

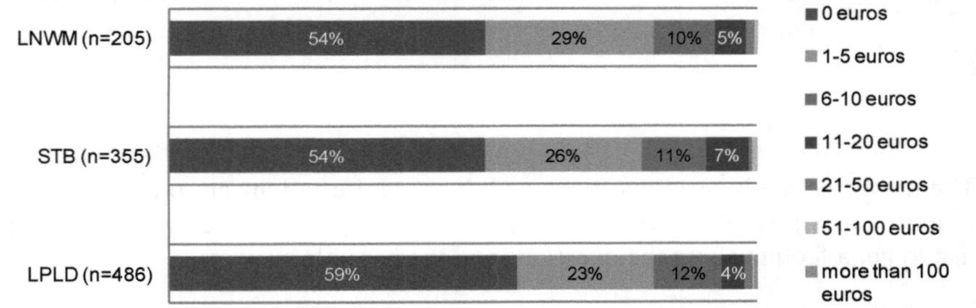

Source: Resident survey, own illustration.

The city landscape could also be found in every region and was therefore asked in each region. The percentages in each region show some similarities. Compared to the water landscape before, fewer people are willing to give any money for the preservation, protection and design of the landscape (66 % for the region LWM to 72 % for the region of STB, Fig. 5-2). Most willing to pay is the region of LNWM: about 20 % are willing to give 1 to 5 euros monthly and additional to the taxes for the landscape preservation. At least 14 % of the sample in this region is willing to pay more than 5 euros monthly. The other two reasons are less willing to pay: 17 % (STB) and 18 % (LPLD) are willing to pay 1 to 5 euros for the landscape protection for their regional city landscapes, while 11 % of the respondents of STB and 12 % of these of LPLD are willing to give more than 5 euros for the city landscape.

Fig. 5-2: Valuation of the city landscape (in percent)

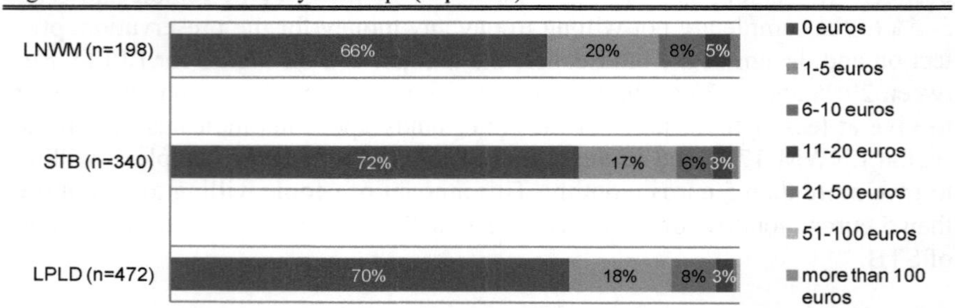

Source: Resident survey, own illustration.

The WTP for the village landscape is a bit higher than the WTP for the city landscape and a bit lower than for the water landscape. In the region of LNWM is 66 % of the sample not willing to pay any money for the village landscape, while this amount is lower for the other two regions (Fig. 5-3). For these two regions, 58 % respectively 57 % of the sample is willing to pay. Differences can also be seen in the amount of respondents who answered that they would be willing to pay 1 to 5 euros: While it is for the region of STB and LPL 25 % each, in the sample for LNWM only 19 % would be willing to pay this amount. Consequently, the amount of people willing to pay more than 5 euros monthly for the village landscapes in their region is higher for these two regions (17 % for the region STB and 18 % for LPLD). For the region of LNWM, 15 % is willing to give a higher amount of money than 5 euros monthly.

Fig. 5-3: Valuation of the village landscape (in percent)

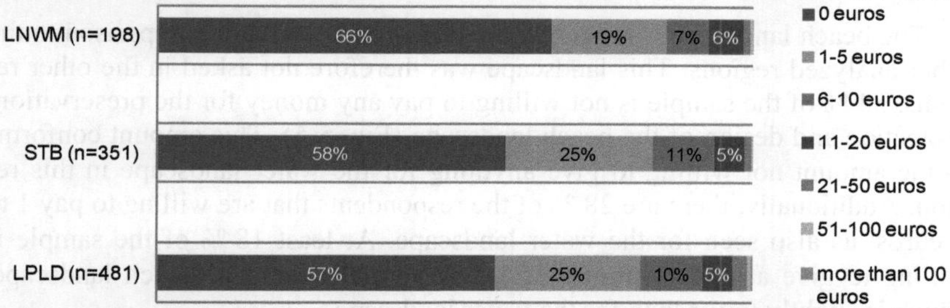

Source: Resident survey, own illustration.

The landscapes covered by forests, hedges and groves were identified to be only important in the regions of LNWM and LPLD. Therefore, Fig. 5-4 does not display any values for the region of STB. It can be seen that the region of LNWM showed a smaller amount of respondents not willing to pay any money for this landscape (52 % LNWM, 57 % LPLD). The amount of people willing to pay 1 to 5 euros is also higher for the region of LNWM with 33 % compared to LPLD with 26 %. Even it seems like the sample of LNWM is willing to give more for this landscape type, it displays that the respondents of LPLD are willing to pay slightly higher amount of money that is higher than 5 euros monthly (17 %) than the sample of LNWM answered (15 %).

Fig. 5-4: Valuation of the forests, hedges and groves (in percent)

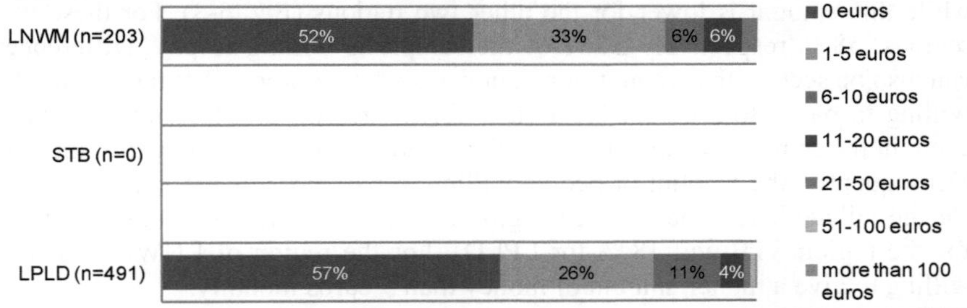

Source: Resident survey, own illustration.

The beach landscape is unique to the region of LNWM in comparison to the other analyzed regions. This landscape was therefore not asked in the other regions. 54 % of the sample is not willing to pay any money for the preservation, protection and design of the beach landscape (Fig. 5-5). This amount conforms to the amount not willing to give anything for the water landscape in this region. Additionally, there are 28 % of the respondents that are willing to pay 1 to 5 euros, as also seen for the water landscape. At least 18 % of the sample is willing to give a higher amount of 5 euros for the regional beach landscape, which is slightly lower than for the water landscape.

Fig. 5-5: Valuation of the beach landscape (in percent)

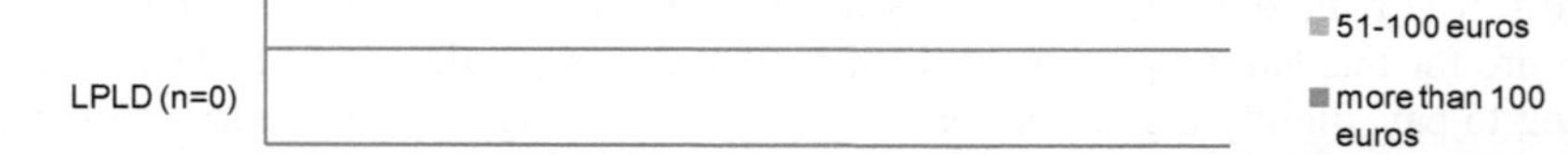

Source: Resident survey, own illustration.

The farmhouse landscape can be found uniquely in the region of STB and was therefore not part of the questionnaire in the other regions. It can be observed that 64 % of the sample of this region is not willing to pay any money for the preservation, protection and design for this landscape type (Fig. 5-6). 19 % of the respondent of the region STB are willing to give at least 1 to 5 euros monthly for the protection of this unique landscape. Almost the same amount of people willing to give 1 to 5 euros is also willing to give more than 5 euros monthly – in additional to current taxes.

Fig. 5-6: Valuation of the farmhouse landscape (in percent)

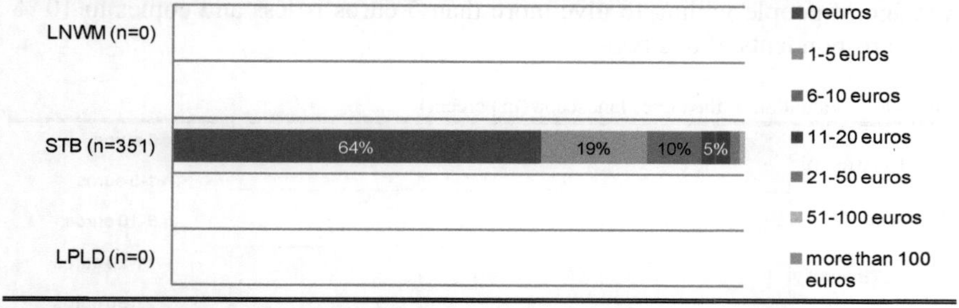

Source: Resident survey, own illustration.

As the farmhouse landscape, the marshland is also unique in the STB and therefore Fig. 5-7 does not show any values for the other two regions. The amount of respondents that are not willing to pay any money is slightly lesser than for the farmhouse yet still 62 % – albeit, more people are willing to give 1 to 5 euros for this landscape type (26 %). With the higher amount of giving 1 to 5 euros, the amount of people giving more money decreases. While it is 17 % of the sample for the farmhouse landscape, it is only 12 % for the marshland.

Fig. 5-7: Valuation of the marsh land (in percent)

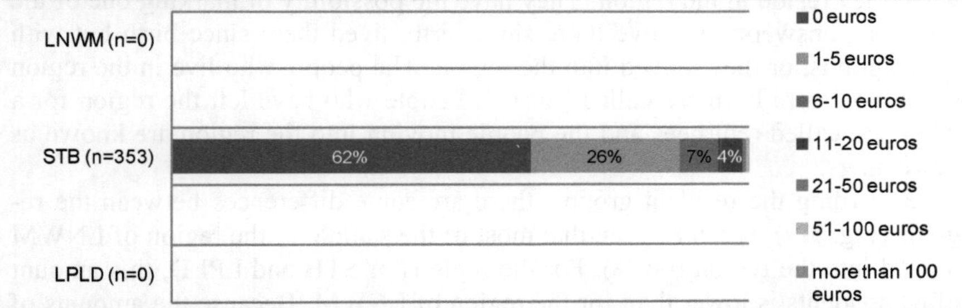

Source: Resident survey, own illustration.

The dunes might be interpreted as kind of beach; but the reason they are not is because the LPLD region does not have direct access to the sea. Therefore, the dunes landscape is unique for the region of LPLD. It can be seen that almost 67 % of the sample in this region is not willing to pay any money for this landscape type (Fig. 5-8). This is the second highest amount in comparison to the other examined landscape in this region. Almost one quarter is willing to give 1 to 5 euros monthly additional to the taxes for the preservation protection and design of this landscape. Due to the two relatively high amounts, the percentage of people willing to give more than 5 euros is less and comes to 10 % of the respondents of this region.

Fig. 5-8: Valuation of the dunes landscape (in percent)

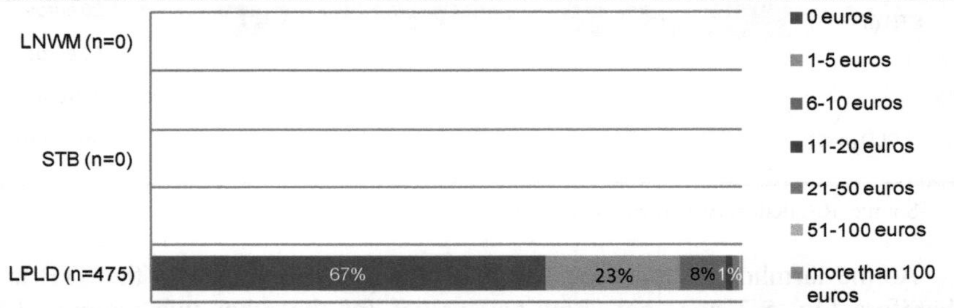

Source: Resident survey, own illustration.

5.2 Newcomers, Returnees and Locals

One of the analyses of the later chapters concentrates on the WTP in relation to different resident groups. The residents groups are divided using *Question 5.13* of the questionnaire (Fig. A 1). In this question the respondent is asked since when they reside in the region. They have the possibility of marking one of the following answers: they live there since birth, lived there since birth but with interruptions, or they moved into the region. The people who live in the region since they were born are called 'locals'. People who have left the region for a while are called returnees and the people moving into the region are known as newcomers.

Regarding the resident groups, there are some differences between the regions (Fig. 5-9). It can be seen that most of the sample of the region of LNWM moved into the region (66 %). For the regions of STB and LPLD, this amount of respondents is lower than for the region of LNWM. Because the amounts of returnees are almost similar in each region, the amounts differ between the rates of locals in the sample. While it is 22 % for the region of LNWM, it is almost 10 % higher for the region of LPLD.

Fig. 5-9: Duration of living in the region (in percent)

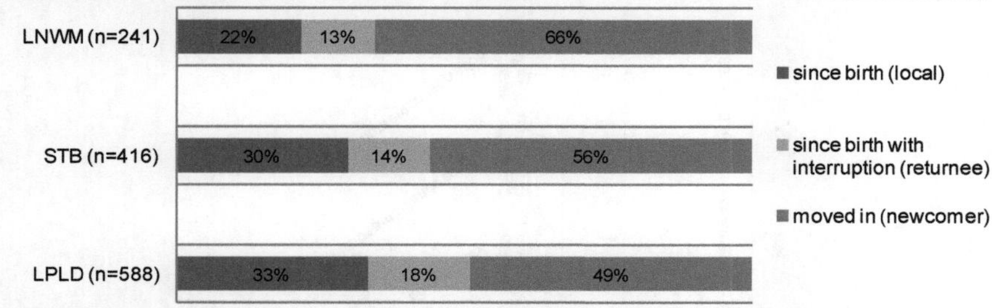

Source: Resident survey, own illustration.

5.3 Volunteer Work

One of the chapters integrates the volunteer work as for of valuation for the landscape. More than one question was asked that are transformed into one so that we present more than one question in this part of the variable description.

The first question that is used is about a future involvement of the respondents in the specific areas of volunteer work (Fig. A 1). The areas of volunteering are animal protection, art/culture/museum, conservation of the environment, education/ kindergarten/ school, fire or ambulance service, local history, monument conservation, place development, politics, religion/ church, society and sports/ exercises. Because of the possibility to have more than one involvement, the respondents had the option to mark more than one area of their involvement.

It can be seen that the most chosen area for volunteering in the future is in every region the society (Fig. 5-10). The respondents of STB show the highest rate of being willing to volunteer in this area, followed by the other two regions with almost the same amount of willing respondents.

The second highest area of volunteering chosen by the region of every region is the area of the conservation of the environment. The area chosen as third option by the respondents is similar for the region of LNWM (20 %) and LPLD (18 %) with animal protection, but for the region of STB a higher amount of respondents would be willing to do some volunteer work in the area of education/ kindergarten/ school (17 % to 16 % for animal protection).

In the three analyzed region, the less favored areas are local history, fire and ambulance services or monument conservation. The ranking is the same for the regions of LNWM and LPLD, but for the respondents of STB, local history and fire or ambulance services need to be changed.

Fig. 5-10: Future involvement of the respondents, sorted by area of volunteering (in percent)

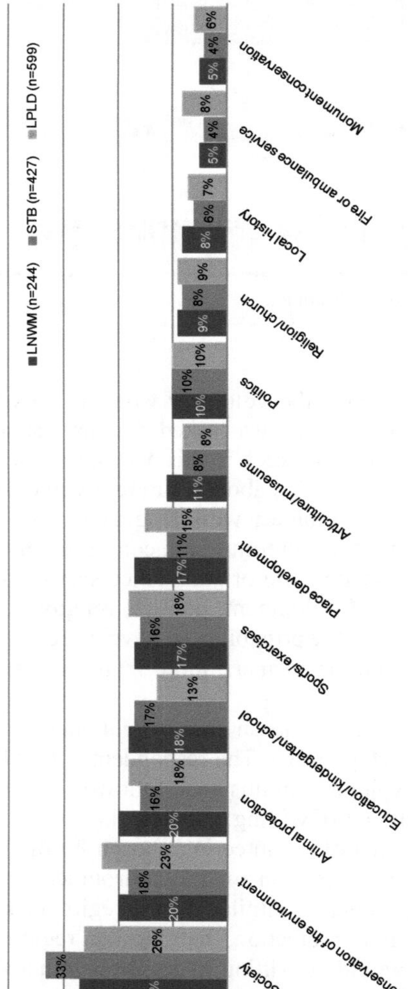

Source: Resident survey, own illustration referring to Matloch et al. (2016).

Question 3.8 is used to match the WTV in one of the named areas with the true willingness of involvement, because this question asked about the hypothetical amount of monthly hours the respondent would like to perform future volunteer work (Fig. A 1). One of the following answers had to be marked by the respondent: 1 to 2 hours, 3 to 5, 6 to 10 hours, 11 to 20, more than 20 hours, or no interest.

In each region, almost 12 % of the respondents had no interest in any future involvement (Fig. 5-11). The differences between the regions are small. The amount of people willing to work 3 to 5 or 6 to 10 hours monthly is about one-third each. Even the respondents who were willing to invest more than ten hours a month for an involvement is about 15 % in each region.

Fig. 5-11: Overview about the possible hours of a future involvement (in percent)

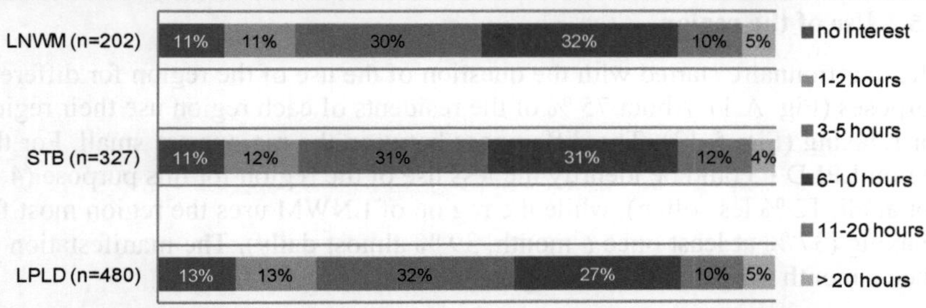

Source: Resident survey, own illustration.

5.4 Life Satisfaction

Life satisfaction was the last question in the questionnaire (Fig. A 1). Most of the residents in each region show a life satisfaction of 7 or higher (Fig. 5-12). The highest satisfaction rate is displayed with an 8, which is relatively high compared to other results. In the case of the Better Life Index of the Organization for Economic Co-operation and Development (OECD), the German population rated their satisfaction on average with 7, which is also higher than the average for all OECD country rankings (Organization for Economic Co-operation and Development (OECD) n.d.). In the SOEP, the sample also rated the life satisfaction of about 7 (SOEP Group 2015b). Only a few respondents said they were not satisfied with their life in the region (about 5 % in each region).

Fig. 5-12: Overview about the life satisfaction (in percent)

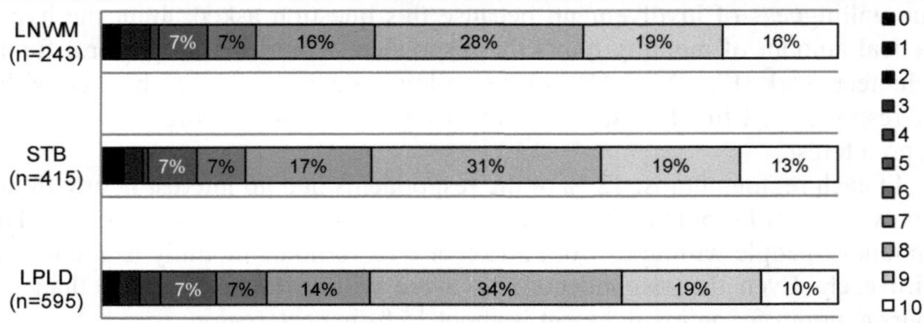

Source: Resident survey, own illustration.

5.5 Use of the region

The questionnaire started with the question of the use of the region for different purposes (Fig. A 1). About 75 % of the residents of each region use their region for relaxing (Fig. 5-13). The differences between the regions are small. For the region LPLD it could be identify the less use of the region for this purpose (4 % not at all, 12 % less often), while the region of LNWM uses the region most for relaxing (37 % at least once a month, 39 % almost daily). The manifestation of once a month is similar in each region.

Fig. 5-13: Overview regarding the use of the region for relaxing (in percent)

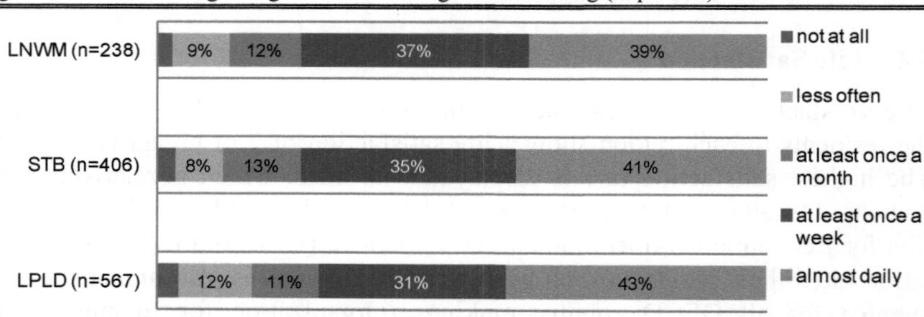

Source: Resident survey, own illustration.

Compared with the relaxing use, the region is not as much used for sports activities (Fig. 5-14). About 60 % in each region marked that they would use the region for sports once a week or more often. It can be observed that the respondents of the region of LNWM do not use the region as much as other regions their region for sportive purposes (15 %, about 10 % in the other regions).

Fig. 5-14: Overview regarding the use of the region for sports (in percent)

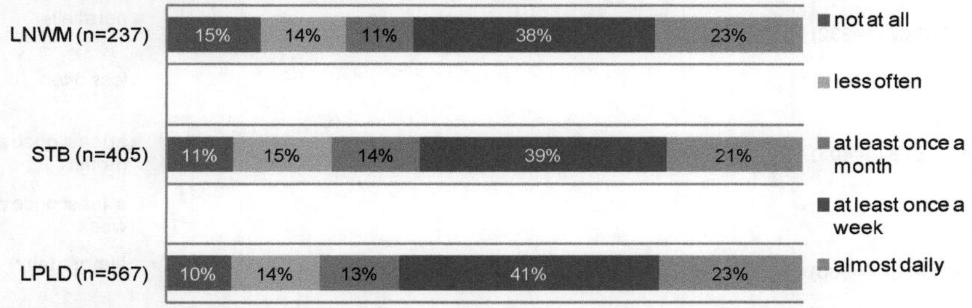

Source: Resident survey, own illustration.

The educational purpose shows that the use could be identified most often in the category of less often. The regular use is not distributed as similar as in the other purposes described above. While the almost daily use is about the same percentage in each region, there are differences between the use of once a week or once a month. The respondents of LNWM and LPLD show similar percentages of use, but the region of LPLD shows the lowest rate. These are almost the same for the use of once a month. Regarding the not using at all option, the percentages are almost equally distributed.

Fig. 5-15: Overview regarding the use of the region for educational purposes (in percent)

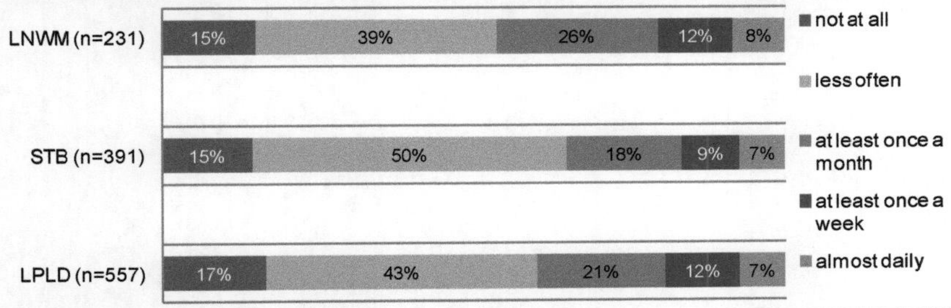

Source: Resident survey, own illustration.

The last use that will be analyzed in one of the chapters later on is the use of cultural purpose (Fig. 5-16). Compared to the other fields of usage, most of the people are using the region for cultural purposes less often. This result is similar to the educational purpose. Less people are using the region at least once a week or almost daily (between 12 % for LNWM and 6 % for STB).

Fig. 5-16: Overview regarding the use of the region for cultural purposes (in percent)

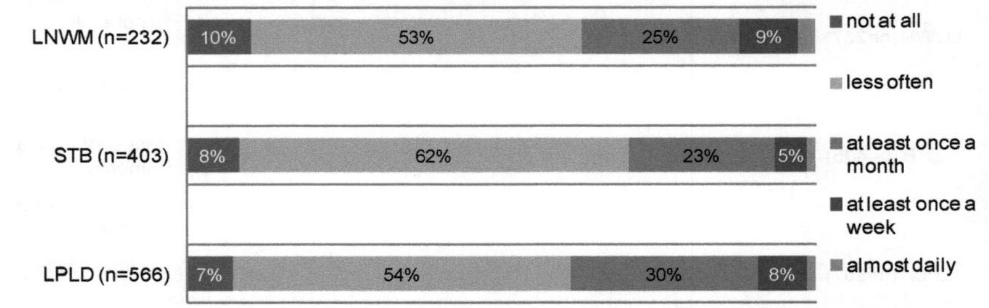

Source: Resident survey, own illustration.

6 Residents´ Perception of Their Everyday Cultural Landscapes – Consistencies and Disparities in Three Regions in Germany

This chapter is about the perceptions of the residents in the examined regions with regard to the literature. Its goal is to identify consistencies and disparities regarding the literature by analyzing the perception of the residents in the selected regions.

6.1 The Importance of Landscape Perception

Landscapes permanently change over time in most cases unconsciously (Antrop, 2005), because the residents experience their regional landscapes day-to-day. These unnoticed shifts in the appearance of landscape particularly concern everyday landscapes that are simply available in many regions (Groth, 1997). Additionally, humans start valuing special places from afar (Szerszynski & Urry, 2006) and that excludes in most cases everyday landscapes. But even if this change might not be recognized in the beginning of the change, the visual character might be change and this could affect the perception of the landscape (de Groot & van den Born, 2003; Tveit, 2009).

Following this development of landscapes sensitivity, this paper analyses the perception of everyday cultural landscapes by the population in their corresponding region. Three questions should be answered: First, how does the resident population value different cultural landscapes in their respective region? Second, what are the differences in valuation for the same landscape types between different sub-regions of the MR HH? Third, which factors have an influence on the perception of the different landscapes in each region?

The approach of this paper will be the comparison of the drivers for perceptions first of each everyday cultural landscape (water, city and village landscapes) in three selected regions and second the comparison of these landscapes across the selected regions. The analysis does not address to find an adequate economic value for landscapes, but uses the concept of ascertain monetary values as an indicator for perception of a landscape. The target groups will be a sample of residents in each region, because the aim will be to learn something about the role of the landscapes in their lives and the perception of their regional landscapes. This is important not only for research but also for the future

J. Matloch, *The Assessment of German Cultural Landscapes*, RaumFragen: Stadt – Region – Landschaft, https://doi.org/10.1007/978-3-658-21416-6_6

development of the landscapes (Antrop, 2005) and how regional planners and managers could ensure an effective regional management of these landscapes (Vos & Meekes, 1999). This understanding could help these planners and managers to learn how the perception of their landscapes could be strengthening (Swanwick et al., 2007) and thereby the region.

The adding to the existing literature will be an analysis of everyday cultural landscapes in Germany that are real and accessible in three different regions. A lot of research used scenarios and not the current landscape. Until now, there was no comparison of similar everyday landscapes in different regions found that could deepen the understanding of the drivers for the valuation of these landscapes. The preferences[16] for different landscapes of residents is the object of interest in this paper, because these people are living next to these landscapes and are those who use and shape them in their everyday life.

The most interesting variables of the examination are the variables that are not often used in the literature before but these might be important indicators of the perception of the landscape of an individual[17]. One could assume that people with a strong bond to nature are also more involved in the support of landscapes because this support should protect the nature and the individual perceive the landscape different than people with a lower bond because of this bond. Another analyzed and interesting variable is the bond to region. The perception of the landscapes and the identity of individuals are close to each other and if the landscapes are part of the individual's identity the perception of the landscape might be even higher. Additionally, a more close-to-nature living environment could be important to the perception of the landscapes around the individual. Likewise, the perception of the landscapes could differ among people who moved into the region or moved back into the region, since both groups have seen different landscapes and the landscape might be a reason for moving.

In the following chapter, the theory of the landscape perception is provided. Next, the findings of the literature are presented and the developed hypotheses, derived from the conceptual part, are described. Following is the description of the variables and the methodology. This part also contains some descriptive analyses. The statistical analysis is done using a logistic regression model because of the indicator character of the variable of WTP. The empirical results are described next and the hypotheses are tested of approval or rejection. It ends with a discussion of the findings and a short conclusion about potential future research.

---

16 Landscape preference means that an individual likes one landscape better than another one (Swanwick, 2009).

17 See Chapter 6.2.3 for an explanation why these variables are important.

6.2 Everyday Cultural Landscapes and Its Perception

First there is a short description of the theoretical background for landscape perceptions as well as findings in the literature concerning the perception of cultural landscapes. These constructs are important to deduce the hypotheses in the last part of this chapter.

6.2.1 Perception of Landscapes

Landscape quality assessment[18] can be divided into two dominated approaches: the expert and the perception-based approach (Daniel, 2001). While the expert approach is highly used in environmental management practices, the perception-based approach is mainly used in research (Daniel, 2001). Because of the described aim of this research above and the use of quantitative data, some theories of the perception-based approach will be described.

In general, landscape perception refers to “the psychology of seeing and attaching meaning to landscape” (Swanwick et al., 2007, p. 4) with the use of every sense the individual has (Tuan, 1990). Interests of landscape perceptions exist in various disciplines (e.g., psychology, geography, environmental science) and all disciplines have developed their own concepts and theories about the perception (Swanwick, 2009; Upham et al., 2009; Zube, Sell & Taylor, 1982).

On the psychological and sociological perspective, the construct of attitudes seems to be an important construct to explain preferences (Upham et al., 2009) but it is complex and uses interpretations by each individual (Swanwick, 2009; Swanwick et al., 2007). Usually, this construct is hypothetical and involves knowledge, emotions and behavior, and varies in its intensity. It can change through information gaining and forms of learning (Upham et al., 2009, Zajonc, 1980). Attitudes might therefore be helpful to explain differences in perceptions by individuals (Upham et al., 2009). In addition to the attitudes, there are those who label attachment, environmental values and the ideas of morality as important for the interpretation of the landscape and how its perceived (Upham et al., 2009). In the following paragraphs, a short description of some selected perception-based approaches is given.

Zube et al. (1982) identified four paradigms for landscape perception that could be used for a categorization of these concepts and theories. The expert paradigm uses the landscape valuation by trained observer while the psycholog-

18 Daniel (2001) distinguishes between assessment and valuation. He describes assessment more as guess of a number of elements (for instance trees in the forest) and the valuation more as a concrete price. Because the price is used as an indicator, this research uses these terms synonymously.

ical paradigm is about the perception of the general public or selected groups of the population. The cognitive paradigm tries to identify a human meaning of the landscape and the last paradigm is the experimental one that consists of the landscape values based on the interaction of human and landscape and its experience. All the identified paradigms could be found in concepts of humans as well as in landscapes and could therefore help to anticipate the outcome of the interaction (Zube et al., 1982).

Landscape preference theories are more general ones and used as basis for more specific theories (Zube et al., 1982). Normally, landscape preferences are about why people like one landscape better than another one and also about the reason for this liking (Swanwick et al., 2007). These theories itself are about the relationship and interaction of humans and their landscapes (Kaplan, 1987; Tveit, 2009) and the expected outcome of this interaction (Zube et al., 1982). It includes individual contexts like past experiences or socio-cultural aspects (Zube et al., 1982). Generally, perceptions of landscapes are formed by visual stimuli, especially by non-experts in valuing landscapes (Appleton, 1975; Zube et al., 1982). But these stimuli could differ from individual to individual (Aoki, 1999; Kaplan, 1987). Most of the research using this general theory tried to find some indices of preferences (Kaplan, 1987). Identified drivers for preferences are gender (e.g., Macia, 1979), age (Balling & Falk, 1982) and the academic background (e.g., Kent, 1993), but also the familiarity of the landscape (Kaplan, 1987; Kent, 1993). This theory could be assigned to all paradigms, depending on the object of examination. More specific theories of landscape preferences are the prospect-refuge theory (Appelton, 1975), the landscape aesthetic theory (Ulrich, 1983) and information processing theory (Kaplan, 1987).

The prospect-refuge-theory was first developed to find preferences for landscapes in paintings (Appelton, 1984). Because of the argument that all individuals observe the environment in a similar way, the prospect-refuge theory was derived from the animal predator behavior (Aoki, 1999). The theory itself was developed using concepts of behavior (Appelton, 1984). Prospect within this theory refers to an overview of the landscape or the environment and refuge refers to a save heaven (Aoki, 1999; Kaplan, 1987). Environmental perceptions were identified as important factor for appropriate behavior in the landscape. Therefore the developed theory were modified and found application in usual landscape perceptions (Appleton, 1984). The prospect-refuge-theory could be assigned to the cognitive paradigm (Zube et al., 1982).

For the landscape aesthetic theory, affect or emotion seems to be the central factor for experience with and behavior in the environment for an individual and thus for the landscape preferences (Ulrich, 1983; Zajonc, 1980). The aim of this theory is explaining why a landscape causes certain emotions in an individual (Ulrich, 1983) and is to some extent an extension of the theory of Appleton (Lothian, 2014). The first affect by recognizing a landscape is a general emotion like interest of the landscape and needs only a little information for evolving these emotions (Ulrich, 1983; Zajonc, 1980). It depends on how strong the first affect is – whether an individual is familiar with the landscape or not (Zajonc, 1980). Because of this first moment and its emotions and the following evaluation, the individual behaves in the landscape or describe it to others (Ulrich, 1983; Zajonc, 1980) and therefore observes it. It was assumed that complexity might be influence the first affect of the individual, because complexity arouses interest. But analyses found low preferences for landscapes with high complexity and concluded that this could not explain preferences for different landscapes (Kaplan, Kaplan & Wendt, 1972; Ulrich, 1983). Additionally to complexity, landscapes with natural elements generally tend to be more preferred than man-made landscapes without any vegetation (Ulrich, 1983). The landscape aesthetic theory could be placed into the psychological, cognitive and experimental paradigm of Zube et al. (1982).

The focus of the information processing theory is the gain of information and the possibility of having an informational disadvantage by not gaining all information about the landscape that could be available (Kaplan, 1987). It includes the idea that individuals need to make sense of the environment around them and therefore need to interpret it (Stamps, 2004; Zube et al., 1982) and uses the learning process changing attitudes described by Upham et al. (2009). Processing the information concludes with the preference judgment of the individual (Kaplan, 1987). As people are usually pursuing to additional information (Appleton, 1975) an information gain could be used to influence the perception of the landscapes (Zube et al., 1982). In general, people tend to prefer natural environments over build environments (Kaplan et al., 1972), as mentioned before by the landscape aesthetic theory. This theory could be assigned to the cognitive paradigm (Zube et al., 1982). A review, made by Stamps (2004) did find heterogeneity of the viewed results and therefore suggest more research to prove the theory.

A newer theory by Nohl (2009) describes that in future sustainability would be influencing the perception of landscapes. Constructing on this, the author identifies four different modes that determine the perceptions of landscapes: beauty, fascination, interest, and prosaic experience. These modes are to some extent co-determined by the type of the landscape (Nohl, 2009). The mode beauty is expressed by the balanced and harmonic order of landscapes (Ritter, 1962) and has a meaningful structure. The individual is able to comprehend the landscape quickly as for instance in traditional cultural landscapes (Nohl, 2009). This description of the beauty modus is similar to the making sense that is stated by Kaplan (1987) to build landscape preferences (Nohl, 2009). The modus of the fascination is about the nature that is not changed by any human impacts (Nohl, 2009). The modus of interest depict landscapes with architectonical elements and natural elements and is often confusing for the observer (Nohl, 2009). These landscapes are usually incomprehensible as a whole – even if the individual knows that there is coherence within the processes and exactly this is attractive for the individual (Nohl, 2009). This modus is often found in periphery landscape where the architectonical structure and natural elements come together (Nohl, 2009). Both experiences modes can also be find rudimentarily by Kaplan and Kaplan (1989). Landscapes with intensive agriculture are defined as landscapes that are prosaic experienced (Nohl, 2009). These landscapes are often monotonous structured and the natural elements are almost gone (Nohl, 2009).

Most of the theoretical work is done in the research field of environmental or landscape planning. Only the theories of Appleton (1975), Ulrich (1983) and Kaplan (1987) are integrated in psychology research. But even if these theories are located in a different research all authors are showing their interface to environmental science. Appleton (1975) started in the field of art, but his theory found also application in psychology and environmental science. The same research areas are served Kaplan (1987). The theory by Ulrich (1983) added the area of economics to the existing field. Therefore, all theories are adequate for the theoretical background of landscape perceptions.

6.2.2 Findings in the Literature

Some studies already looked into the valuation and perception of landscapes, for instance in Ireland (Campbell, 2007; Howley et al., 2012), Slovenia (Marangon & Visintin, 2007; Verbič & Slabe-Erker, 2009), the Scandinavia (Haugen, 2015; Kaltenborn & Bjerke, 2002), South Europe (Marangon & Visintin, 2007; Molina et al., 2016) and also in Germany (Hartje et al., 2003; Job, 2003; Job & Knies, 2001; Kämmerer et al., 1996; Krakow & Gronemann, 2005). The target groups of these examinations differ: some studies compare tourists and residents and some just analyze the residents or compare the valuation of land owners to the general public (Haugen, 2015; Vanderheyden et al., 2013; Verbič & Slabe-Erker, 2009). This paper is focusing on the valuation of residents in the selected region LNWM, STB and LPLD.

Many papers refer to experiments of potential future developments of the landscape to value different development scenarios for the same landscape in comparison (Hartje et al., 2003; Kaltenborn & Bjerke, 2002). The analysis does not have the aim of figuring out which type of future development of a landscape is more preferable to its inhabitants. Instead, real landscapes are used that are accessible for everyone, which is not common in the literature (Aoki, 1999).

Additionally, the already examined landscapes differ from paper to paper. More often the existing papers examined symbolic landscapes. Most research examined the valuation of this landscape by analyzing protected areas (Molina et al., 2016) or vineyards (Job & Knies, 2001; Marangon & Visintin, 2007). Vanderheyden et al. (2013) used everyday landscapes to determine preferences. Examples for the analyses of these landscapes in the literature are agriculture land (Kämmerer et al., 1996; Karkow & Gronemann, 2005) or forests (Haugen, 2015). This analysis is in line with these papers, examining the perception of everyday cultural landscapes.

The comparison of different landscapes within Ireland was used from Campbell (2007). Marangon and Visintin (2007) compared the same landscape in two different regions while Kämmerer et al. (1996) looked at the valuation of a landscape using a rural and an urban area. Yet, there was no paper found that delved into similar landscapes in different regions. This academic void was also identified from Mayer and Job (2014), Morrison and Dowell (2013) and Schmitz et al. (2003). This is because the valuation of a landscape could be affected by specific background variables of the areas and therefore the existing studies are not comparable with each other (Verbič & Slabe-Erker, 2009).

6.2.3 *Hypotheses*

Regarding the conceptual part and the findings of the literature, it could be identified that it would be interesting to analyze different landscape types (water, city and village landscapes) in different regions to learn how individual characteristics influence the perception of these landscape types. This part of the paper is about the landscape evaluation that means to bring the perceptions together as a response to an aesthetic landscape quality (Swanwick et al., 2007; Unwin, 1975). Regarding the paradigms of Zube et al. (1982), introduced earlier, the analyses could be assigned into the psychological paradigm, primarily. This could be stated because Zube et al. (1982) argued that this kind of research is mostly a problem-related one and uses the valuation of the general public.

Kaplan (1987) stated that if there would be no preferences for the environments, it would not matter what landscapes are preserved and which ones are not. Therefore, it is not surprising, that some research found high preferences for their analyzed landscapes (e.g., Campbell, 2007; Howley et al., 2012; Kaltenborn & Bjerke, 2002; Marangon & Visintin, 2007). Some of these studies established that the preferences of the used samples are for water landscapes. Kent (1993), Kaplan et al. (1972) and Kaplan (1987) concluded that closed-nature landscape and more historical sites are preferred over landscapes with houses. Ulrich (1981) argued that especially water is preferred over other landscape types in general because individuals have more positive emotions (e.g., relaxing with the sound of running water or blowing winds) regarding water. Furthermore, regarding the evolution, this type of landscape is more likely to survive for an individual, as explained by the prospect-refuge-theory. Likewise, Sauer (1962, p. 45) described the water as "best opportunity to eat, settle, increase, and learn" for past generations of people and would therefore explain preferences. It might also be possible, that the residents are valuing higher close-nature landscapes higher than these of the city or village because they think the government is responsible for the preservation of inhabited landscapes[19]. All these finding leads to the hypothesis that in all of the three examined regions water landscapes are preferred over the other landscapes (H1).

It could be the case that locals (people who have never left the region) developed an identity with their region, which would lead to the fact that locals value their landscapes higher than tourists do (Fritz-Vietta et al., 2015, Job, 2003; Job & Knies, 2001, Kaltenborn & Bjerke, 2002). The explanation for this phenomenon could be that with increasing distance to the examined good the valuation of it is decreasing (Kim et al., 2015). This result could not be proven by Haugen (2015), who found that forest owners usually value a forest landscape lower than the general public. But forest owners typically have a lower distance to the landscape than the general public and therefore this finding disagrees with the result of Kim et al. (2015). Tuan (1990) stated that especially visitors and locals have different perceptions of the landscape because of their different viewpoints. While visitors build their perception by using their senses, locals include aspects as memories or traditions in their perception of the landscape. This makes for that a visitor observes aspects of the landscape that a local does not see anymore because it is part of the everyday life and therefore, the visitors viewpoint is often valid (Tuan, 1990). Von Reichert, Cromartie and Arthun (2012) delved into reasons for coming back to a rural American community and found that returnees valued the landscapes higher than the locals. Especially newcomers are usually free to choose a region and it might be the case that they chose the region because of the landscapes. The authors found that the newcomers viewed the landscape differently than the locals. Szerszynski and Urry (2006) examined the landscape perception of newcomers and locals and found differences of their perception: while locals viewed the more practical aspects, the newcomers view the landscape character and compare the landscape with this of other places. Therefore there might be a difference between locals, returnees and newcomers. Because the regions are different re-

19 This argument was often stated within the development of the questionnaire from our project partners working in and for the examined regions.

garding their economic power and structure, the comparison could show the impact of these reasons in each region. Out of the previous results, there could be developed another hypothesis: Locals have a lower valuation for their landscapes than newcomers and returnees (H2).

The previous findings regarding the impact of the living environment were inconsistent regarding the place of living. While Kämmerer et al. (1999) found that people who live in the city are valuing their environment (the city in this case) higher than people living in the countryside; Howley et al. (2012) concluded the opposite, that people living in the city value the landscapes that are not close to them higher than landscapes found around their living area. This interaction was also analyzed from Kim et al. (2015), using a metropolitan area in South Korea. They concluded that the valuation of the landscape decreases by increasing distance of the landscape, which would confirm the results of Kämmerer et al. (1999). In general, Purcell (1992) found preferences for an environment that is outside of the environment the individual lives in. In the current case, it would be stated, that people, who live in cities perceives the village and water landscapes more than people who live in the countryside. It might be the case that this result is depending on the examined region and that is what we could test using three relatively different region. Since there are no reliable results of the impact of the living surrounding, the stated hypothesis is that a residence in an urban environment influences the perception for water and villages more positively than a residence in a rural environment (H3).

De Groot and van den Born (2003) and Bauer, Wallner and Hunziker (2008) found that the perception of the landscape is to some extent dependable on how an individual sees the nature and feel in it. The authors used different types of individuals to explain differences in perceptions. Other studies concluded that people with a strong relation to the nature values the landscapes higher than people with a small relation to it (e.g., Carson et al., 2001; Howley et al., 2012; Kämmerer et al., 1996; Karkow & Gronemann, 2005; Morrison & Dowell, 2013). This influence can be explained by the internal image of the landscapes that include, for instance, the emotional relationship, while the external image integrates the natural structure of the landscape (Ipsen, 2006; Meier, Bucher & Hagenbuch, 2010). Upham et al. (2009) used the term environmental values to explain differences in distinct perceptions. The fourth hypothesis out of these results is: relation to nature has a positive influence on the valuation of the analyzed landscapes in the regions (H4).

Additionally to the visual stimuli described in the theoretical part there are other aspects affecting the perception of a landscape for instance to feel like home. This feeling refers to all "memories, ideas, feelings, attitudes, values, preferences, meanings, and conceptions of behavior and experience" (Proshansky, Fabian & Kaminoff, 1982, p. 59) the individual has with the region (Upham et al., 2009). It is bound with emotional security and familiarity (Nohl, 2006). Kaplan (1987) identified familiarity as an important factor for building preferences and Kent (1993) found a positive influence of familiarity on the landscape perception. The so-called sense of place[20] could be emerging if an individual has an unimpaired social environment in the region (Nohl, 2006). The landscapes could be seen as part of the symbolism of the region and its social attachment to the region (Nohl, 2006) and built regional identity (Fritz-Vietta et al., 2015; Morrison & Dowell, 2013). Kaltenborn (1998) conclude that people with a strong bond to region valuing the regional surrounding more positive than people with a lower sense of place, but Dale et al. (2008) describe negative aspects of sense of place, for instance resilience. The results lead to the following hypothesis: relation to the region has a positive impact on the perception of the regional landscapes (H5).

6.3 Data and Statistical Methodology

This chapter is about the data used to answer the research questions. The first part is about the variables that are important for the valuation of the landscapes. Then, the variables of interest (see hypotheses) and control variables are shown. The last part of this chapter presents the methodology of the examination.

6.3.1 Variables for the Valuation of Landscapes

Most of the existing studies asked for the monetary value of the landscapes by using the value per annum (Campbell, 2007; Hackl & Pruckner, 1999; Hartje et al., 2003; Kämmerer et al., 1996), but this paper analyzes the amount of money that would be paid by the participants monthly. This approach was also used from Kim et al. (2015).

The used questionnaire (Fig. A 1) asked in each region for five typical landscapes in the corresponding region imaged by photographs[21]. Every region has its symbolic landscapes: While LNWM has the beach as an unusual landscape, STB has a lot of old farmhouses and the marshland and LPLD healthy land. Because of the differences in the symbolic landscapes, it seems to be meaningful to compare the everyday landscape. The three different everyday cultural

20 Jorgensen and Stedman (2001) identified three constructs with different attitudes: place attachment has an affective component, place identity has a cognitive component and place dependence has a conative component. Soini et al. (2012) described that place attachment might be a synonym for sense of place although there are three more constructs (place satisfaction, dependence and identity) that complete the term sense of place. We use the terms identity, attachment, bond to region and sense of place synonymously.

21 Chapter 5.1 describes the results of this question in more detail.

landscapes that are analyzed are water, city and village. The selection was made because of the differences of these landscapes and because of their presence in all three regions under research. While the water landscape is the most natural landscape, the city landscape is mostly shaped from humans. The village landscapes are an intermediate type between highly structured and natural landscapes. Another justification for this kind of landscapes is that almost every other region has at least two of the landscapes under examination and therefore could use the results for itself.

The photographs of the landscapes covered with water show waterfronts or banks at one place in each region. Around the water there are some trees and grasslands and every photo offers an unrestricted view. The pictures of the cities contain mostly old houses. Two of the pictures showed special buildings like a church or a city hall in the background while the last one highlights a terrace. The last landscape type under examination displays a typical village in each region. The pictures show some houses with trees and grasslands around or a pathway. The sky in each pictures is blue or with some white clouds.

Non-bidders or protest bidders are separated and deleted from the data set for the analyses that is proposed by Spash (2008). In every region almost 12 % did not answer the question and were deleted as in other studies (Hackl & Pruckner, 1999).

The survey asked about a general contribution for the preservation, protection and design of the everyday landscape and for an individual contribution. The motivation for this procedure is that there might be reasons why a person is not willing to give some money in general for the preservation, protection and design for landscape: For example, one would like to support a special kind of landscape and fears that a general contribution does not benefit the preferred landscape.

The first question in the survey was about a general direct financial contribution about preservation, protection and design of landscapes, cities, villages and monuments for everyone in general (Tab. 6-1). Overall, every fifth resident of the three analyzed regions is willing to give a financial contribution in general for the protection, preservation and design of landscapes, but there are differences between the regions. While every fourth person is willing to give money for preservation, protection and design in LNWM, only every sixth resident is willing to pay a financial contribution in general in STB. To compare the results of the general and individual financial contribution, almost 80 % of the sample was not willing to give a financial contribution in general, but wanted to give some money for water, city or village.

Tab. 6-1: Questions of the survey and the manifestations in the analyses

Question in the survey	Possible answers	Transformation of the variable	Name of the variable in the data set
In your opinion, should the whole population pay a directly financial contribution for preservation, protection and design of landscapes, cities and villages as well as historical sites, additionally to the already paid taxes?	yes no	not used in the data set, only for descriptive statistics	
What amount of money are you personally willing to pay monthly and additionally to the already paid taxes for preservation, protection and design of the landscapes that are shown on the following photographes?	0 euro 1 - 5 euros 6 - 10 euros 11 - 20 euros 21 - 50 euros 51 - 100 euros more than 100 euros	transformed into a binary variable with 1 = is willing to give some money for the landscape and 0 = is not willing to give some money for the landscapes	Because of using three landscapes, there are three landscape: Water, City and Village
In what kind of environment is your house located?	in a city in the city limits in a village outside of a city or a village	transformed into a binary variable with 1 = living in a village or outside a city or village and 0 = living in a city or in the vity limits	rural environment
Since when do you live in your region?	Since birth Since birth with interruptions I moved to the region	transformed into binary variables with 1 = yes and 0 = no	local returnee newcomer
Please evaluate the following statements concerning your bond to nature: (1) I try to be as often as possible to be in the nature. (2) Nature has only a minor part in my live. (3) I feel a strong bond to the nature and the landscapes of my region. (4) I think human beings are benefit from an in working order nature. (5) I think it is an obligation to preserve and protect the nature. (6) I feel like I am part of the nature.	All statements could be rated from -- = radical refusal over 0 = irresolute to ++ = unreserved approval	transformed into a binary variable with 1= yes (everyone who marked statements 1, 3 to 6 with ++ or + and statement 2 with - or --) and 0 = no (otherwise)	bond to nature
Please evaluate the following statements concerning your bond to your region: (1) I feel really like home in my region. (2) With my region, i have a lot of personal memories. (3) My personal future is strongly linked to the region.	All statements could be rated from -- = radical refusal over 0 = irresolute to ++ = unreserved approval	transformed into a binary variable with 1 = yes (everyone who marked every statements with ++ or +) and 0 = no (otherwise)	bond to region
For what prupose and how often do you use your region?	(almost) daily at least once a week at least once a month less often not at all	transformed into a binary variable with 1 = yes (everyone who marked relaxing as daily) and 0 = no (otherwise)	daily relaxing

Question in the survey	Possible answers	Transformation of the variable	Name of the variable in the data set
		transformed into a binary variable with 1 = yes (everyone who marked sports as daily) and 0 = no (otherwise)	daily sports
I use for occuptional purpose and work as...	Farmer Official other occuption	transformed into one binary variable with 1 = yes (everyone who marked sports as daily or at least once a week) and 0 = no (otherwise)	occupational use
What is your gender?	female male	with 0 = male and 1 = female	gender
How old you are?	0 - 17 years 18 - 24 years 25 - 34 years 35 - 44 years 45 - 54 years 55 - 64 years 65 - 74 years older than 75 years	manifestations of the categorizes are displayed by the mean of each class	age
What is your highest school graduation?	No school graduation Main school Middle school High school College	used with numbers from 0 to 4	education
How many children do you have?	open answer	amount of children	children
What is your your net income for all members of your household (monthly)?	lower than 1,000 euros 1,001-1,500 euros 1,501-2,000 euros 2,001-2,500 euros 2,501-3,000 euros 3,001-4,000 euros 4,001-5,000 euros more than 5,000 euros	manifestations of the categorizes are displayed by the mean of each class	income

Note: Displayed are the questions in the questionnaire and the appropriate transformation and the used name in the analyses for it.
Source: Resident survey.

These results show that the separation of a general and individual financial contribution was meaningful. Usually, one would except that an individual is more willing to pay generally for landscapes or at least the same amount of people, but that does not seem to be the case, at least in the selected regions in Germany. The residents show a higher individual valuation and thereby higher perceptions of the chosen landscapes than for landscapes in general.

For the water landscape, almost 50 % of the respondents in each region are willing to spend additional money for preservation, protection and design for this landscape type (Tab. 6-2). This figure is perceived as rather high, in particular because Hartje et al. (2003) found only 22 % of their sample willing to give some money for the water landscape of the Elbe. It can be seen, that in more rural areas the residents are more willing to give some additional money for the preservation, protection and design of villages than in more urban regions. No research that could verify this result was found.

Tab. 6-2: Amounts of money that would be paid for each landscape type in each region

	Water			City			Village		
	LNWM	STB	LPLD	LNWM	STB	LPLD	LNWM	STB	LPLD
0 euros	53,66%	53,80%	59,05%	66,16%	71,76%	69,92%	66,16%	58,40%	56,76%
1-5 euros	28,78%	26,48%	23,46%	19,70%	17,06%	18,22%	19,19%	24,79%	24,53%
6-10 euros	10,24%	1,10%	11,52%	8,08%	6,47%	7,63%	7,07%	10,54%	10,40%
11-20 euros	5,37%	7,32%	3,91%	4,55%	3,24%	3,18%	6,06%	4,84%	4,99%
more than 20 euros	1,95%	1,41%	2,05%	1,52%	1,47%	1,06%	1,52%	1,42%	3,32%

Source: Resident survey.

Almost 70 % were not willing to pay some money for preservation, protection and design of the city landscape (Tab. 6-2). These results are similar to Karkow and Gronemann (2005) and Molino et al. (2016) who found that almost two thirds of the sample did not want to pay for the agricultural land. On the other hand, Howley et al. (2012) found that half of their sample did not want to pay for a farm landscape. These results could be compared to the city landscape, because for both landscapes, there might be somebody who seems to be responsible for the landscape: for agricultural land possibly the owners, while for the city it might be the government.

Those participants in the sample willing to pay something for the landscapes in their region stated most often an amount between 1 and 5 euros per month which would be equivalent to 12 to 60 euros per year. This result is in line with other paper. Kim et al. (2015) found monthly amounts for an ecological recreation project in the urban area conducted in South Korea between 0.5 and 1.50 dollars (approximately the same amount in euros, currently). Data for this analysis came from face-to-face interviews. Howley et al. (2012) did interviews in Ireland and found similar results. They used a payment card for determine the value. The work of Campbell (2007) found a similar amount but on the top end of the current result of this research by using discrete choice experiments of symbolic and everyday landscapes in Ireland by using face-to-face interviews.

While almost 30 % in LNWM are willing to pay 1 to 5 euros for the water landscape, only 24 % in LPLD are willing to give money. The third region lies in the middle with 26 %. Such differences can also be seen for the village: whereas 25 % of STB and LPLD would spend 1 to 5 euros for villages, in LNWM this amount is lower with 19 %. The payments for city landscapes are around 18 % for each landscape. The amounts of people who are willing to give more than 20 euros are really low in each region for each landscape (almost 1.5 %). A small exception is the village landscape in LPLD. 3 % of the sample in this region is valuing this landscape higher than 20 euros per month. It could imply that people who live in more rural areas are more willing to support the close-nature landscapes and the village than the urban ones.

Dummies were created that show whether a participant is in general willing to pay some money for the corresponding landscape (Tab. 6-1). This strategy was chosen because a lot of participants stated that they were willing to pay between 1 and 5 euros and only a few were willing to give higher amounts of money. In LNWM and STB the highest preference is for water landscapes (Tab. 6-3). 46 % of the respondents would pay at least one euro for this landscape. In LPLD, however, the preference for village landscapes is slightly higher than for water landscapes (43% for village landscapes in comparison to 41% for water landscape). While respondents in LNWM appreciate the city and village landscapes in the same way (34 %), LPLD and STB have a higher preference for villages compared to city landscape. Almost 42 % in both regions are willing to pay some money for village landscapes, while in both regions the amount of people willing to pay some money for preservation, protection and design of the city is only 29 %. It can be seen that the preferences for particular landscapes differ among regions as also found by Marangon and Visintin (2007) and Schmitz et al. (2003).

Tab. 6-3: All used variables (in percent)

Variables and its manifestations	Variable type	All	LNWM	LPLD	STB
Valuation of landscapes					
Water	dependent Variable	44,3	47,6	40,1	48,3
City	dependent Variable	29,9	34,3	28,9	28,9
Village	dependent Variable	39,1	33,5	40,4	40,3
Rural Environment	Interest				
Rural Environment		53,8	23,9	63,8	57,4
Duration Region					
Locals	Interest	29,8	21,6	33,2	29,8
Returnee	Interest	15,7	12,9	17,9	14,2
Newcomer	Interest	54,5	65,6	49,0	56,0
Bond to Nature	Interest				
Bond to Nature		54,6	55,6	57,4	50,1
Bond to Region	Interest				
Bond to Region		59,6	54,5	65,0	55,0
Daily relaxing	Control				
Daily relaxing		41,54	39,08	42,86	41,13
Daily sports	Control				
Daily sports		21,49	21,98	21,32	21,43
Occupational use	Control				
Occupational use		14,38	9,05	18,01	12,50
Gender	Control				
male		48,0	49,6	50,2	44,0
Age	Control				
18-24		5,1	4,1	5,1	5,7
25-34		10,0	12,3	10,0	8,8
35-44		11,1	11,1	11,3	10,9
45-54		23,3	23,4	22,4	24,4
55-64		21,5	21,7	21,3	21,8
65-74		16,5	15,2	16,7	17,1
75 and older		12,5	12,3	13,3	11,4
Education	Control				
No School Graduation		0,7	0,0	1,0	0,7
Main School		19,2	15,6	17,8	23,1
Middle School		36,8	36,6	37,4	36,2
High School		16,3	14,3	15,6	18,3
College		27,0	33,6	28,2	21,7
Children	Control				
No children		25,0	29,3	23,4	24,9
Child		19,7	19,6	22,1	16,3
2 children		34,7	36,0	36,9	30,9
3 children		13,8	8,9	11,7	19,6
4 children		4,4	4,4	3,8	5,3
more than 4 children		2,4	1,8	2,2	3,0
Income household	Control				
Lower than 1,000 euros		9,9	7,1	12,2	8,3
1,001-1,500 euros		15,1	16,1	16,8	12,2
1,501-2,000 euros		12,0	11,2	12,4	12,0
2,001-2,500 euros		14,4	13,4	15,1	13,8
2,501-3,000 euros		15,7	13,0	15,3	18,0
3,001-4,000 euros		16,6	21,9	14,2	16,9
4,001-5,000 euros		8,4	9,4	7,2	9,6
more than 5,000 euros		7,8	8,0	6,8	9,1

Source: Resident survey.

6.3.2 *Variables of Interest*

There was a question about the living environment of each respondent (Tab. 6-1). It can be seen that almost every third respondent of LNWM lives in urban structures (Tab. 6-3). Especially the suburbs seem to be more important in LNWM than in both other regions. In the sample of LPLD and STB, almost 60 % lives in a village or at least out-of-town. These numbers confirm the statistical data of the regions, presented above. It can be seen that the living environment of an individual might have an influence on the perception of landscapes (Fig. 6-1).

Fig. 6-1: WTP sorted by urban and rural regions (in percent)

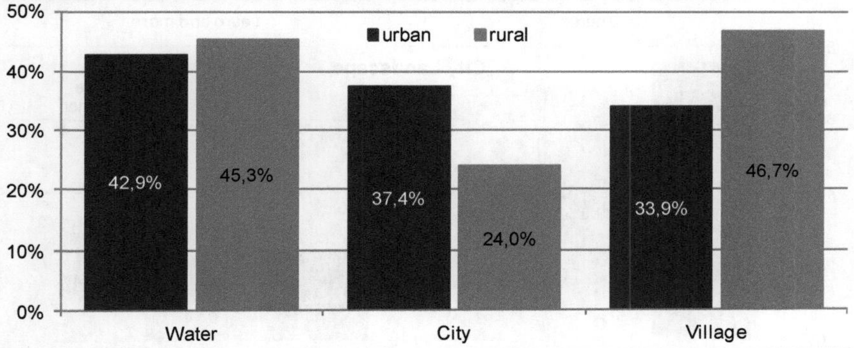

Source: Resident survey, own illustration.

The questionnaire also asked about the duration of being in the region (Tab. 6-1). About 50 % of the residents in each region are newcomers (Tab. 6-3). The amount of newcomer into LPLD and STB is smaller than in LNWM, but instead the quote of returnees (locals living in the region without interruption) is a bit higher in these regions compared with LNWM. In all of the regions only a few participants are returnees. Regarding the valuation of each group, the picture is not clear (Fig. 6-2). On average the locals are willing to pay almost the same amount as newcomers and returnees for villages, but less for city and water.

Fig. 6-2: Appreciation of the different landscapes sorted by residence (in percent)

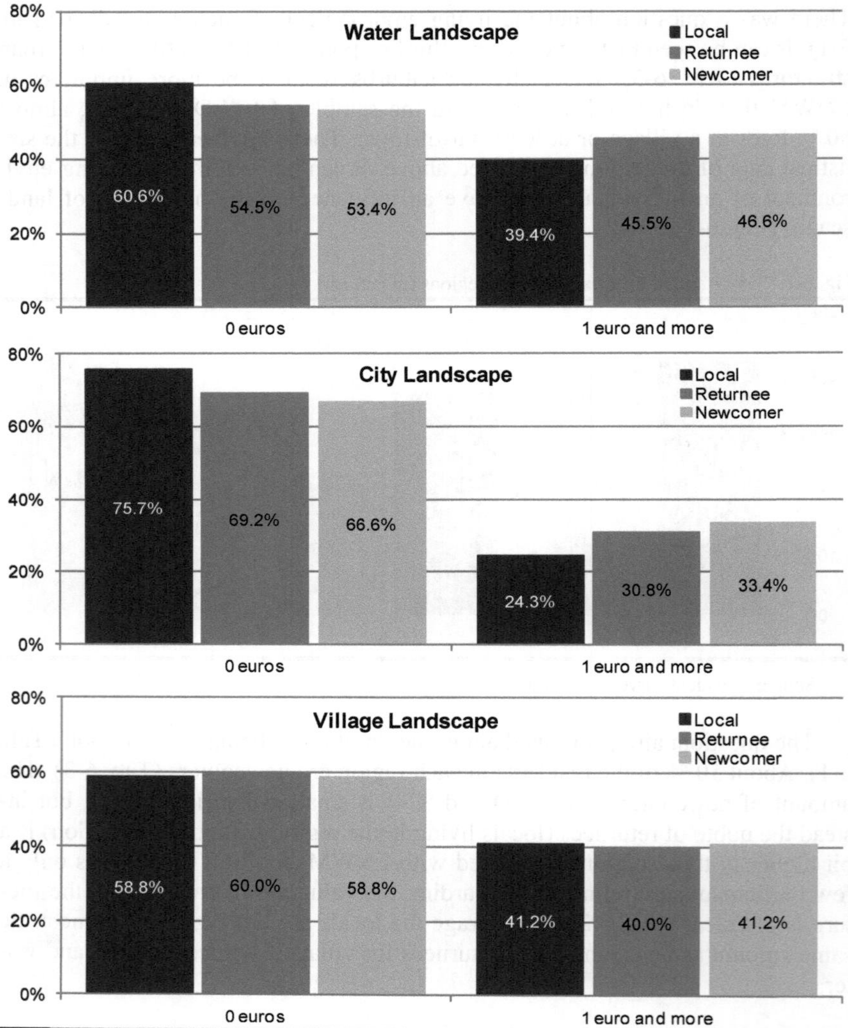

Source: Resident survey, own illustration.

The question used has different statements from which the variable bond to nature is created (Tab. 6-1). These statements were based on a study about the awareness of nature in Germany (BMUB, 2014). Because we just wanted to have one variable indicating the relationship to nature, we transformed the statements into one variable. Tab. 6-3 shows that almost every second participant answered that they would have a bond to nature. The bond is highest described by the respondents of LPLD (57.4 %) and lowest by the participants of STB (50.1 %). Since de Groot and van den Born (2003) found that the relationship to nature might be dependable on education and age, a test was performed. The correlation with age was higher than for the educational level (0.2 and 0.05), but still low. There was also a test about the bond to nature and the living environment because it might also be correlated. The correlation coefficient was about 0.07 and therefore, there was no correlation found. Additionally, a correlation measurement between bond to nature and the WTP for the landscapes was conducted. The data showed correlations between the WTP of village landscapes (0.13 with significance on the level of 0.01) and the water landscape (0.09 on the same level of significance), but no correlation for the city landscape.

Bond to region was created similarly to bond of nature (Tab. 6-1) and was used because of the identity that could influence the value of a landscape (Morrison & Dowell 2013). The statements itself were developed referring to Lalli (1992), Weichhart, Weiske and Werlen (2006) and Soini et al. (2012). Almost 60 % of the sample has a bond to region – even in LPLD there are 65 % participants. Because Kühne (2011) described that identity to a region might be dependable on time, there might be a correlation between the variables of the residents and bond to region. It was checked and it could be found a low correlation. Regarding the frequency, it could be found that more locals and returnees have a bond to region than newcomers.

It was tested if the variables bond to nature and bond to region measure the same, because Farnum, Hall and Kruger (2005) found that people who identified themselves with their places are also higher concerned about the environment. The correlation of these variables was medium high. Only 37 % of the sample is attached to the region and to the nature (Tab. 6-4). Almost 22 % have a bond to either region or nature, while 17 % do not have a relationship to nature nor region.

Tab. 6-4: Relationship between bond to nature and bond to region (in percent)

	no bond to region	bond to region
no bond to nature	25,38	21,98
bond to nature	17,47	35,16

Source: Resident survey, own calculations.

6.3.3 Control Variables

The socioeconomic variables consist of gender, age and education which are common in these kind of examinations (e.g., Howley et al., 2012; Mitchell & Carson, 1989; Morrison & Dowell, 2013; Kaltenborn & Bjerke, 2002; Kämmerer et al., 1996; Karkow & Gronemann, 2005; Kim et al., 2015; Molina et al., 2016; Schmitz et al., 2003) and were also find to have an influence on the perception of a landscape (e.g., Balling & Falk, 1982; Kent, 1993; Macia, 1979,). The whole sample contains almost 48 % male and 52 % female participants (Tab. 6-3). In each region, the distribution of the gender is equal to the statistical data on the significance level of 5 %[22]. Therefore, for all regions it can be said that the theoretical validity of the sample as described by Venkatachalam (2004) and Völkner (2006) are given. The quality of age might raise some doubts: only the distribution of the population in LPLD presents an equal age distribution than the sample. All other regions have no significance on the level of 5 %. Nevertheless, the sample displays an overrepresentation of the range 45 to 74 years in each region. The younger ages are underrepresented, which is also common.

Most of the people in the sample graduated from middle school (almost 37 %) (Tab. 6-3). The second most common graduation is college but with some differences between the regions (LNWM: 34 %, LPLD: 28 %, STB: 22 %). Main and high school are almost equal in each region between 15 % and 18 %. The only expectation is STB with 23 %.

On average, the participants in LNWM and STB have between 2,000 and 3,000 euros monthly per household while in LPLD the averaged income lies between 2,000 and 2,500 euros a month (Tab. 6-3). As stated by Carson et al. (2001), the income elasticity is positive, but more than one means the good seems to be an inferior good[23]. The income elasticity might be biased because almost 10 % of the sample did not answer the question about income. It could be that the question about income is too sensitive for these respondents and for that reason they did not answer it (Spash, 2008). Heyne et al. (2009) also had the phenomenon that some people did not fill in the income question.

Almost 25 % of the residents in STB and LPLD do not have any children (Tab. 6-3). This amount is slightly higher (about 30 %) for the respondents of LNWM. STB is the region with respondents that have on average the lowest amount of children (16.3 % have one child) and LPLD have the highest amount

22 The data of the sample (variables gender and age) was compared with official data of the regions (Effective [12-31-2013]) using a chi test. Only data of the defined municipalities were taken into account to compare sample and statistical data.

23 An inferior good is a good whose demand gets lower with the increase of income, because with an increase of income, other goods are in demand.

of respondents with a child (22 %). On average, the respondents of each region have mostly two children (about 35 %). The number of respondents decreases with the increasing amount of children. This could be correlated to the age; the data displays medium high correlation coefficients.

Nohl (2009) described that the intensity of land use might be reasons for different perceptions even if Tuan (1990) argued that especially in modern societies the contact with the environment is just indirect or special to some occasions which would indicate that the use might not be that important. Therefore relaxing and sports in the landscape as well as the occupational use of the land were integrated as control variables. Almost 40 % of the participants of every region are using their region daily for relaxing reasons (Tab. 6-3). Only one out of four of the respondents use the region daily for sports. The occupational use displays the economic structure of the regions. In LPLD, the highest amount of respondents' works in or with the landscapes of the region (18 %) while the slightest amount of people working in or with the landscape could be found for the participants of LNWM (9 %). The region of STB is with almost 13 % located in between.

6.3.4 Statistical Methodology

The questions about the influences of the valuation of the landscape should be answered by using a binary logit model like Hartje et al. (2003) and Marangon and Visintin (2007) did. The following model is used:

$$WTP_{l,r,n} = f(X_n, E_n, N_n)$$

where $WTP_{r,n}$ denotes if a respondent is willing to give some amount of money for landscape l (water, city, village landscape) in region r (*LNWM, LPLD, STB, All*). X_n is a matrix of the socioeconomic variables (*sex, age, education*) of person n. E_n displays a matrix of the environment (*children, income, daily relaxing, daily sports, occupational use, rural environment, returnee or newcomer*), and N_n includes the concern about the nature (*bond to nature*) and the region (*bond to region*). The model is first estimated for all regions and afterwards for each region separately. The variances of the influences can be seen using a step-by-step extension of the variable groups.

Subsequent to the described model, there will be some statistical tests to test the results of the data. Additionally to the logistic regression, a linear probability model (LPM) is also estimated as a robustness check. The estimation models are tested using likelihood ratio and Wald test. For the interpretation of the effects of the results are used marginal effects by means of the variables.

6.4 Empirical Results

The empirical analysis of the data starts with the estimation of a model that includes only all socioeconomic variables. This model is subsequently expanded in two steps by adding the variables regarding living circumstances and the relationship with nature and region[24].

6.4.1 Results of Each Landscape Type in the Analyzed Regions

The first examined landscape was water. The results for the whole sample suggest the importance of age, income, the occupational use of the region, the fact to be newcomer and the bond to nature for the appreciation of water landscape (Tab. 6-5)[25]. Highly significant is the bond to nature and therefore it could be concluded that the valuation of water increases if the person has a relationship to nature. Also the fact to be a newcomer has a positive influence on the valuation for water. Additionally, it can be said that the older the respondent is, the lower the valuation of water. This result is highly significant, but its effect is close to zero. That implies that the variable is not a critical factor for the perception of the landscape. The occupational use has a negative impact on the perception of the water landscape and implies that people working in or with the landscape valuing the landscapes lower than people who are not working in or with the landscapes. Its effect is medium high[26]. The variable income displays statistically significant on a low level, but its effect is zero. That could be interpreted that the income does not have any impact on the valuation of the water landscape.

In comparison of the regions there are some differences towards the result of the overall sample. Not every significant variable of the overall sample seems to be important in each region. A positive and significant influence on the WTP has the variable newcomer for the region LPLD. The effect of this variable is medium-high. In the other regions, this variable does not seem to explain the valuation well. Bond to nature is significant on the highest level for the region STB and has a positive influence on the WTP. The effect for STB is almost as double as for the overall sample and therefore influences the valuation with a high effect. Likewise, the occupational use has a high effect on the

24 Because the differences of the coefficients are marginal, there are only presented the results of the model with all variables. The likelihood ratio test and the Wald test show that the model fit increases by adding these variables to the first model. Furthermore, the LPM shows almost the same variables as significant coefficients and a similar effect of the variables in each model.

25 The coefficients and the values of the pseudo R^2 are shown in Tab. A.1.

26 Following, the marginal effect of 0.000 to 0.05 is described as low effect, an effect between 0.051 and 0.2 is presented as a medium high effect and the higher effects are used as high effect.

valuation. Its influence is negative and would imply people working in or with the landscape have a lower valuation of the landscape than other people. On a lower significance level than the overall result, the variable age seems to be important for LNWM and STB for the valuation of the population. As in the result of the whole sample, this impact is negative and is still close to zero.

Tab. 6-5: Marginal effects of the logistic regression for water landscapes (margins, standard errors in parentheses)

	All		LNWM		LPLD		STB	
Gender	-0,017		0,074		-0,014		-0,099	
	(-0.035)		(0.083)		(0.051)		(0.066)	
Age	**-0,005**	***	**-0,008**	***	-0,002		**-0,007**	***
	(-0.001)		(0.003)		(0.002)		(0.002)	
Education	0,013		0,013		0,009		0,020	
	(-0.017)		(0.040)		(0.025)		(0.030)	
Children	0,008		0,018		0,009		-0,001	
	(-0.014)		(0.035)		(0.021)		(0.024)	
Income	**0,000**	*	0,000		0,000		0,000	
	(0.000)		(0.000)		(0.000)		(0.000)	
Daily relaxing	0,026		0,024		0,051		0,022	
	(-0.037)		(0.091)		(0.052)		(0.067)	
Daily sports	0,037		-0,013		0,065		-0,015	
	(0.043)		(0.105)		(0.064)		(0.077)	
Occupational use	**-0,096**	**	0,134		-0,084		**-0,205**	**
	(0.047)		(0.129)		(0.063)		(0.093)	
Rural environment	0,033		0,018		0,056		0,045	
	(0.034)		(0.091)		(0.052)		(0.063)	
Returnee	0,055		0,095		0,075		0,030	
	(0.053)		(0.139)		(0.073)		(0.102)	
Newcomer	**0,106**	**	0,136		**0,148**	**	0,002	
	(0.042)		(0.100)		(0.059)		(0.080)	
Bond to nature	**0,117**	***	0,022		0,084		**0,235**	***
	(0.037)		(0.084)		(0.054)		(0.069)	
Bond to region	0,028		0,138		0,060		-0,080	
	(0.038)		(0.087)		(0.056)		(0.069)	
N	921		179		438		304	

Note: Stars indicate significance levels: * $p<0.1$, ** $p<0.05$, *** $p<0.01$. N displays the number of cases.
Source: Resident survey, own calculations.

The most important variables for the WTP of cities are the rural environment around the place of living, the fact to be a newcomer and bond to nature (Tab. 6-6)[27]. All these variables are highly significant and have medium-high effects on the valuation of this landscape type. On a lower significance level, the fact to be a returnee and the occupational use are important. The influence of these variables on the valuation is also medium-high, with 0.119 for returnees and 0.087 for occupational use. The variables of age and income are signif-

27 Tab. A 2 shows the coefficients of the estimation as well as the pseudo R^2.

icant on the lowest level of significance and show only a low effect on the valuation.

Tab. 6-6: Marginal effects of the logistic regression for city landscapes (margins, standard errors in parentheses)

	All		LNWM		LPLD		STB	
Gender	-0,022		0,085		-0,010		**-0,107**	*
	(0.035)		(0.086)		(0.051)		(0.063)	
Age	**-0,002**	*	-0,004		-0,001		-0,003	
	(0.001)		(0.003)		(0.002)		(0.002)	
Education	0,006		0,033		-0,008		0,012	
	(0.017)		(0.040)		(0.024)		(0.029)	
Children	0,014		**0,073**	*	0,011		-0,007	
	(0.014)		(0.038)		(0.020)		(0.024)	
Income	**0,000**	*	0,000		0,000		0,000	
	(0.000)		(0.000)		(0.000)		(0.000)	
Daily relaxing	0,010		0,043		0,078		-0,083	
	(0.035)		(0.089)		(0.050)		(0.065)	
Daily sports	0,057		0,092		0,042		0,018	
	(0.041)		(0.101)		(0.060)		(0.074)	
Occupational use	**-0,087**	*	0,145		-0,097		**-0,238**	**
	(0.048)		(0.123)		(0.063)		(0.104)	
Rural environment	**-0,120**	***	-0,146		**-0,134**	*	-0,081	
	(0.034)		(0.095)		(0.051)		(0.060)	
Returnee	**0,119**	**	**0,258**	*	**0,150**	**	-0,006	
	(0.052)		(0.143)		(0.070)		(0.103)	
Newcomer	**0,127**	***	0,144		**0,191**	***	-0,003	
	(0.041)		(0.108)		(0.059)		(0.076)	
Bond to nature	**0,106**	***	0,054		0,059		**0,215**	***
	(0.036)		(0.085)		(0.053)		(0.066)	
Bond to region	0,007		0,025		**0,093**	*	**-0,128**	**
	(0.037)		(0.088)		(0.055)		(0.064)	
N	921		179		438		304	

Note: Stars indicate significance levels: * p<0.1, ** p<0.05, *** p<0.01. N displays the number of cases.
Source: Resident survey, own calculations.

The picture for each region is again some kind of different to the overall result. The significant variable rural environment is only important for the valuation of the city landscape of the region LPLD. Its effect is medium-high on the highest significant level. As seen in the overall result, the variables returnee and newcomer are positive and significant for the region LPLD. Newcomers have a higher effect on the valuation than returnees but still both effects could be categorized as medium-high. For the region LNWM returnee is positive and the marginal effect of this variable is almost twice as big as the coefficient for LPLD. Bond to nature and bond to region are relevant for the region STB. While bond to nature is highly significant and positive with a high effect of 0.215 on the valuation of the landscapes in STB, bond to region is significant on a lower level and with a negative and medium-high effect. This result is surprising, because that would imply, a person with a low relationship to the region value the city higher than a person with a strong relationship. A different result is shown from the model for the region LPLD: bond to region displays positive on the lowest significance level. Its effect is medium-high. The occupational use of the region has again a negative impact on the valuation, but its effect is high even if the significance level is low. The number of children is only significant and with a medium-high, positive effect for LNWM.

The last examined landscape – the village landscape – displays three highly significant variables. Highly significant is bond to nature and its effect on the valuation is medium high (Tab. 6-7)[28]. People with a relation to nature are valuing the village landscape higher than people without a bond to nature. Age is – like in the model for water – negative, but its effect is again very close to zero. That indicates that age have only a minor effect on the valuation. Rural environment has a high significance and the variable influences the valuation in a positive direction. Its effect is about medium-high. The variable newcomer has a positive impact and a medium high effect for the valuation. Income is significant, but its effect is zero that would imply no effect on the valuation.

The rural environment is positive with a medium-high effect on the valuation of this landscape type, but only for the region STB. Bond of nature for STB is more than twice as big as the marginal effect of the model for LPLD and also significantly higher. This variable seems to be more important for the landscape located in STB. For the valuation of the landscapes of STB, the occupational use of the landscape displays again significant and negative. This time, the effect is lower than for each of the other landscapes. Age is negative in region STB and LPLD on the highest significance level and for LNWM on a lower level, but all margins imply a minor effect on the valuation. Although the number of children is not significant for the overall result however it is for LNWM. This variable has a positively medium-high effect on the valuation. That would imply that with an increasing number of children the valuation of the village landscape also increases. The income is significant for LNWM, but again with an effect that is zero.

28 The coefficients of the estimation and the pseudo R^2 can be found in Tab. A 3.

Tab. 6-7: Marginal effects of the logistic regression for village landscapes (margins, standard errors in parentheses)

	All		LNWM		LPLD		STB	
Gender	-0,032		0,051		-0,037		**-0,114**	*
	(0.035)		(0.085)		(0.051)		(0.067)	
Age	**-0,005**	***	**-0,005**	*	**-0,005**	***	**-0,007**	***
	(0.001)		(0.003)		(0.002)		(0.002)	
Education	-0,013		-0,017		-0,032		0,018	
	(0.017)		(0.040)		(0.025)		(0.030)	
Children	0,017		**0,074**	**	0,021		-0,009	
	(0.015)		(0.036)		(0.022)		(0.025)	
Income	**0,000**	**	**0,000**	*	0,000		0,000	
	(0.000)		(0.000)		(0.000)		(0.000)	
Daily relaxing	0,003		0,107		0,018		-0,046	
	(0.037)		(0.089)		(0.053)		(0.067)	
Daily sports	0,046		0,046		0,069		-0,014	
	(0.044)		(0.101)		(0.065)		(0.080)	
Occupational use	-0,060		0,162		-0,076		**-0,168**	*
	(0.047)		(0.122)		(0.063)		(0.096)	
Rural environment	**0,113**	***	0,010		0,084		**0,194**	***
	(0.035)		(0.090)		(0.053)		(0.064)	
Returnee	0,048		0,084		0,051		0,019	
	(0.053)		(0.136)		(0.074)		(0.103)	
Newcomer	**0,077**	*	0,042		0,079		0,084	
	(0.042)		(0.101)		(0.060)		(0.080)	
Bond to nature	**0,148**	***	0,028		**0,115**	**	**0,288**	***
	-0,037		(0.085)		(0.054)		(0.069)	
Bond to region	-0,027		0,039		0,070		-0,057	
	(0.038)		(0.088)		(0.056)		(0.069)	
N	921		179		438		304	

Note: Stars indicate significance levels: * p<0.1, ** p<0.05, *** p<0.01. N displays the number of cases.
Source: Resident survey, own calculations.

6.4.2 Comparison of the Results between the Landscape Types and Regions

Overall, comparing the results of each region, there are differences between the valuations – but for each landscape or in each region, the influences of some variables on the perception are similar to some extent. This part of the paper will show the consistencies and disparities of the landscapes and the regions and will answer the hypotheses.

The first hypothesis was about preferences for water landscapes in each region. This could only be partially proven. Regarding higher preferences for this landscape type, Fig. 6-1 shows that the participants of LNWM and STB have higher preferences for water landscapes, but the residents of LPLD show higher preferences for village landscapes.

H2 assumed that locals are less valuing their landscapes compared to returnees and newcomers. The fact to be a returnee is relevant for the city landscape only (Tab. 6-6). Its effect differs between medium high for region LPLD and the overall model, and high effect for region LNWM. In comparison to every analyzed region, it could be stated that this variable seems to have a positive effect on the valuation of the residents of LNWM and LPLD on each landscape, except the city landscape of the region STB (Tab. 6-8).

Newcomers are more important for the valuation of water and city (Tab. 6-5, Tab. 6-6) than for the village landscape. The effect for the water landscape is smaller, but still medium-high. Newcomers are especially relevant for the city landscapes in LPLD because the effect and the level of significance are highest there. In comparison of the landscape types and the regions, it could be concluded that newcomers value each landscape type higher than locals in LNWM and LPLD. For the region of STB it could be only stated that the water and village landscape is valued higher if the participant moved into the region. The other landscape type shows a negative influence of newcomers. Additionally to the statistical models, Fig. 6-2 shows that newcomers and returnees value water landscape and city landscape higher than locals while villages are valued highest by the group of locals. On average, locals have higher preferences for villages, but returnees and newcomers show higher preferences for water landscapes. The results could only be approved the hypothesis for LNWM and LPLD. For region STB this could only be stated for the water and village landscape if newcomer valued this landscape, but not for each landscape. Therefore, it can be concluded that being a returnee or a newcomer might depend on the examined region and on the landscape types.

The following hypothesis assumed that people living in an urban environment value the village and water landscapes higher than people living in the countryside. The rural environment influences the valuation on different ways: While the marginal effect is positive for village and the water landscape, it is negative for city landscapes (Tab. 6-5, Tab. 6-6, Tab. 6-7). The LPM further indicates a positive influence on rural environment for villages and water landscapes and a negative coefficient for city. This proves the results of the used logistic regression model. Additionally, these results were also found analyzing the amount of money people would spend on the protection, preservation and design of the landscape.

People living in an urban or suburban area value water the highest, for people living in rural environments the village is valued highest (Fig. 6-1).

Tab. 6-8: Comparison of the landscape types in the different regions

	Water Landscape	City Landscape	Village Landscape
Rural Environment			
LNWM	+	-	+
LPLD	+	-***	+
STB	+	-	+***
Returnee			
LNWM	+	+*	+
LPLD	+	+**	+
STB	+	-	+
Incomer			
LNWM	+	+	+
LPLD	+**	+***	+
STB	+	-	+
Bond to nature			
LNWM	+	+	+
LPLD	+	+	+**
STB	+***	+***	+***
Bond to region			
LNWM	+	+	+
LPLD	+	+*	+
STB	-	-**	-
Daily relaxing			
LNWM	+	+	+
LPLD	+	+	+
STB	+	-	-
Daily sports			
LNWM	-	+	+
LPLD	+	+	+
STB	-	+	-
Occupatinal use			
LNWM	+	+	+
LPLD	-	-	-
STB	-**	-**	-*
Gender			
LNWM	+	+	+
LPLD	-	-	-
STB	-	-*	-*
Age			
LNWM	-***	-	-*
LPLD	-	-	-***
STB	-***	-	-***
Level of education			
LNWM	+	+	-
LPLD	+	-	-
STB	+	+	+
Number of children			
LNWM	+	+*	+**
LPLD	+	+	+
STB	-	-	-

	Water Landscape	City Landscape	Village Landscape
Income			
LNWM	-	-	-*
LPLD	-	-	-
STB	-	-	-

Note: Positive or negative influences of each coefficient are shown. Stars indicate significance levels: * $p<0.1$, ** $p<0.05$, *** $p<0.01$.
Source: Resident survey, own calculations.

Due to these results, H3 – that people living in an urban area are valuing close-nature landscapes (in this case water and village landscapes) higher than people living in more rural environments – cannot be proven. This result is in line with those of Kämmerer et al. (1999) but not with the result of Howley et al. (2012) and Purcell (1992). It could be assumed that for this result, the landscape type decides the impact of this variable.

The hypothesis regarding bond to nature stated that bond to nature has a positive impact on the valuation of each landscape. Even if the coefficients are not statistical significant, all coefficients are positive. For the whole sample, the highest effect is displayed by the village landscape (Tab. 6-6). The results of the analysis show that H4 could be verified. In addition, it could be said that bond to nature is an important aspect in valuing landscapes because its impact does not differ between the landscape types or regions.

The last hypothesis in this analysis assumed that bond to region has a positive impact on the valuation of the landscape types. Having a look on the margins without significances, a positive influence of this variable on the valuation of each landscape type for the regions LNWM and LPLD can be observed, but the influence on each landscape valuation is negative for STB. This result is surprising, and indicates that bond to region and its influence on the valuation depends on the landscape type under research. Because of these results, H5 could only be partially proven.

6.5 Discussion

This part of the paper discusses the results, further variables, the suitability of the pictures of landscapes and its generalizability. It should be noted that there are different types of values[29], the questionnaire asked the participant only to value the landscape types monetarily and then was done a transformation into an indicator. This indicator is not able to distinguish between different value types of the landscape and therefore it is possible that the respondents have dif-

29 Brown and Brabyn (2012) examined eight different value types: scenic/aesthetic, recreation, economic, ecological/ life sustaining, native flora and fauna, social, historical/ cultural and wilderness.

ferent values of each analyzed landscape. But it is usual to have different values for one landscape type (Brown & Brabyn, 2012).

6.5.1 Empirical Results

An explanation why the participants of LNWM and STB prefer water landscapes over others might be their location. In both regions, the residents are confronted with water: LNWM because it is located at the Baltic Sea while STB has an intensive drainage system to make the land usable. The region LPLD has the river Elbe, but it might be the case that the river is not as present as in the other regions. Additionally, it might be possible that because of the most rural area of the selected regions the residents recognized that the closest landscape is the most valuable one. Generally speaking, landscapes with natural elements tend to be preferred over man-made landscapes (Ulrich, 1983). Additionally, Kaplan (1987) stated that close-nature landscapes are higher valued because the individual is able to learn in this landscape type. That would mean that the highest preferences are for the water landscape, following the village – and the lowest preferences are for the city landscape. This displays exactly the results of the current analyses.

The fact to be a newcomer or returnee had a negative impact on the valuation of the city landscape of the participants of STB. That might be explained by the effort of dewatering the landscape. Returnees and newcomers might not know that the dewatering needs a lot of effort that would not be anticipated in advance. Because locals have never lived without this effort of dewatering, it might be usual and make for higher valuation of the landscape. For LPLD and LNWM, there is not that much effort to preserve the landscape as in STB and therefore they valuing the landscape higher. This could be underlying the argument of not recognizing the everyday landscapes anymore by the locals and the argument that both other groups have a higher perception of it, except of region STB. Because of these results, there is the assumption that the perception of landscapes might be dependable on the history and the experiences with other landscapes of the individual that values the landscapes. But, the variables of newcomer and returnee might be affected by positive selection because these respondents chose to move back or into the region. Therefore, the perception of the landscapes might be different than for locals (Von Reichert et al., 2012) because they might have moved because of the landscapes. The high significance of the coefficient for the city landscape of LPLD might be explainable by the resistance movement for active waste and a lot of creative people coming in this region to participate on this resistance movement[30]. On the other hand, it might be the case that locals wanted to move, but had because of lack of money or strong ties to the other residents. There was a test done if the landscapes are the reason why the participant came into or back to the region but that could not be ascertained. It could not be tested if the participated locals wanted to move but could not with the data available.

The geographic proximity was found to have a negative impact on the valuation (Carson et al., 2001). When Job (2003) asked locals and tourists he concluded that the locals are more willing to pay for their landscape than the tourists. As an explanation he used identity of the locals and the time spending in those landscapes. Also Upham et al. (2009) and Kim et al. (2015) concluded that the geographical proximity have a positive impact on the valuation. These would disprove the argument of geographic proximity stated by Carson et al. (2001). There is the question if the results could have changed over time. The geographic proximity was not part of the questionnaire, but the surrounding of each participant was. This variable is not adequate to determine the distance of the participants and therefore the paper could neither prove nor disprove the results of the authors. But people value the landscape higher that is not within their living surrounding. If the landscape could be seen and experienced every day, it might not be that interesting anymore than landscapes that are outside of the own living surrounding (Groth, 1997). This finding would indicate that the distance might be important for the perception of everyday landscapes in a positive way.

Bond to nature was found to be an important variable for the perception of landscapes. This already showed the correlations between this variable and the WTP for the water and village landscape. People with a strong relationship to nature might try to preserve and protect the landscape more than people without any relationship to it. This was also found in previous research on landscape valuation (Howley et al., 2012; Morrison & Dowell, 2013). It might also be the case that people with a bond to nature are more often in the landscape (e.g., going for a walk) and collect positive memories concerning the landscapes and consequently value the landscapes higher than people without such a relationship (Ipsen, 2006; Meier et al., 2010), but during the analyses it could not be identified as impacting the use of the landscape. Additionally, it was not possible to distinguish between nature-bound types as de Groot and van den Born (2003) or Bauer et al. (2008) did, because there was only one statement for

30 This argument was given from the partners in the region. Additionally, in another project we asked especially artists why they moved into the region, and active waste seemed to be an important reason. But with the used data, we could not answer, if this argument can be proven.

each type, which could distort the results. Nevertheless, bond to nature is the most important variable for the landscape perception in this examination.

The assumed positive bond to region could not be found for the participants of STB: Participants having a bond to region value the landscape lower than people without one. This could be based on different parts of this term (mentioned in footnote 20). Kyle, Graefe, Manning and Bacon (2004) found that all three parts (place identity, place attachment and place dependence) have different perceptions regarding the environmental values. For instance, the authors concluded that people with place identity value the environmental conditions more critical, while people with place dependence are less critical about these conditions but value it more favorably. It could be the case that people living in the region of STB are more identified with the region and therefore are more critical regarding the corresponding environmental conditions than in other regions while, for instance, the respondents of LPLD are more place-dependent and therefore value it higher. Lin and Lockwood (2014) distinguished between an emotional and a functional attachment. The emotional one is very deep and part of the individuals′ identity while the functional attachment includes only physical effects of the landscape. It could also be the case that the respondents of the different regions not have the same type of attachment to the region and therefore have different impacts on the valuation of the landscape. Another explanation could be that these respondents might know that there is enough financial support for the landscapes for their owner and they therefore do not feel responsible for the regional landscapes in STB. This argumentation would also be proven by the information processing theory of Kaplan (1987): residents with more information value the landscape differently from people with less information. Another argument could be that the resilience of the respondents of STB is higher than for the other regions and because of this high resilience bond to region has a negative impact on the perception for the landscapes of STB.

Surprisingly, the socioeconomic variables age and income displays indeed significant variables, but their effect on the perception of landscapes is only really small. It might be the case, that as stated by Tuan (1990), the perception of the used sample does not differ between the ages and is therefore not important. Only two of the used variables – gender and number of children – have an impact on the perception. The gender could be found to be important for the region STB and because of its negative coefficient men are more willing to spend some money for the landscapes city and village. The amount of children seems only be important for the region LNWM and for the same landscape types than the gender for region STB. These two socioeconomic variables do not seem to be used commonly in WTP research, but because of the different manifestations of these two variables it might be interesting to look deeper into the impacts on the landscape perceptions.

Another surprising result is that the use of the landscape of private persons does not have any impact of the landscape perception. It does not matter how often an individual uses the landscape for relaxing or sports, the perception does not change. Additionally, the professional use of the landscape does not be important except for region of STB, where the use of the landscape in occupation decreases the perception of all landscape types.

6.5.2 Further Variables

Culture does not have an influence on the preferences because emotions are similar in every culture (Ulrich 1983). Evidence suggests that environmental values and attitudes are similar between different countries, but are still bound with culture and the subjective background (Schultz & Zelezny, 1999; Yu, 1995). However, Swanwick et al. (2007) found that the culture could be a factor for the landscape valuation. Upham et al. (2009) said that some authors were not sure whether the attitudes were held by individuals or created by social construction. Tuan (1990) argued that individual perspectives were always bound with culture and therefore each individual has different perceptions. But the author also stated that each individual might to some extent have similar perceptions of the world. If culture might have an impact on the valuation of the landscapes it could be an explanation for the differences of the valuation between the regions. Even if culture does not have an impact on the perception of the landscape, it formed the attitudes and has at least an indirect effect (Tuan, 1990; Upham et al., 2009). A direct test of the influence of culture in the selected regions in the used models was not possible, because culture is a complex construct with a lot of relevant factors to include (Upham et al., 2009).

6.5.3 Pictures of the Landscapes

One problem of these analyses might be the choice of landscapes: presence of man-made elements, animals and visible water (Arriaza, Cañas-Ortega, Cañas-Madueño, & Ruiz-Aviles, 2004; Kent, 1993; Ulrich, 1981) was found to dive preferences and might be another explanation for difference in the valuation of the landscapes in the regions. For instance, Tveit (2009) found that the general public (also the examined group in this case) preferred pictures with a wider view of the landscape than what is only represented by a water landscape photograph and therefore are probably the most preferred in STB and LNWM. He also discussed that this result might not be in line with the prospect-refuge or the information processing theory, because these theories suggest a preference for half-open landscapes (Appleton, 1984; Kaplan, 1987). Regarding these theories, the preferences would be on the village landscapes, given that these photographs display settlements and nature, which was only the case for LPLD.

Swanwick et al. (2007) found that people′s perceptions to landscapes may depend on what kind of image the landscape is on (e.g., diversity, color). Tuan (1990) and Unwin (1975) agreed with these finding. Tuan (1990) stated that colors are usually important for building preferences: while red tones are associated with dominance or danger, blue tones are known as receding colors or symbolize safety. Zajonc (1980) and Unwin (1975) believed color differences might be explained by the variances in answers and that it is hard to integrate all color aspects into landscape perception analysis. Regarding these results, the colors could cause different preferences of the three tested types of landscapes: Within each landscape type picture there were almost the same amount of tones but between the types, there are differences of the tones. For instance, the city landscape has higher rates of red tones (rooftops) than the water landscape and the water landscape have higher rates of blue tones (water) than red. The village landscapes fall somewhere in between.

Concerning the experience modes described by Nohl (2009), the water landscape could be experienced with fascination because this landscape is almost wild, and the village landscape is experienced more by intrigue. No modus describes exactly the aesthetic experiences for a city landscape, but the closest one might be the prosaic experience. This theory could not answer what kind of landscape would prefer one over another, only in which landscape aesthetic experiences modes could found. It also does not state if an individual experiences the landscape always in one of these modes or if there are exception when an individual perceives landscapes differently. But Nohl (2009) already stated that this theory might not be fully developed and therefore could be interesting to look deeper into.

Zajonc (1980) describes that the attitudes an individual has can change by alternatives. It might be the case that even if the description of the question asks for each landscape to be evaluated separately; some respondents have compared the landscapes and built some kind of rating. This could mean that the results might to some extent be biased because this was already described as a weakness of the CVM (Kahnemann & Knetsch, 1992; Mitchell & Carson, 1989). Because of this weakness, the decision was made to use only the monetary value as an indicator for perception of the landscape.

The order of the goods (*order bias*) could be another explanation why the city and the village landscape are less valued than the one with water. This bias describes that the first good on a list will be valued higher than one in the last position (Kahnemann & Knetsch, 1992; Mitchell & Carson, 1989). Regarding the analyzed landscapes, water was in the third position and has higher valuations, while the city and the village were on the end of the list.

6.5.4 Generalizability

It might be the case that there is a positive selection of the respondents, because it could be the case that respondents are more interested in the topic of the survey than non-respondents. This was taken into consideration during the conception of the questionnaire. Hence, it started with questions that might be easy to answer by everyone. Additionally, there was a pretest of the questionnaire with people living in the region done and the questions were discussed with the partners in the region. A positive selection could not be tested, but there was a comparison of the results to other findings in the literature to show if the results might be generalizable.

The literature shows a positive influence of education on the valuation for landscapes (e.g., Hackl & Pruckner, 1999; Howley et al., 2012; Kaltenborn & Bjerke, 2002; Kämmerer et al., 1996; Marangon & Visintin, 2007; Verbič & Slabe-Erker, 2009). More specifically, Vanderheyden et al. (2013) found that people with a higher level of education value close-nature landscapes higher than other landscape types. Education was tested for each landscape in this research but the results were not displayed as significant. Some negative impact from education could be found for the LPLD region city and village landscape and for the village landscape of LNWM. All other coefficients were displayed positively. The same could be said for the coefficients of the LPM.

Age has a negative effect on the valuation of water and village landscape as well as the city but with no statistical significant coefficients. This is a common finding in practical valuation research (Carson et al., 2001; Hackl & Pruckner, 1999) and also in preference analyses (Balling & Falk, 1982). The negative impact of the age could be explained by the decreasing senses by the elderly people and the shrinking future for them in the landscape (Tuan, 1990).

Swanwick (2009) found that the experiences with different landscape types are important for later life. A test was not possible but a positive influence could be found for the number of children for each landscape perception, except for the region STB. It could be the case that people with a higher amount of children spend more time in the environment than families with a lower number.

Sex roles could explain different perceptions of the environment; for instance, men generally value wild landscapes higher and women prefer more vegetated landscapes (Tuan, 1990). This could not be found in these analyses – instead, it might be dependent on the region's influence. For LNWM, women generally value the landscape higher, while in the other regions men tend to value each type of landscape higher.

The results with respect to income in the literature usually show a positive influence (e.g., Campbell, 2007; Howley et al., 2012; Kämmerer et al., 1996; Marangon & Visintin, 2007; Verbič & Slabe-Erker, 2009), but one research study found a negative influence (Morrison & Dowell, 2013) and Hackl and Pruckner (1999) did not find any significance for income. An explanation for the negative tendency in the current examination might be that people having more money than the average do not appreciate the village in the same way because they are able to spend holidays in a symbolic landscape and spend money for these landscapes. This would also emphasize the assumption that everyday landscapes are inferior goods, as stated before. But because the effect turned out to be zero, there might not be any effect at all.

Because of the selection of landscapes and regions and the results of the control variable, the results of this paper might be transferrable to other regions. There might be other regions with similar problems as the analyzed regions. For instance, other metropolises might also have strong attractions and the regions in the hinterland identify themselves over the metropolis instead of their own attractions and landscapes.

6.6 Conclusion

The paper had the aim to determine the valuation of three cultural landscapes in three regions of the MR HH. The research questions were: First, how the resident population values different cultural landscapes in their respective region; second, what the differences in valuation for the same landscapes between different sub-regions of the MR HH are; and third, which factors have an influence on the valuation of the different landscapes in each region?

The results suggest that close-nature landscapes are perceived higher than the built-up landscape of the city. But even if this is the case, less than 50 % of the people in each regional sample are willing to give any money for landscape preservation and protection.

Overall, it can be stated that the examined variables of the bond to nature, bond to region, rural environment, and the fact of being a newcomer or a returnee to the region are important for the perception of the landscapes – even if only the bond to nature and the newcomer factor are significant in every chosen landscape. Surprisingly, the importance of bond to region differs between the regions. By contrast, the use of the landscape does not seems to be important for the perception, but with the exception of STB, where the occupational use negatively influence the perception of landscape. However, the use of the landscape for relaxing or sports does not have any impact on the perception. The socioeconomic variables of age, education and income have almost no effect on the perception of the landscapes.

The results show interesting aspects of the differences between the landscape types and the analyzed regions. Bond to nature is an important and stable aspect to integrate into examinations of landscapes perceptions. Besides the provision of information, this finding is also an important for regional planers if they would like to strengthen the perception of landscapes. It could be shown that the valuation of the close-nature landscapes (water and village landscapes) are valued higher if the participant is living in an urban area. This could imply that the rural living environment depends only on the landscape type but not the analyzed region. Bond to region might be dependable on the region and not on the landscape type in valuation of landscapes. Concluding, the results show the complexity of landscape perception used as arguments about why no comparisons of different landscapes and regions in the literature exist yet (Mayer & Job, 2014; Morrison & Dowell, 2013; Schmitz et al., 2003). But this analysis shows that a comparison indeed produces interesting results.

It could be proven that valuation of landscapes differs between different regions and landscape types. Because of the results of this work, it can be said the valuation of landscapes is partly a subjective measure as stated by Campbell (2007) and Job and Knies (2001). But, it could also be influenced by the values and norms in the region and the resources that are available in these regions (Paasi, 2002), thus showing the subjectivity of the region and not only of the individual. Individuals living in the same cultural context share the same understanding of their landscapes and therefore could value it in similar ways than people outside the specific cultural context (Butler, 2016, Stephenson, 2008). This was shown especially for bond to region and the occupational use of the landscape as influence factors for valuation.

The results of this analysis show the complexity of the valuation of landscapes and the used method also has some difficulties[31]. It might be interesting to integrate other constructs that could be used as indicators for perception – for instance, volunteering in or for a landscape. Individuals who are volunteering for the region and its landscapes also invest something into the protection, preservation and design of landscapes and might not be willing to give some money for the landscapes. Some researchers already used this combination of the monetary and volunteer valuation, but for developing countries (e.g., Echasseh et al., 1997; Vásquez, 2014). Another aspect that could be interesting for future research might be analyzing the differences between locals, newcomer and returnee. These groups showed statistical significant values in this analy-

31 For an overview of the limitations see for instance Venkatachalam (2004) or Mitchell and Carson (1989).

sis, but dependable on the region or the landscape type, so that the valuation of each group could also be explained by different factors.

7 Does Migration Biography Affect the Perception of Landscapes? Empirical Evidence for Newcomers, Returnees and Locals from Steinburger Elbmarschen, Germany

The following chapter analyzes the perception of symbolic and everyday landscapes of newcomers, returnees and locals in the region of STB. The cultural landscapes are split into symbolic and everyday landscapes.

7.1 The Importance of Migration Biography for the Landscape Perception

The term landscape can be used for every type of landscape (Schein, 1997), because in general landscapes could be interpreted as a result of practices, relationships and interaction of natural and/or human factors (Butler, 2016; Council of Europe, 2000). They are a valuable source of understanding the past of a culture or a region (Kuechler, 1993). This implies that a landscape is not only valued by the past culture, but at the same time with the cultural context of each individual (Stephenson, 2008). It can be suggested that people living in the same cultural context value a landscape in similar ways (Paasi, 2002). Halfacree and Rivera (2011), for instance, stated that past experiences are important for developing a view of the landscape in question and that these can change over time because of more or different experiences. Therefore, people have to value a landscape subjectively, depending on their individual experience with and in the landscape (Butler, 2016; Meinig, 1979; Stephenson, 2008).

De Haan (1999) argued that the triggers for migration are determined by the social and cultural institutions that in individual has and at the same time by the local customs and ideologies. Because of this, the literature suggests dividing groups of newcomers and locals because the social and cultural values are different from each other (Brennan & Cooper, 2008). Long-term residents seem to have other practices working with the landscape (practices regarding productivity) than people moved to a region because of the amenities[32]. These differences between the two groups are arising because of the different experiences the

32 Organization for Economic Co-operation and Development (OECD, 1994) defines an amenity as specific landscapes, natural resources or human activities in the land that are providing benefits for the people by consuming it.

J. Matloch, *The Assessment of German Cultural Landscapes*, RaumFragen: Stadt – Region – Landschaft, https://doi.org/10.1007/978-3-658-21416-6_7

people have: while locals stayed in this one region, newcomers have at least experiences two different regions and its landscapes. The amenity-orientation of the newcomer might lead to more regard to the preservation of the landscape (Abrams, Gosnell, Gill & Klepeis, 2012; Hurley & Halfacree, 2011; Mendham & Curtis, 2010) and therefore could lead to different valuation of the landscapes itself. Fortmann and Kusel (1990) argued that the values might be different between newcomers and locals, but the values of the newcomers might already exist in the local community, the newcomers only express them visible. Returnees share or shared some cultural context with locals but also with newcomers, but its context might be different because of different experiences (Butler, 2016; Meinig, 1979; Zäch, Schulz, Waltert & Pütz, 2015). Therefore, the perception of returnees might be on the one hand close to these of the newcomers and on the other hand to locals.

The migration literature tells a lot about the characteristics and motives of the movers (Barcus & Brunn, 2010), but there was no research found about the perception of the landscape and the differences between these groups. It might be the case that amenity-orientated people could lead to more preservation and protection of the landscape because of their knowledge about the landscapes (Abrams et al., 2012; Deller et al., 2001; McElhinny 2006). If there are different landscape perceptions this could have impacts on the institutional government of the region. Lorah (2000) concluded that regions with natural amenity (especially wilderness areas) have higher levels of population growth and could therefore help to understand future development in the regions. There might be some other indications in the general landscape preference literature that these groups have different preferences (Tveit, 2009).

The aim of this paper is to identify the impact of migration biography on the perception of landscape. In this context, the paper will answer two research questions. First, how does the perception of landscapes in general differ between newcomers, returnees and locals? Second, is the perception of different kinds of landscape affected differently by migration biography? The sample is small so that this analysis could be understood as explorative examination, because there was not identified any analysis using these groups and the landscape types at once before.

Following this introduction, there is describing the theory about different migration theories, including an overview about previous results in the literature. Subsequently, the hypotheses are formulated. A description about the region of STB and its regional landscapes is the next part of this analysis. This chapter should help imagine the appearance of the landscapes in the region. Afterward the used variables and the methodology are described. In this part, there are also some descriptive statistics to have a look on the manifestations and get some ideas about the data. Hereafter there are the results of the empirical analyses presented. These results and possible difficulties as well as further research are discussed subsequently. The research ends with a conclusion.

7.2 Migration Biographies and Landscape Perception

This chapter describes the differences between the newcomers, returnees and locals based on the theoretical background, and connects these differences with the landscape perception. The second part shows some findings of the literature between different groups of residents (including second homeowners and tourists). Followed is the part that describes the differences between the analyzed landscapes and a categorization of these into landscape types. The chapter ends with the formulation of hypotheses for the following analyses.

7.2.1 Migration Theories

There are a lot of theories of migration or at least research in the economic literature (Ritchey, 1976). The neoclassical theories are available at the macro and micro levels (Massey et al., 1993)[33] and focus mostly on the earnings of an individual (Ritchey, 1976). On the macro level, the theories try to explain migration initiated by geographic differences in labor supply and demand, while theories on the micro level integrate the cost-benefit assessment into their decision to migrate (Massey et al., 1993). Because this work looks into the differences between migrant and locals these theories do not seem to be adequate to explain some possible differences. The "new economies of migration" integrates further aspects like the opinion of related people or the minimization of risks (Massey et al., 1993). This theory is on the micro level, but the focus is still on economic decision rather than on environmental ones and is not about why individuals differ from each other. Therefore, these theories do not seem to be adequate to characterize migrants from locals.

The Dual Labor Market Theory explains migration by using the intrinsic labor demand, especially in more modern societies (Massey et al., 1993). As argued for the other theories, the focus is again on the labor market rather than on other further aspects why an individual would migrate or what a migrant differ from a local person and therefore such theories are not useful for this analysis. The last kind of theories categorized by Massey et al. (1993) is called World Systems Theory. In these theories the development of the conditions for labor is

33 The classifications of the theories differ by each corresponding author. For instance, Ritchey (1976) named the theories of migration as labor mobility studies and social demographic studies. King (2002) differentiates theories by process (the act of migration itself) and the product (e.g., cultural characteristics, integration analyses).

in focus. These theories are more broadly orientated but do not give any indication of the differences between the groups of newcomers and locals and their characteristics examined in this analysis. The authors describe four more theories (network theory, institutional theory, cumulative theory, migration system theory) that could interpret as broader theories that include more soft aspects for migration. These theories could help in understanding why migration is easier in the future – but they do not help in deciphering the differences between migrants and locals. It might be the case that psychology orientated theories might be better suitable towards understanding differences in the characteristics of locals and (re)migrants.

The other part of existing theories is about structural and social-psychological analysis of individuals (Ritchey, 1976). The more structural theories are about the position of the individual in society while the social-psychological theories are about the internal attributes (e.g., motives, values, norms) (Ritchey, 1976). Mabogunje (1970) argued that people moving from a rural area to an urban one transform their attitudes and motivation to get more involved in the urban contexts. This shows that the characteristics can change over time, because they are depending on the economic, political and cultural circumstances of each individual (de Haan, 1999). The research suggests that migrants are often younger, higher educated and without strong ties to the community through their spouse or child (Lansing & Mueller, 1967; an overview of results is, e.g., shown by Ritchey, 1976), but that does not mean that every migrant shows these characteristics (King, 2002). In addition to these characteristics, the institutional agencies are also an important factor for the migration decision (Mabogunje, 1970). This part of the migration research seems to be more adequate in explaining differences between migrants and locals. Therefore, some chosen theories are shown in the following.

One of the earliest and most common theories of migration in the so-called social demographic theories might be the theory of pull and push factors (e.g., Heberle, 1938; Lee, 1966; Massey et al., 1993; Petersen, 1958; Ritchey, 1976; Toren, 1976). Pull factors are factors that are located in the new place that are able to tempt a person to this place while push factors are these that are located in the place of residence that an individual is dissatisfied with (Lee, 1966; Toren, 1976). The push factors are associated with negative aspects of the country of origin while the pull factors are showing the positive aspects of the new place (Heberle, 1938; Lee, 1966; Toren, 1976). Lee (1966) completed the model of push and pull factors with intervening obstacles (e.g., distance, physical barriers) and personal factors (e.g., new job, marriage). The personal factors include structural and social psychological aspects (Ritchey, 1976). Therefore, this theory might be adequate to explain differences in migration in this case.

This model comes closer to the aim of the following analyses because it already integrates personal factors of the individual.

Petersen (1958) criticized the approach of the push and pull factors because in his opinion it is based on the social causation hypotheses[34] rather than on the motives of the migrants, concluding that the factor theory might therefore not be relevant for all layers of the population. Because of this, the author tried to bring the results for international and internal migration together into one more general typology of migration. Such a typology was also made for return migrants from Gmelch (1980). King (2002) formed a general typology of migrants for the "current" migration phenomenon. While the theory of the push and pull factors might be applicable to this examination, the typologies might not be. But this typology might be helpful in other cases.

In earlier research, wanderlust was named as the central motive of migration, but Petersen (1958) argued that this could not be the only reason why people move. Based on this argument and on other research before (Toren, 1976; Williams & Sofranko, 1979), Boneva and Frieze (2001) developed a model about the motives of emigrants to leave the country. This theory is based on the personality of the migrants and could therefore be defined as a psychological theory. The authors found that people who want to leave are more work-orientated and usually have higher scores of achievement and power motivation, while the affiliation motivation and the family centrality is low comparing with the people who want to stay. They justify the high scores with the argument that people with a high achievement and power motivation tend to look for more challenging and are more willing to take risks than people with a low score. The authors concluded that these arguments went along with a higher work orientation of emigrants. Previous research found that job transfers or a new job were the most important reasons for migrants to move in general (Price & Sikes, 1975; Shaw, 1975). Additionally, Ritchey (1976) argued that the motivation to migrate because of work-orientation is lower for older people than for younger once. The low affiliation motivation and family centrality are explained by a lower importance of networks and (family) relationships to others than for those people who want to stay (Boneva & Frieze, 2001). This argumentation is in line with other research (e.g., Lansing & Mueller, 1967) or theories of other research areas for instance regarding the activity in volunteering (Carstensen, 1995). Newer research confirmed that people living in places with fewer kin are more willing to move than people having kin in their neighborhood (Kolk, 2017; Spring, Ackert, Crowder & South, 2017). Additionally to

34 Some events (e.g., migration) happen because of social disadvantages like unemployment of some layers of the population rather than because of internal motives of the individual (Petersen, 1958).

these, Boneva and Frieze (2001) concluded that the migration behavior is depending on the actual opportunities for resettle and some environmental factors of both the sending and the entering country. Other psychological factors are also important for migration decisions. But besides these individual characteristics, the background of the individual might be important as well (Mabogunje, 1970). Even if this model is applied to emigrants, it might also be true for people who want to migrate to different regions within a country and it makes it possible to find common characteristics between newcomers and locals.

Halfacree and Boyle (1993) used a different approach with the integration of the biography of each migrant for the decision to migrate. For instance, as indicated by Williams and McMillen (1983) before and by Kupiszewski (1996) later, the past of the individual regarding migration is important for future decisions of migration: Individuals who migrate once show higher motivations to migrate again. Halfacree and Boyle (1993) argued that the biographies of the migrants indicate the reasons for moving. The authors stated that Wolpert (1965) had already taken the first steps into a biographical approach because he concluded that the decision to move depended on the past of the individual and on the future the individual hopes to have, which indicates the importance of the biography. The biographical approach might be part of the psychology of the individual, but also of the human geography, more concrete population geography. As in the landscape preferences[35] and landscape perception[36] theories, it integrates the individual contexts and characteristics (Swanwick, 2009; Swanwick et al., 2007; Upham et al., 2009; Zube et al., 1982). It could also be used for the explanation of the current analysis because the link between population geography and social psychology seems to be beneficial (Halfacree & Boyle, 1993). This might be even made clearer by the work of Warnes (1992), who argued in his biographical approach that the social and environmental conditions have the same rate as the personal circumstances in which the individual lives for the migration. This approach could show differences in the analyzed groups in this paper and might therefore be adequate to understand different perceptions of landscapes.

7.2.2 Previous Findings of Differences in the Migration Biographies

Brennan and Cooper (2008) used different groups of newcomers, separated by the years of living in the region of interest. They found that usually locals are younger, less educated than newcomers, and newcomers were less likely to be

35 Landscape preferences are about why people like one landscape better than another one and also about the reason for liking (Swanwick et al., 2007).

36 Landscape perception refers to "the psychology of seeing and attaching meaning to landscape" (Swanwick et al., 2007, p. 4) with the use of every sense the individual has (Tuan, 1990).

employed and have children less than 18 years of age. Because of these results, the authors argued that there might be differences in the attitudes and values of these two groups (Brennan & Cooper, 2008). The authors found that locals are valuing the heritage higher than the newcomers and that the locals perceive the in-migration and the development of the region more negatively than the newcomers do (Brennan & Cooper, 2008). Additionally, Brennan and Cooper (2008) found that the percentage of newcomers and locals that perceive the cultural heritage as important is almost equal to all groups while more newcomers find it unimportant and more locals find it very important.

Kondo et al. (2012) looked into the different perceptions regarding the environment of locals and second homeowners. The authors were able to show that second homeowners use their second home to get away from the everyday live and as a chance to be isolated. While the literature of newcomers shows that this group tends to support environmental protection because of its preservation aspects (Hansen et al., 2002; Rudzitis, 1999), second homeowners are more supportive of guidelines to restrict the access to environmental areas for a broader public (Kondo et al., 2012).

Ní Laoire (2007) looked at the motives of Irish return migrants to rural areas. She argued that returnees are in some characteristics similar to newcomer but also similar to locals because they have the inside and outside perspective for the region. The people who left the region might have a higher esteem of the landscape in the region they left because they have different experience as well as have seen other landscapes and regions. This argumentation was confirmed from von Reichert et al. (2014).

Kaltenborn and Williams (2002) analyzed the differences between tourists and locals and found that for the local residents the natural environment and the regional social network are important factors for the landscape perception. These two aspects are usually closely related to their experiences in the region and the cultural history and its landscapes. Job and Knies (2001) also looked into these two groups and found that residents value the landscapes in their surrounding higher than tourists. As Kaltenborn and Williams (2002), the authors explain the differences in the perception with a higher degree of identification with the region.

Gottlieb (1994) examined the importance of amenities for firm locations and attractiveness of jobs. The employees would have a higher benefit of the amenity itself while the firms want to benefit of qualified employees who search for jobs close to natural amenities. Therefore, this connection analyzed by Gottlieb (1994) is more an indirect connection because the firm chose the location and the employees follow or not. Amenities did not seem to have a strong influence on the choice of the firm location. The author argues that an amenity-related firm location might make sense if the firm is working on the region market if the community in the region is needed.

7.2.3 The Landscape and Its Types

For the analyses, the landscapes were divided into two types: symbolic and everyday ones. This differentiation of landscapes was used in the literature before: As described by Antrop (2005) a symbolic landscape involves old landmarks or special places with history value for the region. For instance, Molina et al. (2016) delved into protected areas and Marangon and Visintin (2007) as well as Job and Knies (2001) used vineyards. This landscape type embraces special landscapes that could not be found in every other region and could therefore be summarized as symbolic landscapes. These landscapes are valuable for the identity of the people living in the region and therefore they are conserved already (Antrop 2005).

An everyday landscape is originated over time by the cultural background and could be experienced by almost everyone who is in the region (Groth, 1997). The landscape type was, for instance, analyzed by Vanderheyden et al. (2013), who examined different landscapes closer to nature. Tweed and Sutherland (2007) and Bamert et al. (2016) analyzed typical old buildings and its symbolic character for the people living in this type of landscape. Other research looked at agriculture land (e.g., Kämmerer et al., 1996; Karkow & Gronemann, 2005) or forests (e.g., Haugen, 2015). These landscapes could be found in almost every other region and could therefore be resumed as everyday landscape. Because of this variety of landscape in the examined region, this region was defined as an amenity-rich one, as described by Abrams et al. (2012).

Boll, von Haaren and Albert (2014) determined the reaction of changes in specific landscapes in the MR HH. One landscape was the Elbe Marsh, but with a different portion than analyzed in this elaboration. The Elbe Marsh is usually covered by meadows and open land and the authors wanted to find out how changes in this landscape are accepted. They found that buildings show the highest rate of rejection, followed by more fields in the expense of the meadows. The highest rate of acceptance was found for hedges and shrubs. Also the increase of the amount of tourists or a higher rate of maintenance was seen as a more negative change than in the other symbolic landscapes of the MR HH.

7.2.4 Hypotheses

As seen in the parts before, there is some research that looked into differences between groups or between landscapes. This research tried to put together the different groups and their perceptions of different landscapes so that in the end there would be a comparison of the groups for different landscape types. None of the previous works brought together these both construct. This comparison might be useful because identified differences between the resident groups and different landscape types might be helpful for the region. It might be the case that the perceptions of different landscape types also differ and not only the perception of the groups itself. To analyze these differences, the following seven hypotheses out of previous findings of the literature were formulated.

Because of different experience and different cultural contexts in the past (Butler, 2016; Meinig, 1979), there might be different valuations of the landscape. The assumption is that people who have never left the region have a lower valuation of the symbolic landscape of marshlands and farmhouses. People who have left the region for a while value these landscapes higher, because they realized that these landscapes are very special to the region. This was already identified from von Reichert et al. (2012). They examined reasons for returning to a rural community and found that returnees valued landscapes higher than locals. This phenomenon was also found by Haugen (2015), who concluded that landowners value the landscape lower than the general public, because the general public identified the land as special and for the owners it was an everyday one. This might be going back to the argument that the recognition of the landscape is different if the individual has seen different landscapes. Similar aspects could be true for newcomers who might not have seen these landscape types before they moved to the region and therefore value them higher than locals. In summary, the first hypothesis says that newcomers **and** returnees value the symbolic landscape higher than the locals (H1). This hypothesis should indicate if the recognition of landscapes is higher because of his or her experience with other landscapes than for locals.

Some studies already concluded landscape preferences in WTP research (e.g., Campbell, 2007; Howley et al., 2012; Kaltenborn & Bjerke, 2002). Some of these examinations and theories identified the everyday landscape of water as preferred landscapes. For instance, Deller et al. (2001) found that water seems to be important for economic growth and therefore the valuation of the landscape might be higher for this landscape type than for other landscapes. Therefore the research will test if the preference of all three groups is for the water landscape (H2). This hypothesis is important for planners, for instance, because it could indicate what everyday landscape types might be valuable and not only symbolic ones.

Bond to region might be an important factor for landscape valuation, as argued by Vaske and Kobrin (2001), who found a high correlation between bond to the region and the behavior in the landscape. The bond to region is described by Gustafson (2006) as the place attachment – that means the connection between people and a specific place and all the connecting aspects of this place (e.g., feeling, knowledge or actions). The bond is developed by the identity with the region that has been build over the years (Fritz-Vietta et al., 2015; Morrison & Dowell, 2013). Kaltenborn and Williams (2002) concluded that a high attachment to a region have a positive impact on the values of the environmental surrounding. Because locals had more time to build a bond to region, it might be stronger than for returnees or newcomers (Anton & Lawrence, 2014). These two groups might start building a new bond if they move to the region even if Barcus and Brunn (2010) argued that it is not relevant if a person is still living at the place for the bond to region. Bond to region has a positive impact on the WTP for locals and a negative impact on the returnees and newcomers (H3). This hypothesis will underlie the importance of the relationship to the region that could be help the planners to identify an important approach for the attractiveness of the region.

Activities in nature are relevant indicators for the perception of environmental surroundings (von Reichert et al., 2014). It seems more than plausible that people who are often using the nature for activities value it higher than people who do not. For this explanation it does not seem to be a difference between the three groups analyzed. Only a difference between the landscape types could be assumed: while everyday landscapes could be used by almost everyone for free, the symbolic landscapes are usually on private land. Therefore, the hypothesis is formulated as follows: the daily use of the region for relaxing or sports has a positive impact on the valuation of everyday landscapes but not on the symbolic ones (H4). This hypothesis could help planners to understand what they should do to support their landscapes in the right way.

Children were identified in the migration literature as an important factor for moving (Ní Laoire, 2007; von Reichert et al., 2014). This would result in the hypothesis that having children will influence the perception of an everyday landscape in a positive way. This could arise from using the landscape for walks with the children or for playgrounds. Because the symbolic landscapes are agriculture land in the current case meadows or the ground is private property, the valuation of this type of landscape might be lower if people have children. Regarding these results the hypothesis is that children have a positive impact on the valuation of everyday landscape but a negative influence on the valuation of symbolic landscapes (H5).

Level of education might also play a role in the valuation of landscapes. For instance, Rudzitis (1999) and Johnson (2000) found that migrants are more educated than people living in the regions. Von Reichert et al. (2014) concluded that return migrants usually move back because they graduated. In previous

WTP analyses, the educational level had a positive impact on the valuation of landscapes (Hackl & Pruckner, 1999; Howley et al., 2012; Kaltenborn & Bjerke, 2002; Kämmerer et al., 1996; Marangon & Visintin, 2007; Verbič & Slabe-Erker, 2009). It might be the case that the higher educated people learned more about nature and its preservation – in particular about special landscapes (symbolic ones). The results of these studies would mean that people returned or moved to the region have a higher level of education than the locals and might to know more about the idiosyncrasies of German landscapes. Even if it might be the case that popular landscapes are always perceived the best, the regional symbolic landscapes in this case are not popular but still special to the region. Hence, there is the hypothesis that the level of education has a positive influence on the WTP for newcomers and returnees; this effect might be higher for symbolic landscapes than for the everyday ones (H6).

People who returned to a region are more involved in social networks and communities than people who never left the region (von Reichert et al., 2014). Newcomers and returnees who are volunteering might have lower valuation of the landscapes in terms of financial support because they already support the landscape by the volunteer work. Therefore, there is the hypothesis that volunteering of returnees and newcomers has a negative influence on the valuation of the landscapes (H7).

7.3 Landscapes of the Steinburger Elbmarschen

The region STB is located in the hinterland of the MR HH and is part of the county of Steinburg (see Fig. 7-1). Around the county are the counties of Dithmarschen, Pinneberg, Segeberg and Rendsburg-Eckernfoerde and, of course, the Elbe (river). For this research, the cities Itzehoe, Wilster and Glückstadt and the administrative departments of Wilstermarsch, Krempermarsch and Horst-Herzhorn are used.

The region of STB has more than half its fields covered with meadows and pastures (Kowalewski & Schulze, 2010) – especially in the Elbe marsh sector. It is very close to sea level and therefore the protection of the land is one of the most important things for the region (Scheer, 2014). The history[37] indicates that the region had help from Dutch experts when building the drainages of the land. Still today, the dyke building, building calas for the drainage and building of locks is very important for the preservation of the land in the region (Scheer, 2014).

37 Naudiet, Arlt, Jansen and Maiwald (1994) give a good overview of the historical developments of the region of STB.

Fig. 7-1: Location of the Steinburger Elbmarschen in the metropolitan area of Hamburg

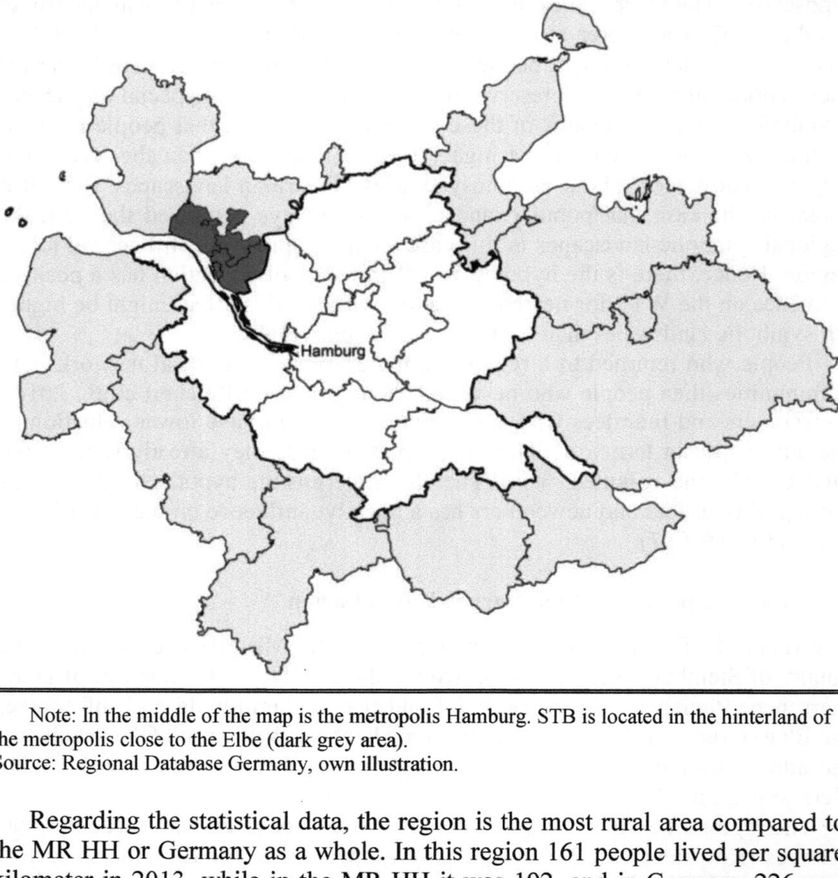

Note: In the middle of the map is the metropolis Hamburg. STB is located in the hinterland of the metropolis close to the Elbe (dark grey area).
Source: Regional Database Germany, own illustration.

Regarding the statistical data, the region is the most rural area compared to the MR HH or Germany as a whole. In this region 161 people lived per square kilometer in 2013, while in the MR HH it was 192, and in Germany 226 persons per square kilometer on average (Fig. 7-1). The amount of the residents younger than 15 years of age is slightly lower in the analyzed region than for the MR HH or Germany. In line with this, the amount of people older than 65 years is a bit higher than in the other regions in comparison. While the region STB has a negative demographic development, for Germany the negative development is lower than for STB and for the MR HH it is even positive. For the BPI that is 100 for the whole of Germany, it can be seen that it is lower than 100 for STB (lower BPI by the residents of this region) than for the MR HH as a whole. The job density is similar in each area. The working population in the agricultural and mining sector is highest for the region of STB and lowest for the MR HH. The sector of manufacturing is highest for Germany as a whole and again lowest for the MR HH. For the MR HH, the service sector is highest. Germany as a whole and the region of STB display similar values in this sector.

Tab. 7-1: Regional data of the region Steinburg in comparison with the metropolitan area of Hamburg and Germany

	STB	Metropolitan area of Hamburg	Germany
Residents per square kilometer (2013)[1]	161	192	226
Residents younger than 15 years (2013)[1]	12,9%	13,3%	13,1%
Residents older than 65 years (2013)[1]	22,4%	21,0%	20,8%
Demographic development per 1,000 residents (2008-2012)[1,2]	-14,5	7,3	-3,1
Buying Power Index per resident (2012)[3]	96,6	104,3	100,0
Job density (2013)[1,4]	55,3%	55,3%	55,5%
Working Population Primary Production (District, 2012)[5]	5,5%	3,9%	4,3%
Working Population Secondary Production (District, 2012)[5]	24,7%	18,1%	27,1%
Working Population Tertiary Production (District, 2012)[5]	69,8%	78,0%	68,7%

Source: 1 Regional Database Germany, 2 Landesamt für Statistik Niedersachsen, Statistisches Amt Hamburg und Schleswig Holstein, Statistisches Amt Mecklenburg Vorpommern, 3 GfK (Buyer Power Index D=100), 4 Bundesagentur für Arbeit, 5 VGR der Länder.

All landscapes used in this paper are photographes of actual accessable landsapes. The everyday landscapes are covered with water, cities and villages. The water landscape shows a part of the Stör river, some green meadows around its banks and the circuitousness of the landscape. Some examinations already showed the attractiveness of water landscapes for migration (e.g., Johnson & Beale, 1998). The city landscape displays some old half-timbered house and a church spire. The village is illustrated by a line of house with an avenue in front of them, and cobbelstone line the streets.

The symbolic landscapes are a marsh landscape as well as a landscape of old farmhouses. The region of STB is characterized by the area of Wilstermarsch and Krempermarsch that are divided by the Stör. The landscapes of the marshs are a relatively young landscape optimal for growing plants (Scheer, 2014). The picture of the marsh landscape in this analysis shows the meadows and that these are slightly elevated to help the water run off. In the background of this picture is a small city with a church. The farmhouses of Wilstermarsch and Krempermarsch differ to some degree (Scheer, 2014): In Krempermarsch, the “Fachhallenhaus” could be find with a hall that exends the whole house. From this hall, there branch off all other rooms and the stables. In the middle of the 19th century, the farmers started to separate the living area from the economical areas. Some farmers built some manor houses out of the "Fachhallenhaus". The so-called "Husmannshus" of the Wilstermarsch had a basement created for dairy processing. These houses segregated between living and work space since the beginning. The picture with the landscape of old farmhouses points out the house next to each other. In the foreground are sheeps and a brown field. In front of the houses are old trees.

7.4 Sample and Methodology

This chapter is about the data used to answer the research questions above. In the first part the data is presented in detail. It describes the variables used for the resident groups, followed by a description of used variables for determining perception. Then, the control variables are shown. The last part of this chapter presents the methodology of the examination.

7.4.1 Newcomers, Returnees and Locals

The sample was asked since when they are lived in the regions and they had three possibilities to answer the question (Tab. 7-2)[38]. These categories were renamed to locals, returnees and newcomers.

Out of the responses, almost 30 % were local, 14% returnees, and most of the sample was newcomers to the region (almost 54 %). The rest of the sample did not answer this question.

A lot of people (almost 36 %) did not need to answer the question about the year of return or income or did not state any year even if they had to answer the question (Tab. 7-2, Tab. 7-3). Most of the people returned to the region or moved in between 1980 and 1999 (39 % of the newcomers, 37 % of the returnees). The second most common years are since 2000 (36 % of the newcomers, 37 % of the returnees). Between 1940 and 1960, the fewest people of the sample moved into the region.

38 The variable is described in more detail in Chapter 5.2.

Tab. 7-2: Overview of the questions

Question in the survey	Possible answers	Transformation of the variable	Name of the variable in the data set	Using as...
Since when do you live in your region?	Since birth Since birth with interruptions I moved to the region	transformed into binary variables with 1 = yes and 0 = no	local returnee newcomer	Variable of interest
In what year did you return / moved into the region	open question	not used in the data set, only for descriptive statistics		
What are the reason why you did not move, did return or did move into the region	job-related reasons family-related reasons neighboorhood-related reasons financial-related reasons other reasons	not used in the data set, only for descriptive statistics		
In your opinion, should the whole population pay a directly financial contribution for preservation, protection and design of landscapes, cities and villages as well as historical sites, additionally to the already paid taxes?	yes no	not used in the data set, only for descriptive statistics		
What amount of money are you personally willing to pay monthly and additionally to the already paid taxes for preservation, protection and design of the landscapes that are shown on the following photographes?	0 euros 1 - 5 euros 6 - 10 euros 11 - 20 euros 21 - 50 euros 51 - 100 euros more than 100 euros	transformed into a binary variable with 1 = is willing to give some money for the landscape and 0 = is not willing to give some money for the landscapes	Because of using five landscapes, there are five landscape: Water, City, Village, Farmhouse, Marshland	Dependen t variable
What is your gender?	female male	with 0 = male and 1 = female	gender	Control variable
How old you are?	0 - 17 years 18 - 24 years 25 - 34 years 35 - 44 years 45 - 54 years 55 - 64 years 65 - 74 years older than 75 years	transformed into three groups where 1 = part of the group and 0 = not part of the group	under 25 years 25 to 65 years over 65 years	Control variable
What is your highest school graduation?	No school graduation Main school Middle school High school College	transformed into one binary variable with 1 = yes (High School or College qualification) and 0 = no (lower education qualification)	higher education	Control variable
How many children do you have?	open answer	amount of children	children	Control variable

Question in the survey	Possible answers	Transformation of the variable	Name of the variable in the data set	Using as...
Please evaluate the following statements concerning your bond to your region: (1) I feel really like home in my region. (2) With my region, i have a lot of personal memories. (3) My personal future is strongly linked to the region.	All statements could be rated from -- = radical refusal over 0 = irresolute to ++ = unreserved approval	transformed into a binary variable with 1 = yes (everyone who marked every statements with ++ or +) and 0 = no (otherwise)	bond to region	Control variable
For what purpose and how often do you use your region?	(almost) daily at least once a week at least once a month less often not at all	transformed into a binary variable with 1 = yes (everyone who marked using the nature daily) and 0 = no (otherwise)	use of nature	Control variable
Do you work on a voluntary basis?	yes no		volunteering	Control variable

Source: Resident survey.

Tab. 7-3: Overview of the year of moving in or return

Year of return/ moving in	Lived in the region since birth with disruption	moved in
1940 to 1959	0,0%	4,3%
1960 to 1979	18,6%	15,5%
1980 to 1999	37,3%	39,5%
2000 to 2019	27,1%	36,5%

Source: Resident survey.

The question about the reasons of staying, returning or moving displays that the reasons for choosing the region STB are similar for each group (Tab. 7-2, Fig. 7-2). Most of the sample said they chose STB due to family-related reasons (73 % of the locals, 64 % of the returnees, 42 % of the newcomers). This reason is often given from people returning to the region: their family still lives there (e.g., Campbell, Strangler & Dailey, 1977). But even people living in neighborhood without any kin show motivations to move into regions where families live (Spring et al., 2017). The least common reasons for the region were financial-related (15 % of the locals and 10 % of each returnees and newcomers). Job-related and neighborhood-related reasons are less important than the family-related ones, and more important than the financial-related reasons. Job-related reasons were also identified in previous literature, and the neighborhood was not that important for the sample (Boneva & Frieze, 2001; Shaw, 1975).

Fig. 7-2: Reasons for staying in the region, returning or moving in

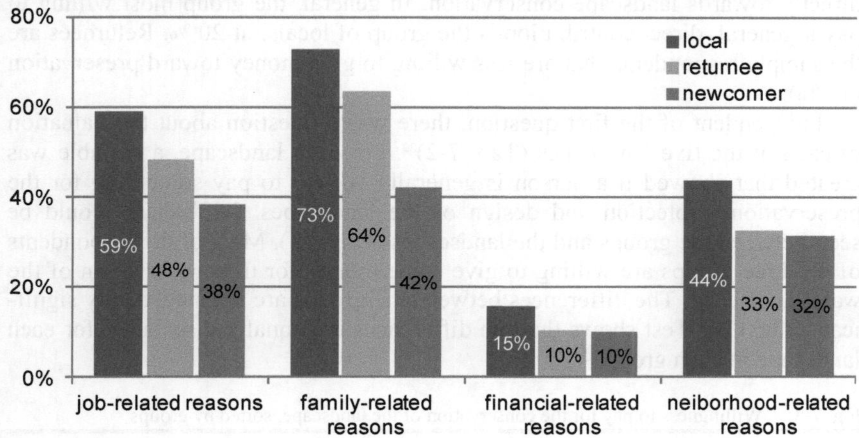

Note: The question allowed multiple answers.
Source: Resident survey, own illustration.

This question was researched before by the context of amenity migration. Von Reichert et al. (2014) found that children are very important for the decision of returning (family-related reasons). Similar reasons for returning were found by Ní Laoire (2007). This could be confirmed by the current data. Johnson (2000) and Von Reichert et al. (2014) found that most of the migrants moved due to economic reasons (job-related/financial). Nevertheless, Rudzitis (1999) found the opposite: economic reason might not be an important reason for moving. The neighborhood was also identified as important reasons for returning to the region (Ní Laoire 2007; von Reichert et al. 2014). In both analyses, the neighborhood not only contains the environmental surrounding but also the feeling of small-town safety. Knapp and Graves (1989) already concluded that the decision to migrate hinges on more than one factor, so the results of the survey are not surprising.

7.4.2 *Willingness To Pay*

The first question regarding the valuation of the landscapes was about a general and direct contribution for protection, preservation and design of the landscape (Tab. 7-2). In addition to taxes, the people living in STB have to pay a concrete amount of money for disposal of the rainwater (Schwarck, 2010). This might be an explanation why less than 20 % of the sample is willing to give some money directly towards landscape conservation. In general, the group most willing to pay a general, direct contribution is the group of locals, at 20 %. Returnees are the sample respondents that are less willing to give money toward preservation (14 %).

Independent of the first question, there was a question about the valuation of each of the five landscapes (Tab. 7-2)[39]. For each landscape, a variable was created that showed if a person is generally willing to pay something for the preservation, protection and design of the landscapes. Differences could be seen between the groups and the landscapes (Fig. 7-3). Most of the respondents of all three groups are willing to give some money for the conservation of the water landscape. The differences between the groups are not statistically significant. The Chi² Test shows that the differences are equally distributed for each landscape in each group.

Fig. 7-3: Willingness to pay for the conservation of the landscape, sorted by groups

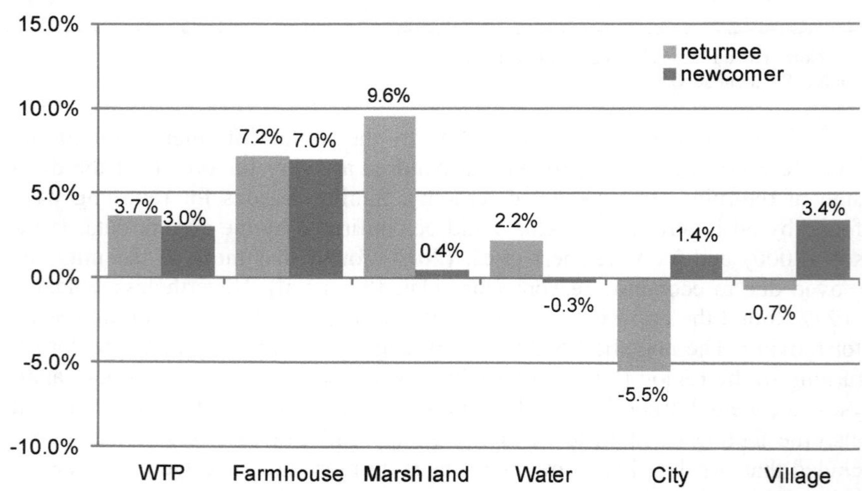

Note: The category of reference is the group of the locals (locals=0%). WTP shows the percentage of people who are willing to pay for at least one of the chosen landscapes.
Source: Resident survey, own calculations and illustration.

But there are also some differences between the landscapes and groups. The landscape of farmhouses is valued highest by the returnees, but the value of the

39 The detailed answers are described in Chapter 5.1.

newcomers is similar. With 29 %, the locals value this type of landscape the least. Marshland is also preferred by returnees. The valuation of the locals and the newcomers are close, with 35 % and 34 %, respectively. The most preferred landscape is the one with water. Almost half of the sample in each group values this landscape as the most protectable landscape.

If a respondent is willing to give some money for the landscapes, most of them are willing to give one to five euros. The willingness to support the different landscapes via money is differently distributed for each group. The highest and the lowest value are within the everyday landscapes. It can be seen that in each group the water landscape is valued highest. The group of the returnees is willing to pay the highest amount of money among the analyzed groups; the lowest amount would be given from newcomers. The locals have valued landscape in between (Fig. 7-3). The confirmation could also be given for the village landscape. The mean of monetary support does not confirm the results for the city landscape. For the symbolic landscape, the results of the means of money are similar to Fig. 7-3. For the landscape of the farmhouse, newcomer and returnees are willing to support the landscape almost the same, the newcomer are on average willing to give higher amounts of money. For marshland, the group of the newcomer is the group with the highest valuation of this landscape type, both giving any money and the amount of money.

7.4.3 Control Variables

Age, gender and education are used in WTP examinations before (e.g., Howley et al., 2012; Kaltenborn & Bjerke, 2002; Kämmerer et al., 1996; Karkow & Gronemann, 2005; Kim et al., 2015; Mitchell & Carson, 1989; Molina et al., 2016; Morrison & Dowell, 2013; Schmitz et al., 2003). These variables are also used in the migration literature to explain moving (e.g., Johnson, 2000; von Reichert et al., 2014; Rudzitis, 1999). These variables were also used (Tab. 7-2). Further control variables are children, bond to region, use of nature and volunteering.

Regarding the gender, the sample consists of more females than men (Tab. 7-4). Only the group of the locals is distributed equally, both other groups have almost 60 % females. In the migration literature, von Reichert et al. (2014) as well as Ritchey (1976) found that the stage of life course might be an important indicator for migration and Johnson (2000) concluded that employment could influence the decision of migration. For this reason, three different variables were defined displaying if a person is less than 25 year, between 25 and 65 years or older than 65 years. The first variable involves young people in the beginning of their work time or still in qualification. The second variable contains the people who have to work and the last is about the people who are retired. Most of the sample is between 25 and 65 years. This group is relatively high represented in the sample and therefore explains that the age is not representative in the region (Fig. 3-2). Because of this result, there were not done any analyses with the group of ages. Regarding education, a variable was generated that shows if a person has a higher education entrance qualification or not. Most of the people in all groups have a lower educational level than the higher education entrance qualification. The amount of people having the asked qualification is highest in the group of returnees (40 %). The percentage of these highly educated people is lesser for the newcomers and the locals, but compared with each other, the members of the group of newcomers are a bit higher.

Tab. 7-4: Overview of control variable, sorted by groups (in percent)

	Locals	Returnees	Newcomers
Male	51	44	41
Under 25 years	13	3	3
25 - 65 years	68	82	77
Over 65 years	19	16	20
0 children	37	18	22
1 child	14	11	22
2 children	25	39	30
3 children	19	26	19
4 children	4	5	4
5 children	2	0	2
6 children	0	0	1
Higher education	27	40	30
Bond to region	74	66	38
Use of nature	53	51	44
Volunteering	42	52	33

Source: Resident survey.

Most of the locals have no children (37 %), while most of the returnees and the newcomers have two children (39 % and 30 %, resp.) (Tab. 7-4). Because of the possibility of multicollinearity between age and the amount of children, this was tested for. Significant negative correlation was only found for those less than 25 years old. All others values do not seem to correlate.

Bond to region was created out of statements concerning the feelings about the region (Tab. 7-2), since the attachment to a place is a complex construct (Anton & Lawrence, 2014). The three used statements (referring to Lalli, 1992; Soini et al., 2012; Weichhart et al., 2006) asked for the emotions about one's home, memories, and future. Only if all three statements were answered by 'very important' or 'important' was there a bond to region. As expected, locals and returnees have a strong bond to region (74 % and 66 %, resp.). The bond to region is weaker for newcomers (38 %) (Tab. 7-4).

From the question on how respondents use nature, a variable for the use of nature was developed (Tab. 7-2). Only if participants use the region almost daily by doing some sports or relaxing in nature did the variable show a positive value. Out of the group of locals and returnees almost half of the respondents use the nature for activities daily (Tab. 7-4). The value for the newcomers is a bit smaller with 44 %.

In the questionnaire there was a question asking about volunteering (Tab. 7-2). The question was used as the respondent marked the question. It can be seen a similar result: half of the returnees are voluntarily involved, while in the group of the locals it is 42 %, and in the newcomers group only 33 % volunteer. Anton and Lawrence (2014) also argued that people with a higher bond to region are getting more involved in social activities in the region. Therefore the correlation between the variables was tested, displaying a coefficient of 0.11, which is low, and thus interpreted as there might only being a low correlation between the two.

7.4.4 Methodology

To analyze the question if newcomer, returnees and locals perceive the landscapes differently, there were used four different logit models, one for each landscape. The first model will use all answers no matter if the individual is a local or somebody who came into the region. The three other models are separated the individuals regarding the fact if they have never left the region (locals), were born in the analyzed county but left it for a while (returnees) or are new to the region (newcomer). This strategy allows a comparison of the influences with the groups and if the influences are special for a group or if it does not matter.

The following function is used to get results for the research question:

$$WTP_{r,l,n} = f(X_n, W_n, R_n)$$

where $WTP_{r,l,n}$ label the WTP of the group r (newcomer, returnee, locals or all) for the asked landscape l. X_n is a matrix of the socioeconomic variables (*sex, age, number of children*), W_n is a matrix of the work-related variables *(academic, full time, part time)* of person n. R_n include the relationship to STB (*bond to region)* and F_n contains the use of the region regarding nature (*use of nature)* and working voluntary for or in the region *(volunteering*). All models are estimated step-by-step to see how the coefficient changes. Subsequent to the described models, there will be the result of the likelihood ratio test described. The interpretation is done using marginal effects by mean.

7.5 Empirical Analyses

This part of the paper presents the empirical results. First, all landscapes are described individually. After this description is following a comparison of the results within the symbolic or everyday landscapes and in the end, the results of the landscape types are summarized. This chapter ends with the answers of the hypotheses.

7.5.1 Results of the Landscape in a General Way by Groups

The WTP for at least one of the examined landscapes for the whole sample displays two significant variables (Tab. 7-5)[40]. The first of these is the variable for the number of children that shows a negative impact on the WTP for at least one of the landscapes. This would imply that people with more children are less willing to pay a financial amount for landscape protection. The coefficient is almost stable with the addition of further variables. It only decreases with the inclusion of the educational level. Its effect is medium-high, with -0.05 h[41]. The second one is higher education. People having at least the higher education entrance qualification perceive the regional landscapes higher than people with a lower educational level. The coefficient is highest for the model without bond to region, use of nature and volunteering, but decreases with the integration of bond to nature. It remains almost stable afterwards. Its marginal effect is high.

The models for the locals did not show any significant variable neither in the models with lesser variables than with more variables.

In the case of the returnees only one variable shows a significant coefficient. The higher educational level displays a positive coefficient. This result implies that people having a higher education are twice as willing to pay for at least one of the landscapes as others. Its impact is a bit smaller for the models without volunteering, and without volunteering and use of nature – but it is still closes to two. The effect of this variable is high.

For the newcomers there are the same two variables significant that are important for the overall sample: number of children and higher education. The coefficient for the number of children is again negative and a bit higher than for the overall sample. It implies that people with an increasing number of children have a lower perception of the nature than people with lesser children. The coefficient remains stable over the addition of more variables. Its effect is medium high. The coefficient for higher education is positive. It decreases with the

40 Tab. A 4 shows the coefficients and the pseudo R^2 of the estimation.

41 Following, the marginal effect of 0.000 to 0.05 is described as low effect, an effect between 0.051 and 0.2 is presented as a medium high effect and the higher effects are used as high effect.

addition of bond to region and remains on this level. It would mean that people with a higher educational level have a higher perception of the landscape than those with a lower degree of education. As seen for the overall sample and the returnees, the effect of the higher education is high. Furthermore, the age between 25 and 65 and bond to region displays some significant coefficients in the models without all variables. The age variable has a positive coefficient, but it loses its significance with the addition of bond to region and decreases then. Bond to region is only significant for the model without use of nature and volunteering. Its impact with all variables – like in the model – is negative.

Tab. 7-5: Marginal effects for the estimation of willingness to pay for at least one of the landscapes, sorted by groups (margins, standard errors in parentheses)

WTP	All		Locals	Returnees		Newcomers	
Gender	0,021		0,039	-0,044		0,035	
	(0.064)		(0.116)	(0.211)		(0.089)	
Under 25 years	0,056		0,159	0,000		0,141	
	(0.155)		(0.251)	(omitted)		(0.258)	
Between 25 and 65 years	0,082		0,052	-0,202		0,158	
	(0.086)		(0.154)	(0.262)		(0.117)	
Number of children	**-0,050**	**	-0,026	-0,056		**-0,084**	**
	(0.025)		(0.045)	(0.102)		(0.035)	
Higher education	**0,224**	***	0,112	**0,503**	**	**0,261**	***
	(0.070)		(0.148)	(0.217)		(0.096)	
Bond to region	-0,063		0,141	-0,078		-0,144	
	(0.063)		(0.135)	(0.207)		(0.089)	
Use of nature	-0,005		-0,017	0,240		-0,082	
	(0.063)		(0.112)	(0.200)		(0.087)	
Volunteering	-0,014		-0,026	0,109		0,036	
	(0.065)		(0.117)	(0.208)		(0.095)	
N	293		92	36		163	

Note. Stars indicate significance levels: * p<0.1, ** p<0.05, *** p<0.01. N displays the number of cases.
Source: Resident survey, own calculations.

The likelihood ratio test indicates an increase of the model fit for the overall sample by the variable of the educational level. The same can be said for the newcomers and returnees. With the integration of bond to region an increase of the model fit can be described for the sample of the locals.

7.5.2 *Results of the Farmhouse Landscape by Groups*

The first, specific landscape analyzed more in depth was the farmhouses. The only significant variable for the whole sample is the variable of higher education (Tab. 7-6)[42]. Its coefficient is positive and indicates that if an individual has a higher education level they value the farmhouse landscape one unit higher

42 The coefficients and the pseudo R^2 are displayed in Tab. A 5.

than individuals with lower levels of education. The coefficient decreases with the involvement of bond to region but then stays almost at one. Its significance is on the highest level of 0.01. The pseudo R-squared increase with the integration of higher education, implying that the model fit does not improve with the involvement of more variables. The marginal effect of it is high.

Tab. 7-6: Marginal effects for the estimation of the farmhouse landscape (symbolic landscape), sorted by groups (margins, standard errors in parentheses)

Farmhouses	All		Locals	Returnees		Newcomers	
Gender	-0,011		-0,072	-0,244		0,066	
	(0.061)		(0.109)	(0.203)		(0.086)	
Under 25 years	0,056		0,205	0,000		0,210	
	(0.142)		(0.226)	(omitted)		(0.246)	
Between 25 and 65 years	0,010		0,028	-0,143		0,026	
	(0.084)		(0.147)	(0.244)		(0.115)	
Number of children	-0,031		-0,013	-0,052		-0,051	
	(0.024)		(0.042)	(0.085)		(0.033)	
Higher education	**0,223**	***	0,084	**0,502**	**	**0,272**	***
	(0.064)		(0.131)	(0.197)		(0.088)	
Bond to region	-0,077		0,146	-0,146		-0,135	
	(0.060)		(0.129)	(0.182)		(0.088)	
Use of nature	-0,020		-0,097	0,195		-0,040	
	(0.060)		(0.102)	(0.178)		(0.084)	
Volunteering	-0,005		-0,011	0,226		-0,002	
	(0.062)		(0.108)	(0.188)		(0.092)	
N	293		92	36		163	

Note. Stars indicate significance levels: * p<0.1, ** p<0.05, *** p<0.01. N displays the number of cases.
Source: Resident survey, own calculations.

Having a look at the locals, there are not any significant variable for the perception of the landscape. The pseudo R-squared increases by the rise of variables but is with 0.05 low.

Higher education is significant on the level of 0.05 for returnees. The marginal effect is as double as in the whole sample and implies that higher education qualification entrance might be really important for returnees concerning the perception of the farmhouse landscape. The coefficient is decreasing with the integration of bond to region and again with the involvement of use of nature, but increases with the involvement of volunteering. The pseudo R-squared indicates the model with all tested variables fits the best for the group of returnee.

Beside higher education, number of children and bond of region display some significant coefficients for the newcomers on the step-by-step estimation, but not in the model with all variables. Number of children is an important factor for newcomers for the model with gender, age and number of children, and decreases minimally with the involvement of higher education. It can be stated that a higher number of children decrease the valuation of this type of landscape by one-quarter unit. With the integration of bond to region, the number of children has no longer has any significant coefficients. Bond to region also has a negative impact on the valuation for newcomers, but only in the model that did use nature and volunteering. Its negative impact is about half-a-unit if an individual has a bond. Higher education is significant on the highest significant level (0.01) in all models where this variable is used in this group. The coefficient decreases again with the involvement of use of nature and volunteering. The effect is a bit higher than for the overall model. Pseudo R-squared shows that the model fit only increases with the involvement of higher education and bond to region.

The likelihood ratio test indicates that for the sample as a whole that the addition of higher education increases the model fit significantly. The same could be said for the group of the returnees and newcomers. The model for the locals does not display a significant increase of model fit in each of the estimated models.

7.5.3 Results of the Marsh Landscape by Groups

The landscape of the marshland shows two significant variables: number of children and higher education. Both are significant on the level of 0.1, but their impacts are different (Tab. 7-7)[43]. While the number of children has a negative impact on the perception of the landscape, higher education has a positive one. The coefficient of number of children is decreasing with the integration of more variables. Its impact is low. The coefficient of higher education is not constant, but sways. With the involvement of bond to region the coefficient increases, but decreases again with the involvement of use of nature and rises with volunteering. Its effect is medium-high, at 0.119. The pseudo R-squared increases with the integration of higher education but after this increase the R-squared is constant.

For the locals, bond to region seems to be an important factor for the perception of the marshland. The impact of it is positive and significant on the level of 0.05. The coefficient remains constant with the integration of use of nature and increases with the model of all variables. The marginal effect displays a high effect. The pseudo R-squared increases with the integration of bond of region and is highest for the last estimated model with 0.09.

43 Details about the coefficients and the pseudo R^2 can be found in Tab. A 6.

Tab. 7-7: Marginal effects for the estimation of the marshland (symbolic landscape), sorted by groups (margins, standard errors in parentheses)

Marshland	All		Locals		Returnees		Newcomers	
Gender	0,086		-0,038		0,095		**0,158**	*
	(0.061)		(0.116)		(0.200)		(0.084)	
Under 25 years	0,114		0,365		0,000		0,210	
	(0.144)		(0.257)		(omitted)		(0.236)	
Between 25 and 65 years	0,044		0,116		-0,085		0,021	
	(0.085)		(0.158)		(0.265)		(0.111)	
Number of children	**-0,040**	*	-0,015		-0,039		**-0,058**	*
	(0.024)		(0.043)		(0.094)		(0.033)	
Higher education	**0,119**	*	0,068		**0,397**	**	0,103	
	(0.065)		(0.145)		(0.199)		(0.085)	
Bond to region	0,055		**0,405**	**	-0,043		-0,059	
	(0.061)		(0.157)		(0.196)		(0.085)	
Use of nature	-0,058		-0,097		0,127		-0,099	
	(0.060)		(0.109)		(0.190)		(0.082)	
Volunteering	-0,058		-0,100		-0,007		-0,042	
	(0.062)		(0.116)		(0.195)		(0.089)	
N	293		92		36		163	

Note. Stars indicate significance levels: * p<0.1, ** p<0.05, *** p<0.01. N displays the number of cases.
Source: Resident survey, own calculations.

The models for returnees show higher education as significant variable (level of 0.05). Its coefficient is positive and increases the valuation of the marshland. The coefficient decreases with the integration of bond to region, use of nature, and volunteering. As in the landscapes before, the effect is high. The pseudo R-squared increases with the involvement of higher education but then almost remains constant at 0.12.

Two different variables are important for the perception of the marshland of newcomers: gender and number of children. In all estimated models, number of children is negatively related to the valuation of this landscape type. The coefficient of this variable is minimal decreasing with the integration of more variables as well as the level of significance decreases from 0.05 to 0.1 if higher education is integrated into the model. Its marginal effect shows a medium-high effect. However, gender is first significant when higher education was added to the variables. Its coefficient is even increasing with the integration of the other variables. This means that the perception of the marshland will be higher if the person is female. Its effect is medium-high. Pseudo R-squared increases with the integration of higher education and again with use of nature, but is still low at 0.07.

The likelihood ratio test shows only significant improvements of model fit of the models of locals and returnees. For the locals, the model fit increases with the addition of bond to region and for the returnees with the involvement of higher education.

7.5.4 Results of the Water Landscape by Groups

The model for the water landscape shows three variables with weak significance level that changes with the integration of more variables into the model. The variables of age (younger than 25 and between 25 and 65 years of age) are positively related to the perception of this landscape type. With the integration of higher education, these variables are not significant anymore. The number of children has a negative influence on the valuation and is significant on the lowest level of significance. Its coefficient is small and decreases with the involvement of higher education. Afterwards it remains constant, but with the integration of volunteering it is not significant anymore. Only one variable (higher education) is highly significant in all models where it is integrated (Tab. 7-8)[44]; the other variables are not significant anymore in the overall model. Higher education has a positive impact on the valuation. If an individual has a higher education entrance qualification, the valuation is significantly higher than with a lower graduation. The coefficient varies in a small range: it is smaller with the integration of bond to region, remains constant with the integration of use of nature and increases with the integration of volunteering. The marginal effect shows a medium high impact. The pseudo R-squared increases with the integration of variables, but is still low with 0.06.

Tab. 7-8: Marginal effects for the estimation of the water landscape (everyday landscape), sorted by groups (margins, standard errors in parentheses)

Water	All		Locals		Returnees		Newcomers	
Gender	-0,001		-0,096		-0,054		0,070	
	(0.063)		(0.121)		(0.202)		(0.088)	
Under 25 years	0,116		0,264		0,000		0,163	
	(0.155)		(0.264)		(omitted)		(0.255)	
Between 25 and 65 years	0,073		0,140		-0,217		0,100	
	(0.086)		(0.164)		(0.257)		(0.116)	
Number of children	-0,041		-0,020		-0,009		**-0,069**	**
	(0.025)		(0.045)		(0.094)		(0.035)	
Higher education	**0,227**	***	**0,257**	*	**0,394**	*	**0,243**	***
	(0.069)		(0.155)		(0.201)		(0.092)	
Bond to region	-0,018		**0,247**	*	-0,137		-0,102	
	(0.063)		(0.146)		(0.199)		(0.089)	
Use of nature	0,008		0,030		0,171		-0,063	
	(0.062)		(0.115)		(0.190)		(0.086)	
Volunteering	-0,051		-0,165		0,002		0,030	
	(0.065)		(0.123)		(0.196)		(0.093)	
N	293		92		36		163	

Note. Stars indicate significance levels: * p<0.1, ** p<0.05, *** p<0.01. N displays the number of cases.
Source: Resident survey, own calculations.

44 Coefficients and the pseudo R^2 are shown in Tab. A 7.

For locals, two variables displays significant on the lowest significance level of 0.1. Higher education has a positive influence on the perception, when all variables are included. Its effect is high. Similar developments can be observed with bond to region. This variable has also a positive influence and increases the valuation. The effect is equivalent to this of higher education. The pseudo R-squared increased with the integration of higher education, bond to region and volunteering, but is still relatively low with 0.07.

Higher education is the only significant variable for returnees. If the individual has a higher education entrance qualification the perception of the water landscape increases. The coefficient decreases with the integration of bond to region and again with the involvement of use of nature, but increases a bit with the model of all variables. Its significance level decreases with the integration of use of nature. Its effect is again high. The pseudo R-squared increased when integrating higher education. With 0.12, it is higher than for the locals and the overall models.

The age between 25 and 65 is important to look at on the valuation of newcomers, but only in the model with gender, age and number of children. If the individual is between 25 and 65 years old, the perception of the water landscape is higher. Its significance is low and the variables lose their significance when higher education is included in the model. Number of children has a negative and significant impact on the valuation of this type of landscape. The significance is lower in the models where bond to region and use of nature come into the estimation, but its significance is the same in all other models. The coefficient is varying lightly between the models. Its effect is medium high. As in the models for returnees, higher education has a positive and highly significant impact. The coefficient sinks with the involvement of more variables but still have a high impact. The involvement of higher education and bond to region make for an increase of the pseudo R-squared but it is again low with 0.09.

The likelihood ratio test indicates that the inclusion on higher education is responsible for a significant better model fit, at least in the model with the whole sample and the group of the returnees and the newcomers. For locals, the extension does not show a significant improvement of the model fit.

7.5.5 Results of the City Landscape by Groups

The valuation of the city landscape is highly dependent of the variable of higher education (Tab. 7-9)[45]. This variable is highly significant in each of the estimated models, but its coefficient is decreasing with the addition of more variables. The last model with all variables only shows a coefficient of 0.82 and a medi-

45 Tab. A 8 shows the coefficients and the pseudo R^2.

um-high effect. Number of children is only significant in the first estimated model with the use of gender, age and number of children. Its influence is negative and only significant on the level of 0.1. With the involvement of higher education this variable is not significant anymore. The pseudo R-squared is increasing with the inclusion of higher education and again with the involvement of use of nature, but decreases with the integration of volunteering. Best model for the whole sample therefore seems to be the model without volunteering.

The group of locals does not show any significant results. The pseudo R-squared increases with the involvement of higher education and again with volunteering, but is with 0.08 relatively low. This implies that the model is not suited for explaining the perception of locals for this type of landscapes.

Tab. 7-9: Marginal effects for the estimation of the city landscape (everyday landscape), sorted by groups (margins, standard errors in parentheses)

City	All		Locals	Returnees	Newcomers	
Gender	-0,045		0,039	-0,133	-0,055	
	(0.057)		(0.109)	(0.130)	(0.078)	
Under 25 years	0,089		0,106	0,000	0,253	
	(0.128)		(0.219)	(omitted)	(0.221)	
Between 25 and 65 years	-0,040		-0,035	-0,008	-0,052	
	(0.077)		(0.145)	(0.191)	(0.104)	
Number of children	-0,032		-0,014	-0,050	-0,033	
	(0.022)		(0.043)	(0.056)	(0.031)	
Higher education	**0,167**	***	0,157	0,180	**0,204**	**
	(0.059)		(0.126)	(0.124)	(0.080)	
Bond to region	-0,080		0,028	-0,015	-0,127	
	(0.056)		(0.122)	(0.120)	(0.083)	
Use of nature	-0,039		-0,068	-0,190	0,002	
	(0.056)		(0.102)	(0.116)	(0.078)	
Volunteering	-0,009		0,123	0,106	-0,074	
	(0.058)		(0.106)	(0.132)	(0.086)	
N	293		92	36	163	

Note. Stars indicate significance levels: * p<0.1, ** p<0.05, *** p<0.01. N displays the number of cases.
Source: Resident survey, own calculations.

Almost the same could be said about the model of returnees for the valuation of the city landscape. None of the used variables displays a significant influence on the perception – not even the constant of the models. The pseudo R-squared increased with inclusion of higher education, again with use of nature, and – like the last time – with the involvement of volunteering. The value of the pseudo R-squared is higher than for locals (0.15). But again, this model does not seem to be adequate specified to explain the valuation.

Highly significant for newcomers is higher education. The variable is significant in each estimated model. If an individual has a higher education entrance qualification, the valuation of the city landscape is higher than for individuals with a lower education. The significance level is decreasing with the inclusion of volunteering. Its effect can be described as high. Bond to region is also significant, but only in the models without volunteering. Its influence is negative what would mean that newcomer with a bond to region are valuing the city landscape lower than newcomers without a bond to region do. The pseudo R-squared is increasing by the inclusion of higher education and bond to region and remains constant with 0.09 afterwards.

The model fit significantly increases with higher education in the models of the whole sample and for newcomers. The other models do not improve the fit significantly.

7.5.6 Results of the Village Landscape by Groups

Number of children and higher education are the variables that are influencing the perception of the village landscape (Tab. 7-10)[46]. Number of children is significant on the level of 0.05, but this level is decreasing with the addition of more variables. The same could be said for its coefficient: It is decreasing with the involvement of more variables. This coefficient for children is negative and would mean that the perception of this landscape type decreases with the increase of children. Its effect is relatively low. Higher education has a positive influence on the valuation of the village landscape. Its significance level is highest. With the addition of bond to region, the coefficient increases slightly, but with the integration of use of nature it decreases again. It remains constant with the integration of volunteering. As for the other landscapes, the effect is high. The pseudo R-squared increased with the involvement of higher education, and again with the addition of use of nature.

For the locals, higher education has only an impact on the valuation of this landscape type if the variables gender, age, number of children, higher education and bond to region are integrated into the model. Its coefficient is only significant on the lowest significance level (0.1). A higher level of significance is displayed in bond to region. Its coefficient remains constant with the addition of use of nature, but decreases with the integration of volunteering. If an individual has a bond to this region his or her perception is increasing with a high effect. The pseudo R-squared increased with the addition of bond to region, and again with the use of nature. The pseudo R-squared is still low with 0.08.

None of the used variables seems to be an important factor for the valuation of the village landscape. The constant is negative and the pseudo R-squared is

46 The coefficients of the estimation as well as the pseudo R^2 can be found in Tab. A 9.

not significant. Its value is increasing slightly with the addition of more variables.

Tab. 7-10: Marginal effects for the estimation of the village landscape (everyday landscape), sorted by groups (margins, standard errors in parentheses)

Village	All		Locals		Returnees	Newcomers	
Gender	0,004		0,055		0,138	-0,047	
	(0.062)		(0.117)		(0.183)	(0.087)	
Under 25 years	0,102		0,161		0,000	0,208	
	(0.149)		(0.246)		(omitted)	(0.253)	
Between 25 and 65 years	0,048		-0,019		0,135	0,074	
	(0.086)		(0.153)		(0.276)	(0.115)	
Number of children	**-0,046**	*	-0,017		-0,023	**-0,066**	*
	(0.025)		(0.044)		(0.084)	(0.034)	
Higher education	**0,210**	***	0,208		0,110	**0,260**	***
	(0.067)		(0.146)		(0.174)	(0.091)	
Bond to region	0,013		**0,302**	**	0,067	-0,072	
	(0.062)		(0.151)		(0.187)	(0.088)	
Use of nature	-0,054		-0,061		-0,001	-0,045	
	(0.061)		(0.110)		(0.172)	(0.085)	
Volunteering	-0,013		0,081		-0,142	0,027	
	(0.063)		(0.115)		(0.177)	(0.093)	
N	293		92		36	163	

Note. Stars indicate significance levels: * p<0.1, ** p<0.05, *** p<0.01. N displays the number of cases.

Source: Resident survey, own calculations.

For the valuation of the village landscape of the newcomers are the variables number of children and higher education are important. Number of children is significant on 0.05 if there are only gender, age and number of children integrated in the model. With the addition of the other variables, the level of significance is decreasing to 0.1. The coefficient itself is negative, so that the valuation of this landscape is decreasing with more children. Its effect is medium high. Higher education is highly significant (0.01) in each of the estimated model. Its coefficient is decreasing with the integration of more variables, except with the involvement of volunteering. In this case the coefficient is slightly increasing. With higher education the valuation is higher than for the higher entrance education qualification. The effect of this variable is high. The pseudo R-squared is highest in the model without volunteering and is decreasing with the integration of this variable. This would mean that the model is better specified without volunteering.

The likelihood ratio test suggests that the models for the overall sample and newcomers have a significant better fit if higher education is adding to the control variables. The fit of the model for locals is significant higher if bond to region is adding to the variable set. The fit of the models for returnees are not significantly better if more variables are integrated.

7.5.7 Comparison of the Results for the Symbolic and Everyday Landscapes

For the perception of the symbolic landscape, education seems to be an important factor (Tab. 7-11). Especially for returnees and newcomers education it has a positive impact. This impact is about as twice as higher for returnees than for newcomers for the farmhouse (Tab. 7-6). For the marshland (Tab. 7-7) it could shown that the educational level is not as important as for the farmhouse landscape – albeit still important. With respect to the everyday landscape, there seems to be two different types of everyday landscapes: education is more important in every analyzed group for the close-to-nature landscapes (in this case water) than for constructed areas like the city and the village (Tab. 7-11). For the constructed landscapes, education is not as important as for the valuation for every group. Only for the valuation of these two landscapes of newcomers, education is important. These results support the results of previous WTP studies (Howley et al., 2012; Marangon & Visintin, 2007) and the findings that people with higher education might return or move into the previous or other regions (Johnson, 2000; von Reichert et al, 2014). The importance regarding education is highest for newcomers (highly significant values of farmhouse, village and water landscapes). The lowest importance within the researched groups could be found amongst locals (lowest significance level only for water landscapes).

These results could be confirmed by the estimation of the WTP for at least one of the landscape (Tab. 7-11). The level of education has a positive impact on the perception of the regional landscapes. The coefficients for the overall sample, the newcomers and returnees show significance, but the coefficient for the locals is at least also positive.

Number of children seems to have a negative impact for the perception of newcomers of the symbolic landscape (Tab. 7-11). None of the step-by-step models of the other groups showed significant influences. Having a look on the everyday landscapes, number of children has also a negative impact on the valuation, at least for water and village landscapes. The city landscape has no significance values for this variable in none of the groups. Number of children has at least for the region STB a negative influence on the valuation of all landscapes, except the city landscape. Even if the values for the other analyzed groups are not significant, the coefficients are also negative.

Tab. 7-11: Overview of the results, sorted by groups and landscapes

	WTP				Farmhouses				Marsh Land				Water				City				Village			
	All	L	R	N	All	L	R	N	All	L	R	N	All	L	R	N	All	L	R	N	All	L	R	N
Gender	+	+	-	+	-	-	-	+	+	-	+	+*	-	-	-	+	-	+	-	-	+	+	+	-
Under 25 years	+	+	+	+	+	+	o	+	+	+	o	+	+	+	o	+	+	+	o	+	+	+	o	+
Between 25 and 65 years	+	+	-	+	+	+	-	+	+	+	-	+	+	+	-	+	-	-	-	-	+	-	+	+
Number of children	-**	-	-	-**	-	-	-	-	-*	-	-	-*	-	-	-	-**	-	-	-	-	-*	-	-	-*
Higher education	+***	+	+**	+***	+***	+	+**	+***	+*	+	+**	+	+***	+*	+*	+***	+***	+	+	+**	+***	+	+	+***
Bond to region	-	+	-	-	-	+	-	-	+	+**	-	-	-	+*	-	-	-	+	-	-	+	+**	+	-
Use of nature	-	-	+	-	-	-	+	-	-	-	+	-	+	+	+	-	-	-	-	+	-	-	-	-
Volunteering	-	-	+	+	-	-	+	-	-	-	-	-	-	-	+	+	-	+	+	-	-	+	-	+

Note. + indicates a positive direction of the coefficient, - indicates a negative direction of the coefficient, o indicates an omitted variable, stars indicate significance levels: * p<0.1, ** p<0.05, *** p<0.01. WTP= implies the WTP for at least one of the analyzed landscapes, L= Locals, R=Returnees, N= Newcomers.

Source: Resident survey, own calculations.

While the impact of the number of children is not always significant for the newcomers, this variable seems to be important for the fact it provides a financial contribution from this group for at least one of the landscapes (Tab. 7-11). The coefficient shows a significant and negative impact on the perception of the regional landscape.

Gender does not seem to be an important variable for the valuation of the landscapes (Tab. 7-11). Even if only a few coefficients are significant, it can be observed that female newcomers value symbolic landscapes higher, while female locals value it lower than men. The returnees are in between these groups: men value the marshland higher, females the farmhouse landscape. For the everyday landscapes, the coefficients indicate that female newcomers and male locals and returnees value water landscape higher. Females value the village higher if they are returnees or locals. For newcomers the coefficient is negative and therefore indicates that men value the village landscape higher. The city landscape is valued higher by male returnees and newcomers. The valuation of the locals is not obvious, given that during the step-by-step estimation the coefficient changed from negative to positive. This variable does not seem to be important for the fact to value a regional landscape. For the decision to give something to at least one of the landscapes this variable does not seem to make a big impact.

Bond to region has a positive impact on the perception of locals (Tab. 7-11). The coefficient is significant for farmhouse, water and village landscape. There is not a difference between the symbolic and the everyday landscape type. For newcomers (significant negative coefficients for city and farmhouse landscape within the step-by-step estimation), bond to region might have a negative impact. For returnees, the similar effects are observed except for villages. For this landscape type the coefficient of bond to region is positive and indicates a positive impact. Again, there does not seem to be difference between the landscape types. But this variable has no direct influence on the WTP for at least one of the regional landscapes.

7.5.8 Results Regarding the Hypotheses

The first hypothesis assumed that newcomers and returnees value the symbolic landscapes more than locals. This hypothesis can be proven using Fig. 7-3 for the farmhouses for the group of newcomers. Returnees and locals perceive it on the same level. For this landscape, the hypothesis cannot be fully validated. For the marshland returnees value this landscape highest, followed by locals. Therefore, the hypothesis can only by proved for this landscape for returnees and needs to be rejected for newcomers.

The following assumption is about a preference for the water landscape by each group. This can also be proved with Fig. 7-3. In comparison with the other landscapes all groups are most willing to pay for the water landscape. These preferences are followed by the village landscape for the newcomers and the locals and the marshland for the returnees. Another observation is common between the groups: The lowest amount in each group is willing to pay for the city landscape.

A positive impact of bond to region is assumed for the locals and a negative one for the returnees and newcomers. The positive impact for the locals can be proved with significant coefficients for the landscape marshland, water and village landscape (Tab. 7-7, Tab. 7-8, Tab. 7-9). The coefficients for the other landscape types show a positive direction, but the coefficients are not significant and therefore the positive impact could only be assumed for the farmhouse and city landscape. The variable has a negative influence on the WTP for newcomers, but the coefficients are not significant and therefore the influence cannot be proved with reliable results. For the group of the returnees, it could only be assumed that this fact has a negative impact on the symbolic landscapes as well as the water and city landscape. The direction of the coefficient changes for the village landscape so that these results for the group are also not robust and cannot be proved with this work.

Hypothesis 4 assumed that the daily use of the region for relaxing or sports has a positive influence of the valuation of the everyday landscape but not on the symbolic ones. The variable use of nature did not show any significant coefficient in none of the landscapes. There might be a pattern for the symbolic landscapes (Tab. 7-6, Tab. 7-7): It might be the case that the daily use of nature might have a negative impact on the symbolic landscapes for the locals and newcomers but a positive one for the returnees. However, because the coefficients are not significant, the hypothesis could not be proven. For the everyday landscapes there were not found any pattern.

The fifth hypothesis is about a positive impact of the number of children on the everyday landscapes and a negative one on the symbolic ones. It could be observed that the coefficients of number of children has a negative direction for the symbolic landscape in each of the groups, but only for the marsh land the groups of the newcomers show a significant coefficient (Tab. 7-7). Further, the coefficients for the water and village landscape (Tab. 7-8, Tab. 7-10) show significant and negative coefficients. This might indicate that for newcomers the number of children is not important to distinguish the perception of symbolic and everyday landscapes. The same might be true for both of the other groups because these coefficients are also negative, even if these are not significant. Therefore the hypothesis could only be partly proved (for the symbolic landscape) but rejected (for the everyday landscape) for newcomers, but not for both of the other groups.

Higher education was assumed to have a positive impact on the perception in each landscape type and resident group. This can be proved for almost every landscape. The coefficient for the newcomers is significant for the farmhouse landscape as well as for the water, city and village landscape (Tab. 7-6, Tab. 7-8, Tab. 7-9, Tab. 7-10). Even if the coefficient for the marshland is not significant, its direction is also positive. The highest coefficient can be observed for the farmhouse, followed by the village landscape and in the end the water and city landscape at the same level. This leads to evidence for the group of the newcomers. For the returnees, the coefficients for the farmhouse, marshland and water landscape display significantly (Tab. 7-6, Tab. 7-7, Tab. 7-8). For the other landscapes, the coefficients are also positive but not significant. The highest coefficient is for the farmhouse landscape, the coefficients for the other two landscapes are on a similar level. This might also be evidence for the assumption that the level of education might have a higher impact on the symbolic landscapes, especially for the returnees. The coefficients for the group of the locals show only one significant coefficient (Tab. 7-8). Indeed, all other coefficients are positive, but not significant, so that the hypothesis for the locals cannot be proven. In comparison of the groups, the results identify the importance of the higher level of education for the water landscape.

The last hypothesis assumed that the variable volunteering has a negative impact on the WTP for the landscapes. This cannot be proved with the used data – none of the coefficients are significant. Even the directions of the coefficients are changing between the landscape types. For instance, returnees show a positive coefficient for the farmhouse, but a negative one for the marshland (Tab. 7-6, Tab. 7-7), and newcomers have a positive coefficient for water and village landscapes but a negative one for the city (Tab. 7-8, Tab. 7-9, Tab. 7-10). The only pattern that could be identified for locals: with exception of the village landscape, they showed negative coefficients for the landscapes.

7.6 Discussion

The following part discusses the empirical results regarding previous literature and the generalizability of the results. Additionally to this, there is a discussion of further variables for research interesting in a similar setting of analyses. In this part there is also shown future topics for research.

7.6.1 *Empirical Results*

The symbolic landscape of the farmhouses is valued higher by the newcomers and the marshlands by returnees. This might be the case because locals have been grown up in these landscapes and because they have never left the region, if might be the case that they do not know about the uniqueness of these landscape in comparison to other regions. The newcomers and returnees have seen other regions and might therefore identify the landscape as being more special. Therefore the perception of the groups of newcomers and returnees is better, but not for both landscapes. It might be the case that newcomers perceive the farmhouses as more special than the marshlands and for the returnees it is the other way around.

The second hypothesis was based on the theories about the landscape perception. In regard of the everyday landscapes, the water landscape was preferred by each of the groups. This is not surprising because these landscapes have the aspects called 'prospect' and 'refuge' used by Appelton (1984). This type of landscape is also the least man-made landscape of the three everyday ones. Therefore the result is also in line with Ulrich (1983) and Kaplan (1987).

For newcomers the bond to region has a negative impact and for locals a positive impact. In the results it could be seen that returnees are in some cases similar to newcomers and in others for locals because the direction of their impact is not clear. Ní Laoire (2007) also concluded that returnees are in some cases similar to newcomers and in other cases to locals. Brennan and Cooper (2008) and Kaltenborn and Williams (2002) explained it in the way that returnees and especially newcomers might have a lower bond to region because of, for instance, lower insight on the culture of the region they moved, or because of a lack of social connections in the region. This might be true of newcomers or returnees who just moved to the region, but the question is what it is like for the newcomers and returnees who already been there a long time?

The research could not identify any significant coefficients for the use of nature but could assume a negative impact for newcomers and locals for the symbolic landscapes. This could be explained using the arguments of Green (2001). He stated that some people just liking to have the opportunity to use the nature or try to preserve it for further generations even if they do not use it themselves. For the everyday landscapes, no impact was noted, hence it could be assumed that using the nature on a regular basis is not relevant to landscape perception.

The impact from the number of children for newcomers in the symbolic landscapes could be identified in these results. This could indicate that people without any children perceive these landscape types higher than people with children. Even the newcomers might move to the region because of these symbolic landscapes. The returnees and locals might know the symbolic landscapes already, and for them the number of children might not matter in their perception of the landscape.

The educational level has a positive impact for the newcomers and returnees – even if the influences are more important for the returnees in cases of the symbolic landscape type and the everyday landscape more so for the newcomers. It might be the case that these people have learnt more about the landscapes because of their experience in different regions. It could also be the case that these two groups are wealthier than the locals, so that another of Green's (2001) arguments comes into focus. He suggested that people who are getting wealthier become more aware of the nature around them. A comparison of the means of the income of these groups shows that locals have the lowest means, which could indicate that this argument might be true for this sample.

In the data no impact from volunteer work could be found. Even if Halfacree and Rivera (2011) argued that people moving to a region are often involved in volunteer activities with social interaction to develop bonds to the region and its community, the importance for the perception of the landscape could not be verified. Differences – or at least influences – between the resident groups or the landscape types could not be identified.

7.6.2 Difficulties Regarding the Data

One of the difficulties coming along with the data is the question at what time newcomers or returnees are similar to locals regarding their surrounding (Ní Laoire, 2007). This research used directly a question that asked if a person is a local, returnee, or newcomer (Tab. 7-3). It might be the case that some returnees or newcomers already felt like they have never been somewhere else, but needed to answer the question as newcomers and returnees. It might also be that people that just moved to the region were included. Halfacree and Rivera (2011) concluded that migration to a region is a long process and does not end with the move. Therefore, the relatively new newcomers or returnees might have different perceptions of the landscape than people living in the region for some years – even if they moved to the region or came back to it. To have a solution for this phenomenon, Brennan and Cooper (2008) used the years of residence for the categorization of the newcomers and locals. But the authors had also trouble because for different ages the time span might be appearing differently.

The origins of movement could not be determined. For instance, Rudzitis (1999) found differences between the regions (urban or suburban) returnees came from. This could also lead to different values regarding the landscape and the perception of it. Even if the size and the surrounding of the origin seem to be matter, there does not seem to be differences between a domestic or international migration (von Reichert et al., 2014).

There might be two types of positive selection in the data: first, it might be the case that only people who are already have perceptions about the landscape answered the questionnaire and second, that locals who wanted to move but could not due to, for instance, family obligations, could be included. These people might have different perceptions than the other locals who stayed because they wanted to stay. Anton and Lawrence (2014) concluded that people who were not moving once get more satisfaction from their current place than other groups. But it might be the case that people who wanted to move but could not are more dissatisfied with their surrounding and therefore the group of the locals might be biased the value by these people.

Finally, the sample is relatively small – even if the response rate was about 15 %. The whole sample includes about 300 answers with almost 55 % newcomers and 30 % locals, respectively. The group of the returnees is consequently even smaller. The results of this examination could therefore be understood as explorative analyses. The results need to be proven by a bigger sample – even if Brennan and Cooper (2008) did their regression with a small sample from about 50 cases of each of their groups as well.

7.6.3 Further Variables

It might be good to integrate a variable of the infrastructure and the accessibility of the landscape in the analyzed region of STB, because Shumway and Otterstrom (2000) found that these could influence the region's economic growth. This sample asked about the quality of life regarding the infrastructure using a car or local public transportation, but there was no information available about accessibility to the landscapes using these methods. We could have also integrated the economic situation of the region itself. Only if the region is able to spend some money for amenities in their landscapes – for instance, better infrastructure to reach the amenities or to create special signs, etc. – could the region grow alongside the amenities. Quite similarly, Green (2001) found that the existence of water and economic growth have a positive relationship. The region of STB has a lot of water, but the questionnaire was addressed to residents, in which precise information about the financial situation, or a list of prior-year direct or indirect regional investments in amenities or landscapes towards economic growth, were not available. As a result, these three aspects could not be integrated into the analyses.

Not only could the financial situation of the region have an impact on the landscapes but also the economic situation of the respondent (Brennan & Cooper, 2008). The authors argue that a variable like this might be similar to the educational level or the income, but the economic situation includes also the living status of an individual and represents therefore the lifestyle of a person. This might go along with the time spent outside. For instance, it could be the case that people with a lower standard of living spend more time outside than people with a relative high living standard, because these people are more often visiting music lessons or other things usually done indoors. This could infer the importance of the nature or landscapes to the individual – but when it comes to income, it could be that people with higher living standards spend their money on other things than the environment. These two aspects could not be integrated in the analyses because there was no question about the lifestyle or living standard in this questionnaire. But this topic might be an interesting one for further research.

Close to the economic situation is the argument of Frey and Liaw (1998) that the movement of people is dependent on some socio-economic variables. The authors identified educational level, income and children as important for the choice of the region. An analysis of these three variables were done, but the aim of this research was to identify the perception of the landscape by different groups of residents and less to explain why an individual is more likely to move to one region than to another one. But these explanations might be interesting – especially for the regional management to understand in what direction the region could develop in the near future. Because of the complexity of the decision to move and plenty of reasons to move, this topic might be better off in further research.

People who just stayed a while in the region and then left it again could not be included. This problem was already described by Gmelch (1980), who examinations the characteristics of returning migrants. Some people tried to integrate themselves in the community; others did not or were dissatisfied with their new environment. Reasons why they have left might be that the live in the new place did not meet their expectations (Brennan & Cooper, 2008; Jobes, 2000) or they moved because of other private reasons, like a new partner or a new job. Close to this explanation was also the argument of Halfacree and Rivera (2011) that each individual has experience from the early childhood and these might be idealized. If these images are not matched with the region then the individual might move again. These indicate again the positive selection described in the previous chapter. But it might be interesting to look into how newcomers try to integrate themselves into the new community and how that works out. Especially for the administration of the region, this might be valuable information because they could use this kind of information to help integrate newcomers into the community.

Most of the research of migration does not integrate the perception of the environment. It usually analyzed the motives of leaving or staying. Some examinations looked at the environmental and the influences to move (e.g., Adams, 2016; Brennan & Cooper, 2008; Rudzitis, 1999), but the combination of migration and the following perception of regional landscapes might be to some extent new in research. The theories explained above only show the reasons for moving, but there is still a missing link between the perception of landscapes and the motives for migration. In future research, this link could be analyzed because there was no information about why an individual came (back) into the region. The reasons for migration into the region are more general and they are less expressive. There might be a link between these aspects.

Movement could be seen as major event in the biography of each individual (Fielding, 1992; Halfacree & Boyle, 1993; Ní Laoire, 2000). The movements in a biography are often including emotions about leaving a place and coming to another (Fielding, 1992) so that these emotions could also influence the perception of the environment or the landscape. The reasons for moving were asked but the emotions that went along with the movements could not be analyzed. Also the ties to the family seem to be important for migration (Boneva & Frieze, 2001; Kolk, 2017; Spring et al., 2017), even if it is not clear if this is also important in the perception of landscapes. Furthermore, if there were more information about the emotions since moving these were only available about the newcomers and the returnees who are still in the region. Therefore these emotions could not be examined in the analyses.

Anton and Lawrence (2014) found that owning an own house in the region might be a predictor for long-term stay in the region and that these individuals try harder to get a bond to the region than people living in the region for rent. The questionnaire included a question about having their own house or apartment but a lot of the sample stated that they live in their own house. There might be a bias (80 % of the sample stated to owning a house) because it might be the case that the people owning a house are more sensible to nature and therefore answered the questionnaire. Because of this, the fact of owning a house in the region or not is not included. Further, using the argument of Anton and Lawrence (2014) there might be a correlation between owning a house and bond to region.

7.7 Conclusion

This examination focused on the impact of migration biography on the perception of landscape. There were two questions in the focus of the analyses: First, how the perception of landscapes in general differs between newcomers, returnees and locals; and second, whether the perception of different kinds of landscape is affected differently by migration biography.

Differences between newcomers and locals regarding their perception of landscapes were identified. There are some differences between the perceptions of each landscape, but every group perceives the water landscape as highest. While for the groups of locals the bond to region has a positive impact on the perception, for the newcomers, it influences is negative. An increasing number of children have especially for the newcomers a negative impact. The impact for the returnees is not clearly, because as argued before, the returnees are in between the two other groups. In some case, this group is more like the newcomers and in others more like locals. But, the groups also have similarities: a higher educational level influenced the perception of the landscapes in all landscapes positive. Therefore, it could be said that the biography of each group might have an impact on the perception of the landscapes.

The differences between the types of landscapes are not clear. In some cases, the coefficient displays significant in the one type of landscape, but at the same time it is significant for the other type. But, it this is the case, often the landscapes within a type differ in its types. In summary, there are not as many differences between the landscapes than expected.

For the regional management, this would imply that the higher education is the most important factor to improve the perception of the landscape. It might good to offer more information about the regional landscapes and its features. Especially newcomers with more than one child could be more incited to get more information (on a playful basis) about their environmental surrounding with their children. For instance, in the analyzed region there are a lot of old houses. Someone could offer afternoons where some places of the houses could be used to play hide and seek or do other things in the landscapes. This might be strengthening the feature of the landscape for the region. In regard to the locals, the regional management could offer regular tables with environmental topics or actions for these people who attend.

The sample was relatively small so that this analysis could be interpreted as an explorative analysis about the differences of newcomers, returnees and locals. Especially for the regional management it might be interesting to get to know why newcomers or returnees left the region after a few years instead of staying. But also the emotions of movement might be important for the management and could be some potential for further research. The manager could try to make their regions more attractive for newcomers by having this information handy.

8 Willingness to Pay or Willingness to Volunteer: Who Gets Involved in Regional Landscape Protection in What Way?

The chapter shows the analyses of a combination of WTP and WTV regarding the support of regional cultural landscape and shows individual characteristics that are responsible if an individual will support the landscape in general, in terms of money or volunteer work or in a combination of both forms.

8.1 Why Examine Willingness to Pay and Willingness to Volunteer in Combination?

The protection, preservation and design of regional landscapes are often challenged by financial constraints (Asah & Blahna, 2012; Asah, Lenentine & Blahna, 2014); but it seems to be necessary to find a solution concerning the overcoming of these constraints for the landscapes to maintain it for further generations. Volunteering seems to have high potential for the solution of these constraints. Hence, this paper looks into the individual characteristics that examine how people support the preservation, protection and design of regional landscapes (in terms of money, volunteer work or in a combination of both). By now, it is common to analyze the value of a public good using a monetary assessment (Marangon & Visintin, 2007; Verbič & Slabe-Erker, 2009) and it is also common to use volunteer work for the valuation of these goods, yet especially in developing countries (Anderson & Zimmerer, 2003; Stone et al., 2008). Volunteer work is used for valuation, because volunteers and their work are believed to be an important factor for the maintenance of landscapes (Measham & Barnett, 2008; Overdevest, Orr & Stepenuck, 2004) and to be an economic benefit for the society (Houle, Sagarin & Kaplan, 2005). But there are only a few analyses (e.g., Lankia et al., 2014) that are used both constructs at the same time in western countries. This is the strategy of this research.

Combining the constructs of the monetary valuation and volunteer work could be lead to a broader understanding of landscape valuation by the residents. Integrating volunteering in the valuation of public goods could be more efficient to display the true valuation and is a more plausible alternative for its valuation (Hardner, 1996; Vondolia, Eggert, Navrud & Stage, 2014). Using only a monetary valuation could lead to underestimations of true values of the public good because the cash constraint varies between regions (Rai & Scar-

J. Matloch, *The Assessment of German Cultural Landscapes*, RaumFragen: Stadt – Region – Landschaft, https://doi.org/10.1007/978-3-658-21416-6_8

borough, 2013, 2014). Monetary value for landscape valuation might not be adequate to find the true valuation for instance Hossack and An (2014) found that the use of in-kind paid results in a higher valuation of the public good than using only a monetary payment. Eom and Larson (2006) stated that the combination of WTV and pay might be a more adequate measure than using only the monetary value. As stated by Rai and Scarborough (2014) and Gibson, Rigby, Polya and Russell (2015) concluded that the alternative of volunteering might be suitable for low-income households as well as people living in rural areas, but these households and its surrounding could not only be found in developing countries.

Additionally, some people might not be familiar to the valuation of a public good by monetary values (Berrens, Jenkins-Smith, Bohara & Silva, 2002; Echessah et al., 1997) and this construct of monetary payment and volunteer work might help value the landscape in a more precise way than the monetary valuation solely (Gibson et al., 2015; Vondolia et al., 2014). Furthermore, the integration of volunteering could solve some critics[47] concerning the hypothetical market for the monetary valuation of a public good (Schiappacasse, Vásquez, Nahuelhual & Echeverría, 2013).

Otherwise, some people already working on a voluntary basis for or in the landscapes and for that reason might not be willing to give some money for protection, preservation and design of the landscapes. For instance, some of the governmental programs rely on volunteer work – especially in countries like the United Kingdom and Australia, where the volunteer rates are about 30 % (Anderson & Zimmerer, 2003; Measham & Barnett, 2008; Salamon, Sokolowski & Haddock, 2011). Some of these committed people might be affected by the argument above and therefore are interpreted by a researcher as having no valuation for the landscape despite their volunteer work because they are not willing to pay financially for their landscapes.

The appreciation of the residents for their regional landscapes in terms of WTP in combination to WTV is examined. WTV will be an addition to the already known and analyzed results of WTP for environmental concerns in western countries. The aim is understand which individual characteristics are responsible whether an individual is only willing to pay some money for the landscape or is only willing to volunteer for the landscape protection or is willing to do both at the same time. This approach submits a comparison between the different forms of contribution and the individual characteristics that are important for the contribution form.

To take volunteering and monetary payment for landscape protection in combination, this work analyzes the determinants that influence the willingness

47 An overview regarding critics and limitations are shown in the Chapters 4.2 and 4.3.

to contribute in terms of money and/or volunteer work in the corresponding region and show differences between the impacts on the willingness´. This approach offers the answers of the following research questions: first, who (in terms of individual characteristics) is willing to contribute for the protection, preservation or design of their regional landscapes and if so, in what kind will the individual support it (volunteering, money, a combination of both). Second, what individual characteristics are influencing each form of contribution in what way?

Following is presented a short overview about motives for contribution and of the concept of WTV as valuation method for a landscape. Findings of the literature and the examined hypotheses are next. Subsequently the variables and the methodology are described. The data and some descriptive analyses are shown in the next section. The statistical results are presented afterwards. The paper ends with a discussion of the findings and a short conclusion.

8.2 Theoretical Constructs of this Analysis

First, the motives for contributions are presented. They are not directly part of this examination but they might help to understand why an individual contributes. To analyze the support for landscape protection by residents the approaches of hypothetical WTP and WTV for the landscape are used. To understand these approaches, a short description of the WTV follows. The constructs and measurements regarding the WTP are described in Chapter 4. Subsequently, some findings of the literature are then shown before the hypotheses under examination are explained.

8.2.1 Motives for Giving Behavior

The motives for involvements are important to understand why one individual is willing to invest into the region and another one is not willing to invest anything in his or her surrounding environment. But the motives are not explicitly objects of this work. Starting with the theories about why an individual is voluntary willing to give something in general. The motives of giving behavior act as basis for explanations of possible differences between groups in the discussion later on.

One of the first authors to analyze the motives of giving behavior was Gouldner (1960), based on the elaboration of Malinowski (1926). Gouldner (1960) found that giving behavior is not only functional, but also have components of reciprocity. This result is based on believes of the individual (1) that giving would be balanced out on average, (2) that individuals only help others of whom they expect to be helped and (3) that people only help others if they do not get certain penalties by helping them. These believe of repaying the benefits someday are anchored in the moral norms of society the individual is living in (Gouldner, 1960). The author already limited his results because he found that children, old people and handicapped people are acting different and exceptions of the reciprocity approach. However, this theory as a whole might not explain why people contribute for the protection of landscape, but the first and third motive could be interesting for the explanation.

One of the first researches about the motives of volunteering was by Titmuss (1971). He analyzed the motives of blood donors and compared their motives in the United States and the United Kingdom. The author divided the motives into three types that are advanced by Arrow (1972) into three classes that are matching the utility theory. These three motives are (1) the individual welfare depends on the own satisfaction and the satisfaction of others, (2) the individual welfare depends on the contributions of the individual to the benefit of others and (3) the individual has egoistic motives that are in line with an implicit social contract. A different approach used Schindler-Rainman and Lippitt (1971). The authors concluded that each individual analyzes the possible benefits he or she gets out of it. The explicit named motivations are service to society, self-actualization, need for power, emotional associations, autonomy and need for mutual support. These theories are more adequate to explain why people contribute for landscape protection. Even if these were developed to explain motives for volunteer work, it could also be used for a financial contribution.

Becker (1974) introduced a theory of social interaction also called altruism in giving behavior, based directly on the utility function in economic theory. The author distinguished between family members and unknown individuals as recipient. While givers to family members are mostly motivated by improvements of the general well-being, givers to unknown recipients have their motivation regularly through the avoidance of scorn or the achievement of social impact (Becker, 1974). This theory assumed that an individual charitably gives something because of selflessness (Schervish & Havens, 1997) and because they receive satisfaction because of the increase of the good (Harbaugh, Mayr & Burghart, 2007). This motive implies that the individual receive higher satisfaction even if the good is supplied through mandatory payments like taxes (Harbaugh et al., 2007). Harbaugh et al. (2007) found evidence that the altruistic motive of giving is present: in their experiment, money was giving by 58 % of the respondents with altruistic motives and only by 31 % of the respondents with other motives. This theory could be useful for this examination because these people give because for instance they want to preserve the landscape for further generations.

An advancement of the theory of altruism is the theory of impure altruism by Andreoni (1989, 1990), which is also called warm-glow giving. The author (1989) stated that people are usually contributed because of a higher demand of the public good (altruistic motive) or benefit from their giving itself (impure altruistic or warm-glow motive). The warm-glow motive implies that the donor may not gain material utility, but from the act of giving by, for instance, positive feelings. These forms of motives were advanced from Andreoni (1990) as social pressure, guilt, and sympathy. A similar approach has Simmons (1991), who described motives for giving behavior in an improvement of self-picture, feelings of more happiness than without giving, empathy and obedience of social norms. The higher level of satisfaction could only be reached by voluntary payments instead of mandatory ones, which implies that people with impure altruistic motives give less if the giving is mandatory (Harbaugh et al., 2007). These theories could also be applied in this analysis for financial and volunteer contributions for landscape protection, since these people are motivated because they get their financial donation back from the government in some form.

A different approach, called socio-emotional selectivity theory and based on psychological constructs, was introduced from Carstensen (1995). Motives in this theory are social regulation, development of the self-concept, and information seeking. It is assumed that the goals of an individual vary within their life span and therefore the involvement of a person. Carstensen (1995) describes that children and older people have stronger motives for emotional regulation and chose therefore to be more with known social partners. Because of this, the involvement is not as intense as it was in adolescence. Information-seeking is highest in the age of adolescence to middle age, and therefore individuals chose to spend time with unknown partners, while the self-concept is the in every age in between the other motives. Beyond this, Carstensen (1995) concluded that people with a limited future were more focused on the feelings and therefore less motivated to get involved – especially with unknown partners. Information seeking is more important as a long-term goal. This theory could help to understand possible differences in the individual characteristics.

Clary et al. (1998) formulated the most used approach (called functional approach) for explaining motives of volunteer based on older work of Katz (1960) and Smith, Bruner and White (1956). Clary et al. (1998) identified six functions that are influencing the motives for volunteering: values, understanding, social, career, protective, and enhancement. The value function of volunteering is that the individual has the opportunity to expresses its values related to others and contributions to others. The understanding function includes the motive of learning and gaining new experiences. The social function is about the fit into the social surrounding of an individual. The career function implies that the volunteer work would provide career-related benefits for the individual. The protective function is a motive that should protect the individual for feeling guilty to be better off than others or for ignoring the own problem. The last function is related to self-esteem and self-improvement or personal growth. At least the value function could be a motive for financial contribution and therefore, this theory might also be important for the explanation of possible differences.

Later on, Clary and Snyder (1999) concluded that this approach can give insights into personal and social processes but that individuals can perform the same volunteer work with different of the named functions or with more than one function at the same time. The author found that the functions values, understanding, and enhancement are more important than career, social and protective functions. Clary and Snyder (1999) found that volunteers whose motivations were matched by the volunteer task and have their own control about the task are more satisfied and are willing to volunteer longer than people with no match of motivation and task. Houle et al. (2005) found that not all tasks are equal and that the level of satisfaction differs by the motives for each task. Individuals tend to be volunteering in task in which they see the highest rate of satisfaction of their motives for volunteer work. Even these theoretical findings could help understanding possible difference in the contribution by the individual characteristics.

The first theories regarding the motives for charitable giving by Goulder (1960) and Titmuss (1971) are based on research in sociology. Later, the theories are more psychological ones, but always with a corresponding research area. The theory of altruism by Becker (1974) is the first of the presented ones with a corresponding research area of economics. Andreoni (1989, 1990) has the other research area in the political and environmental sciences. Clary and colleges (1998, 1999) went back to the sociology as the other research area. Carstensen (1995) is the only one with the focus on the psychology itself. Because all these areas are also touched by these analyses, the theories are mostly adequate to explain the motives of landscape support. The charitable giving in terms of money or volunteer work can be interpreted as a decision that has different components. These are psychological, economical, sociological as well as environmental. As shown, the described theories are all in these research areas located.

8.2.2 Concepts and Measurements of Willingness to Pay and Willingness to Volunteer

The approach of WTP is described in more detail in Chapter 4. A short description of the concept and the measurements of WTV will be provided to better understand the findings of the literature. The WTV is added to get a better understanding why an individual is willing to support the landscape protection especially in terms of voluntary work or financial contribution.

In low-income countries it is not suitable to use a monetary measure for landscape preservation, because the residents in these countries might not have enough money to give some to protect their environment (Ahlheim, Frör, Heinke, Duc & Dinh, 2010; Gibson et al., 2015; Hardner, 1996; Hung, Loomis & Thinh, 2007; Rai & Scarborough, 2014; Schiappacasse et al., 2013; Vondolia et al., 2014). That does not mean they are not interested in the environment, but that they need their money for more essential things like food. Probably the first research using payments in the form of labor was done in Ecuador from Hardner (1996). He decided to use labor contribution, because of the low income in Ecuador the use of money was not an adequate measure. Volunteering can also be seen as an investment in the region rather than monetary investments and as a result could also determine the value of a landscape.

Usually, volunteering can be seen as a classical labor-leisure decision by a constraint of income and time for each individual (Bockstael, Strand & Hanemann, 1987) and could therefore be used for the determination of the value of a public good (Schiappacasse et al., 2013), yet especially in developing countries. But, volunteer work is still a non-market service that could not be converted into an exact monetary value (Gibson et al., 2015; Salamon et al., 2011). But non-profit organizations as well as other institutions with volunteers are still interested in calculating the value of volunteer work by using a monetary value for that work (Anderson & Zimmerer, 2003).

It can be stated that investing time in preservation, protection and designing is an investment in the region and their landscapes. Because of these arguments, volunteer work is used for the analyses as an indicator for the value of the landscape.

8.2.3 *Findings of the Literature*

Stone et al. (2008) looked into the WTV or WTP of user groups for the restoration of mangroves in India. They compared the amount of money resulted of the combination of volunteering and payment in terms of Indian rupees for each group and found differences by user groups and found different motives for the willingness to contribute. Vásquez (2014) examined the willingness of households for improved water services in Guatemala using a combination of volunteer work and money. He found that people prefer municipal service instead of community-managed services. The municipal service showed an increase of the water bill by more than 200 % and 19 hours of volunteering while community-based services had no willingness to contribute. Rai and Scarborough (2014) determined a value for a recreational project in Nepal. Schiappacasse et al. (2013) assessed benefits of ecosystem restoration in Chile by using WTP and WTV and found that most of the respondents preferred payment in form of volunteering. Vondolia et al. (2014) examined the willingness to contribute for a canal in Ghana. They did four different analyses, using only financial contribution, only volunteer work as contribution, and two measures of both where volunteering was transformed into money. Hung et al. (2007) examined the valuation of the Vietnamese by using a payment card with money and working days and found a preference for labor contribution.

Echessah et al. (1997) examined the willingness to contribute (monetary or time) for control of the tsetse fly in a western area of Kenya. The respondents were able to state either money, volunteering in terms of work or both. The authors found that most of the respondents were willing to contribute financially and voluntarily work. By using a Heckman model they examined the factors influencing the decision about the form of contribution (monetary or time payment) as well as the amount of these contributions (days or Kenya Shillings). They found influences of gender, education and income. Moreover, Echessah et al. (1997) also compared different regions in the selected area and found differences of the distribution of WTP and WTV across these regions.

Yet, volunteer work was not included in measuring the valuation of public goods in industrialized countries, only for undeveloped countries, except for an analysis by Lankia et al. (2014). The authors identified a conflict about the preservation, protection and design of forests in Finland. Huge areas of land are private, but used by the recreationists. The landowners decide about the recreation about their land without taking the recreationist into account. To resolve the conflict the authors explored the willingness of the recreationists to support the landowners by money or time. They found that the recreationists are willing to contribute 92 euros per year and 3.5 days a year on average, respectively.

8.2.4 Hypotheses

Out of the motives for involvement in more general ways and the concepts and measurements for both constructs, hypotheses are generated out of the previous findings that might influence whether an individual gets involved. The first hypothesis is a general one. The following four hypotheses are about the impacts on the contribution as a whole while the others are more about what kind of contribution an individual would choose.

Generally, preceding researches show more acceptance of the WTV than the WTP, especially in developing countries. These were found by Schiappacasse et al. (2013) for Chile, by Hung et al. (2007) for Vietnam, by Rai and Scarborough (2013, 2014) for Nepal and by Echessah et al. (1997) for Kenya.

But Gibson et al. (2015) concluded that there is only a small difference in preferences for WTV and WTP, even in developing countries. Because Ahlheim et al. (2010) and Schiappacasse et al. (2013) argued that WTV is more common for poorer countries or regions the research will test the hypothesis whether people are generally more willing to pay for their landscape than they are willing to volunteer for it – even if Lankia et al. (2014) found a higher preference for WTV for Finland.

While Echessah et al. (1997) found that women are more likely to contribute labor and money, Rai and Scarborough (2014) concluded that men more often select a monetary value than women. Rai and Scarborough (2014), Measham and Barnett (2008) and Menchik and Weisbrod (1987) found that women are on average more often volunteering than men but the authors did not define the area of volunteering. Wilson (2000) found differences in the job type that men and women choose for their volunteer work. For Europe this results cannot be confirmed: In some European countries, there are more males volunteering than women but the ratio adjusts within the lifecycle (Wilson, 2000). This could also be confirmed from Simonson, Vogel and Tesch-Römer (2016) in Germany: the authors found that more men than women are volunteering. Orientating on these controversy findings, this paper will use the opportunity to determine if there is a difference between the genders and tests the hypothesis that men are more involved in the landscape protection than women.

While Campbell (2007) indicated a positive influence of the age on the WTP, Carson et al. (2001) and Hackl and Pruckner (1999) found a negative impact of age. Similar influences were found for WTV: A higher amount of the middle-aged respondents were found to more willing to volunteer than give a financial contribution (e.g., Lankia et al., 2014; Menchik & Weisbrod, 1987; Wilson, 2000). Asah et al. (2014) who analyzed the motives for volunteering found that the middle-aged generation is more willing to volunteer because they want to do something meaningful. Adjuvant to this result, they found that these people are less willing to pay. On average, Tidwell and Brunson (2008) found that people who are willing to volunteer are younger than those who are not willing and therefore they concluded that prospective volunteers are younger than the average population.

For volunteering, it can be said that socio-emotional selectivity theory explains using different motives in the lifetime of an individual and why older people are less motivated to get involved in volunteering (Carstensen, 1995). This is because they usually choose to be around already known individuals and that might not be always the case in volunteering. Tselios, Noback, McCann and Dijk (2015) also find that engagement rises in younger years and fall with getting older. Even if the rate of older volunteers is increasing in Germany the rate is still lower as in every other younger age group (Enste, Neumann & Schare, 2012; Simonson et al., 2016). In line with that is the result that retired people are not starting volunteer work because they have a lot of time (Wilson 2000) as one might think. Rather, Schiappacasse et al. (2013) and Vásquez (2014) stated that older people are less interested in contributing for their environment in general because of their health conditions and life expectancy. These results suggest the hypothesis that older people are less willing to support the regional landscapes than younger people and the age therefore needs to have a negative influence on the willingness to contribute.

The positive influence of education could be found in the literature of WTP and WTV (e.g., Ahlheim et al., 2010; Howley et al., 2012; Menchik & Weisbrod, 1987; Verbič & Slabe-Erker, 2009). While Kämmerer et al. (1996) and Howley et al. (2012) have different explanations why education has a positive impact on WTP, in WTV is only a small discussion about the importance of education on WTV. But its importance might be dependable on the type of volunteer work (Wilson, 2000). Some work types have a high degree on intellectual work while other tasks have a higher degree in physical labor and that influence the impact of education. Rai and Scarborough (2014) did not find any significant values neither for WTP than for WTV, but they concluded that education has a positive tendency of the WTP, but a negative one on the WTV. Some other examinations did not find any impact of education on the willingness to contribute at all (e.g., Karkow & Gronemann, 2005; Stone et al., 2008). For Germany, Simonson et al. (2016) found that people with higher education are more volunteering than people with lower education degrees. This paper will therefore test the hypothesis that education has a positive influence on the willingness to contribute for the regional landscapes.

People with a strong relationship to nature value a landscape higher than those who do not have such a relationship to nature concluded different researches (Carson et al., 2001; Howley et al., 2012; Morrison & Dowell, 2013). Verbič & Slabe-Erker (2009) found that the relationship to nature is the most important variable for valuing a landscape. These were also found by examining the motivation of volunteering (e.g., Asah & Blahna, 2012; Bramston, Pretty & Zammit, 2011; Bruyere & Rappe, 2007). Because of previous findings this paper explores if bond to nature has a positive influence on the contribution for landscape protection.

The literature does not analyze the effect of children on the WTP, but it could be argued that children reduce the amount of freely available household income and time. As a consequence, the influence of children on WTP and on the WTV needs to be negative. But, it was found that persons with children volunteer more often than individuals without children (Freeman, 1997). A person with young children is frequently more voluntarily involved than people with older or no children (Menchik & Weisbrod, 1987) but it is not clear if the area of volunteering differs and if this areas might be social areas like the kindergarten or school for instance. These results indicate the hypothesis that children have a negative impact on the WTP, but a positive influence on the WTV.

The economic theory (Hicks, 1939) came up with said that individuals need to decide along income constraints and therefore income is expected to have a positive impact on the WTP. But this result is not completely uniform along analyses of the WTP: While some authors (e.g., Howley et al., 2012; Kämmerer et al., 1996; Marangon & Visintin, 2007) found a positive impact on the WTP, Morrison & Dowell (2013) uncovered a negative influence. On the WTV, income displays to have a positive influence in recent examinations (Schiappacasse et al., 2013; Wilson, 2000). Vohs, Mead & Goode (2006) found that people with a higher income are more likely to work alone and are less willing to volunteer as well as donating money which would prove the argument of Freeman (1997) that the income has a negative impact on the WTV. Analyses combining the constructs of volunteering and payment found that people with higher income are more willing to pay than to volunteer (Echessah et al., 1997; Lankia et al., 2014; Rai & Scarborough, 2014). For Germany, Simonson et al. (2016) found a positive relationship between income and WTV and since this research is conducted in Germany, it will be examined the hypothesis that the income has a positive influence on the WTP and on the WTV.

Newcomers or returnees have different awareness's about the landscapes in their region because they have seen other regions and its landscapes already (von Reichert et al., 2012). This was proven regarding WTV from Tidwell and Brunson (2008) and Bramston et al. (2011). Additionally, it might be the case that newcomers are more often active voluntarily because they try to get to know other people living in the region. This social component is often identified (DiEnno & Thompson, 2013; Wilson, 2000,). Tidwell and Brunson (2008) concluded that people who are new to a region consider the regional landscapes in a different way than the locals and therefore are very valuable for a region concerning volunteer work. Because of these findings the hypothesis is: newcomers and returnees value the landscapes higher and are more willing to contribute in all forms for their landscapes than locals.

Identification is seen as an important factor for giving behavior (Schervish and Havens, 1997) and could therefore be an important factor for valuing environmental goods (Morrison & Dowell 2013). That volunteering can help building an identity was one important finding by DiEnno and Thompson (2013). If a resident has a bond to his or her region, he or she usually knows the people living in the same region and therefore prefer to volunteer than people who did not know others (Lankia et al., 2014). This leads to the hypothesis that a bond to region has a positive influence on all forms of contribution.

Regarding the environment around the place of living, the results are different: Kämmerer et al. (1999) stated that people living in a city are more willing to pay than people living outside the city because they are not able to visit the landscapes everyday because of the distance (Howley et al., 2012). DiEnno and Thompson (2013) stated that people who experienced nature on a regular basis like living in a natural environment outside a city are more likely to support its conservation. The authors had an interviewee who stated that especially when living in the countryside the motivation to get involved in volunteer work might be higher than for people who live in a city. Freeman (1997) concluded that people living in cities are volunteering lesser than people living outside a city, but this influence of the environment could not be applied to every landscape. In the research of Bruyere and Rappe (2007) the reason for volunteering was often that the people are taking the opportunity to be outside, away from home or their work, which is in line with the argument stated before. This paper examined the hypothesis that people living in the countryside are more willing to volunteer for the landscapes, but less willing to pay than people living in the city.

Volunteering in the nature and its landscapes can be seen as similar activity of gardening (Miles et al., 1998). Kaplan (1973) looked into the benefits of gardening and found that gardening requires knowledge as well as provides it. This could activate a learning process and could therefore change attitudes of the landscapes (Kaplan, 1987; Upham et al., 2009). Gardening can therefore offer benefits that are associated with working in the landscape (Miles et al., 1998). Summarizing, it could be concluded that people having their own garden can use their garden for outdoor recreation while people living without an own garden would use the landscapes for outdoor recreation. Out of these, the following hypothesis is assumed: People having their own garden are more willing to contribute financial contribution for protection of the landscape, than work voluntarily in or for the landscape.

Haugen (2015) analyzed differences of landscape supports by landowners and the general public. She found that landowners consider the landscape in a more practical way concerning the production of goods, while the general public includes the soft values like recreational value or the preservation of plants and animals. Other researches examined the differences of people working in or for the landscape professionally and the general public could not be found. But the results might be transferable for people working in or for the landscape. On the one hand, they already invested a lot of time in the landscape and therefore might prefer a financial contribution, and on the other they know how much money is investing in the landscape protection and preservation so that they are more willing to contribute volunteer work than money. For these analyses the assumption is that people using the landscape in their occupation are less willing to contribute volunteer work or money.

8.3 Sample and Methodology

This chapter is about the data used to answer the research questions. The first section describes the used variables for the valuation, including the landscapes. Then, the variables are shown. The last part of this chapter presents the methodology of the analysis.

8.3.1 Variables of Willingness to Pay and Willingness to Volunteer

The topic of the WTP and WTV consisted of three questions in the current survey (Tab. 8-1)[48]. One of the three questions is about the classical theory of WTP and the other two are about the WTV. Protest voters – meaning interviewees that did not answer the question – were deleted from the sample.

The question about the WTP asked the sample about a personal contribution (additionally to the taxes) for different landscapes that were imaged on pictures corresponding to each region (Tab. 8-1). These landscapes are water, city and village landscapes. Especially for the water landscape, almost 50 % of the respondents in each region are willing to spend additional money for preservation, protection and design for this landscape type (Tab. 6-2). This figure is perceived as rather high, in particular because Hartje et al. (2003) found only 22 % of their sample willing to give some money for the water landscape of the Elbe. It can be seen that in more rural areas the residents are more willing to give some additional money for the preservation, protection and design of villages than in more urban regions.

Almost 70 % were not willing to pay some money for preservation, protection and design of the city landscape (Tab. 6-2). These results are similar to Karkow and Gronemann (2005) and Molino et al. (2016) who found that almost two thirds of their sample did not want to pay for agricultural land. Howley et al. (2012), on the other hand, found that half of their sample did not want to pay for a farm landscape. These results could be compared to the city landscape, because for both landscapes, there might be somebody who seems to be responsible for the landscape: for agricultural land it might be the owner, while for the city it might be the government.

48 All questions regarding the WTP and WTV are described in Chapter 5.1 for the WTP and in Chapter 5.3 for the WTV in detail.

Tab. 8-1: Overview about the questions and variables used

Question in the survey	Possible answers	Transformation of the variable	Name of the variable in the data set	Using as...
What amount of money are you personally willing to pay monthly and additionally to the already paid taxes for preser-vation, protection and design of the landscapes that are shown on the following photographs?	0 euros 1 - 5 euros 6 - 10 euros 11 - 20 euros 21 - 50 euros 51 - 100 euros more than 100 euros	transformed into a binary variable with 1 = is willing to give some money for at least one of the landscapes and 0 = is not willing to give any money for the landscapes	financial	
In future, in what area could you imagine to volunteer most likely (for the first time, strengthened or additional)?	place development politics conservation of the environment animal protection fire or ambulance service sports/ exercises education/kinderga rden/school religion/church art/culture/ museums monument conservation local history society other	transformed into a binary variable, that displayed if the person is willing to volunteer for the region (place development, conservation of the environment, animal protection, art/culture/museu m, monument conservation and local history)	Only people with one of the named forms of involvement and a statement about the duration of future involvement are seen as involved people or already worked volunteeringly = involved	
In what area do you volunteer at the moment?	same areas as the question before			
How many hours per month could you imagine to volunteer for your region?	no interest 1-2 hours 3-5 hours 6-10 hours 11-20 hours more than 20 hours	transformed into a binary variable that shows if the respon-dent is willing to volunteer at least for one hour per month		
		Variable that displays if a person is involved in the regional context with the manifestations 0=no involvment 1=only volunteering 2=only financial support 3=both forms	WTP_WTV	Dependent variable
What is your gender?	female male	with 0 = male and 1 = female	gender	Control variable

Question in the survey	Possible answers	Transformation of the variable	Name of the variable in the data set	Using as...
How old are you?	0 - 17 years 18 - 24 years 25 - 34 years 35 - 44 years 45 - 54 years 55 - 64 years 65 - 74 years older than 75 years	manifestations of the categories are displayed by the mean of each class	age	Control variable
How many children do you have?	open answers	amount of children	children	Control variable
What is your highest school graduation?	No school graduation Main school Middle school High school College	transformed into a variable that shows if a respondents have higher education with 1 = yes (High school, College) and 0 = no (no school graduation, main school, middle school)	higher education	Control variable
What is your your net income for all members of your household (monthly)?	lower than 1,000 euros 1,001-1,500 euros 1,501-2,000 euros 2,001-2,500 euros 2,501-3,000 euros 3,001-4,000 euros 4,001-5,000 euros more than 5,000 euros	manifestations of the categories are displayed by the mean of each class	income	Control variable
Since when do you live in your region?	Since birth Since birth with interruptions I moved to the region	transformed into binary variables with 1 = yes and 0 = no	local returnee newcomer	Control variable
Please evaluate the following statements concerning your bond to nature: (1) I try to be as often as possible to be in the nature. (2) Nature has only a minor part in my live. (3) I feel a strong bond to the nature and the landscapes of my region. (4) I think human beings are benefit from an in working order nature. (5) I think it is an obligation to preserve and protect the nature. (6) I feel like I am part of the nature.	All statements could be rated from -- = radical refusal over 0 = irresolute to ++ = unreserved approval	transformed into a binary variable with 1= yes (everyone who marked statements 1, 3 to 6 with ++ or + and statement 2 with - or --) and 0 = no (otherwise)	bond to nature	Control variable

Question in the survey	Possible answers	Transformation of the variable	Name of the variable in the data set	Using as...
Please evaluate the following statements concerning your bond to your region: (1) I feel really like home in my region. (2) With my region, i have a lot of personal memories. (3) My personal future is strongly linked to the region.	All statements could be rated from -- = radical refusal over 0 = irresolute to ++ = unreserved approval	transformed into a binary variable with 1 = yes (everyone who marked every statements with ++ or +) and 0 = no (otherwise)	bond to region	Control variable
In what kind of environment is your house located?	in a city in the city limits in a village outside of a city or a village	transformed into a binary variable with 1 = living in a village or outside a city or village and 0 = living in a city or in the vity limits	rural environment	Control variable
In what kind of house do you live at the moment?	detached house (single family) semi-detached house row house apartment house other	transformed into a variable that show if a respondent might have a garden with 1 = garden (detachted house, semi-detached house, row house) and 0 = no garden (appartment house)	garden	Control variable
I use for occupational purpose and work as...	Farmer Official other occupation	transformed into one binary variable with 1 = yes (everyone who marked sports as daily or at least once a week) and 0 = no (otherwise)	occupational use	Control variable

Source: Resident survey.

Those participants in the sample willing to pay something for the landscapes in their region stated most often an amount between 1 and 5 euros per month which would be equivalent to 12 to 60 euros per year. This result is in line with other paper. Kim et al. (2015) found monthly amounts for an ecological recreation project in an urban area conducted in South Korea between 0.5 and 1.50 dollars (currently, approximately the same amount in euros). Data for this analysis came from face-to-face interviews that could explain the lower amount. Howley et al. (2012) did interviews in Ireland and found also similar results. They used a payment card for determine the value. The work of Campbell (2007) found a similar amounts but using discrete choice experiments of symbolic and everyday landscapes in Ireland by face-to-face interviews.

While almost 30 % in LNWM are willing to pay 1 to 5 euros for the water landscape, only 24 % in LPLD are willing to give money for their water landscape. The third region lies in the middle with 26 %. Such a difference can also be seen for the village: whereas 25 % of the residents of STB and LPLD would spend 1 to 5 euros for villages, in LNWM this amount is lower with 19 %. The payments for city landscapes are around 18 % for each landscape. The amounts of people who are willing to give more than 20 euros are really low in each region for each landscape (almost 1.5 %). A small exception is the village landscape in LPLD. 3 % of the sample in this region would pay more than 20 euros monthly for their landscape. It could imply that people who live in more rural areas are more willing to support the close-nature landscapes and the village than urban ones.

Dummies were created that show whether a participant is in general willing to pay some money for at least one of the corresponding landscapes (Tab. 8-1). This strategy was chosen because a lot of participants stated that they were willing to pay between 1 and 5 euros and only a few were willing to give higher amounts of money. Additionally, it was not possible to distinguish the variables for volunteering in the different types of landscapes.

Two questions address the potential of volunteer work in each region (Tab. 8-1). Because the WTP is a concept of a hypothetical market, the WTV used is also hypothetical. This variable was generated to shows if a person is willing to volunteer in their region. It consists of first, the question if a person is willing to volunteer in or for the region[49] or is already volunteering for the region[50], and second, one about the hours the person would be willing to volunteer.

There are huge differences between the percentage of current volunteer work and the hypothetical ones (Tab. 8-2). For all regions it can be seen that most of the respondents would hypothetically like to volunteer for animal protection. This is also the area (paired with environment conservation) in which the amount of volunteers in Germany increased in the last years (Enste et al., 2012). This can be observed for every region by the huge increase rates, but there are some differences within the next options the respondents would chose to protect the landscapes.

---

49 The question did not asked about volunteer work for or in the region, it asked about different fields of acting. The following fields are indicating volunteer work for the corresponding region: development of the location, conservation, animal protection, culture, cultural heritage preservation, local history.

50 Some people marked that they are already volunteering. It might be the case that people already volunteering have missed the notice about a first, additional or strengthened involvement in the hypothetical question, or they might be willing to stay with the current involvement without any additional work. Therefore we also integrated these people – even if they did not answer the hypothetical question about volunteering.

Tab. 8-2: Current and hypothetical areas of volunteering and the increase per area (in percent)

	All current	All hypothethical	Increase in percent	LNWM current	LNWM hypothethical	Increase in percent	LPLD current	LPLD hypothethical	Increase in percent	STB current	STB hypothethical	Increase in percent
Place development	16,02	18,87	117,79	12,5	22,76	182,08	19,12	19,89	104,03	12,61	15,33	121,57
Conservation of the environment	14,99	26,87	179,25	18,75	26,21	139,79	16,67	28,01	168,03	10,08	25,67	254,66
Animal protection	5,94	24,90	419,19	10,94	27,59	252,19	6,86	24,93	363,41	1,68	23,37	1391,07
Art/ culture/ museum	12,66	10,48	82,78	9,38	12,41	132,30	14,71	10,36	70,43	10,92	9,58	87,73
Monument conservation	5,94	6,95	117,00	4,69	8,28	176,55	5,88	7,28	123,81	6,72	5,75	85,57
Local history	5,43	8,39	154,51	6,25	9,66	154,56	5,88	9,24	157,14	4,2	6,51	155,00
Hypothetical regional volunteering		49,89			54,71			48,84			48,71	

Source: Resident survey.

For STB and LPLD the second highest increase is for the conservation of the environment, followed by local history. For LNWM it is different: the second highest increase is on space development, followed by monument conservation. It does not seem highly important to conserve the environment in this region for the respondents or there might be already enough initiatives protecting the environment.

The mean of the hours asked in each class for the people who would like to volunteer in or for the region were generated. In each region, half of the sample is hypothetically willing to volunteer on average eight hours monthly for or in the region (Tab. 8-3). These results in the examined regions is equal than the hours of involvement found by Simonson et al. (2016) for volunteers in Germany. The authors found a high percentage of people working voluntary for about two hours a week, consequently about eight hours monthly on average.

Tab. 8-3: People who stated they would work voluntary for or in the region stated the following amounts of hours to work (in percent)

Hypothethical monthly hours for volunteering in or for the region	All	LNWM	LPLD	STB
No hours	100,00	100,00	100,00	100,00
On average 1.5 hours	95,26	96,15	88,85	96,53
On average 4 hours	83,52	83,65	77,09	85,42
On average 8 hours	49,89	49,03	43,96	50,70
On average 15.5 hours	18,06	15,38	15,79	18,06
On average 30 hours	5,42	4,80	5,26	4,87

Source: Resident survey.

The hours of potential work are an indicator if the WTV could be a true answer, because if a person indicated a hypothetically willingness to do volunteer work in a field that supports the region does not mean they are willing to do this in reality. Looking into this aspect, at least 2 % of the whole sample are willing to volunteer for the region but has no interest to volunteer (Tab. 8-4). The result for all regions shows that almost 50 % of the sample would work voluntarily. Again, in LNWM more than half of the sample is willing to volunteer for supporting their region. In LPLD and STB the share is lower than 50 %. Furthermore, except for a small percentage in the whole sample, almost every respondent stated that they would be hypothetically willing to volunteer – but not in the defined areas of volunteering for this research.

Tab. 8-4: Comparing volunteering for or in the region and hours to spend for volunteering (in percent)

Volunteering in or for the region	no		yes	
	Willing to spend hours for volunteering		Willing to spend hours for volunteering	
	no	yes	no	yes
All	7,25	42,86	2,31	47,58
LNWM	4,71	40,59	1,76	52,94
LPLD	8,37	42,79	3,02	45,81
STB	7,10	44,19	1,61	47,10

Source: Resident survey, own calculations.

The correlation coefficients of the WTP for regional landscapes and the WTV for regional development show a low correlation of about 0.2. Because of this result, a variable was generated out of the financial contribution and volunteering as contribution form that shows if a person is generally willing to pay a financial contribution for at least one of the landscapes and / or if they are willing to volunteer in the region and its landscapes (hypothetical) (Tab. 8-1). The categories of this variable are that the respondent is only willing to pay and not willing to volunteer, they are willing to volunteer but not to pay, they are willing to do both, or they are not willing to contribute at all. The last category included these people who were neither willing to pay for the regional landscapes nor willing to volunteer for – or in – the landscape[51]. In advance, it was tested whether the individual is supporting the landscape in form of money or in volunteer work but this was only found for a small total of respondents (three people) and could therefore be ignored.

The variable displays the smallest amount for volunteer work only (19 %) and the highest amount for no contribution (31 %) in every region (Tab. 8-5). The financial contribution or both forms of contribution were possible for 22 % and 28 %, respectively. For LNWM, most of the respondents are willing to volunteer and pay for their landscapes (30 %). The next common answer for LNWM is not contributing at all (25 %). For STB the answers are almost two equal groups: either the respondent is willing to pay and volunteer (30 %) or is not willing to contribute for the landscape (31 %). Most of the respondents in GGEW are not willing to volunteer or pay (33 %), while only 26 % of the sample is willing to invest financial and volunteer contribution.

51 This group includes the respondents who were willing to volunteer in other areas that were not defined as areas that are offer volunteering for or in the landscape. This means the last groups comprises, for instance, of volunteers that marked they would work voluntary for the school or the fire department.

Tab. 8-5: Overview about the manifestations of the variables (in percent)

Variables and its manifestations	Variable type	All	LNWM	LPLD	STB
Desicion of support (WTP_WTV)	Dependent				
no involvement		27,6	24,1	28,4	28,4
volunteering		22,4	23,5	23,5	20,3
financial		18,8	18,8	17,9	20,0
both involvements		31,2	33,5	30,2	31,3
Gender	Control				
male		49,8	52,1	52,6	44,5
Age	Control				
18-24		6,4	5,3	6,5	6,8
25-34		12,6	16,5	12,8	10,3
35-44		14,0	15,3	13,7	13,6
45-54		25,4	24,1	24,7	27,1
55-64		22,4	22,4	22,1	22,9
65-74		12,2	11,8	12,6	11,9
75 and older		7,0	4,7	7,7	7,4
Children	Control				
No children		27,2	32,3	25,1	27,3
Child		22,1	24,0	24,1	18,3
2 children		32,1	31,7	34,6	29,0
3 children		12,3	6,6	9,7	19,0
4 children		4,4	4,2	4,6	4,3
more than 4 children		1,8	1,2	1,9	2,0
Higher education	Control				
higher education		48,0	54,1	46,7	46,5
Income household	Control				
Lower than 1,000 euros		8,7	6,9	11,0	6,6
1,001-1,500 euros		12,8	13,2	15,0	9,7
1,501-2,000 euros		11,1	12,6	11,0	10,4
2,001-2,500 euros		13,3	9,4	15,0	13,2
2,501-3,000 euros		16,4	13,8	15,7	18,8
3,001-4,000 euros		18,6	23,3	16,2	19,4
4,001-5,000 euros		9,9	10,7	8,2	11,8
more than 5,000 euros		9,1	10,1	8,0	10,1
Duration Region					
Locals	Control	32,6	23,1	36,9	31,9
Returnee	Control	15,7	14,8	18,5	12,3
Newcomer	Control	51,7	62,1	44,6	55,8
Bond to Nature	Control				
Bond to Nature		52,6	55,3	54,4	48,7
Bond to Region	Control				
Bond to Region		57,1	52,9	62,8	51,6
Rural Environment	Control				
Rural Environment		54,6	26,5	63,7	56,1
Garden					
Garden		77,7	70,0	76,5	83,6
Occupational use	Control				
Occupational use		15,6	11,2	19,3	12,9

Source: Resident survey.

It is obvious, that the answers are separated in the most extreme answers in every region: do not contribute at all or contribute in both forms (volunteering and financial support). This result is different from the research of Echasseh et al. (1997) who found that a lot of respondents are willing to pay or/and willing to volunteer, and the result of Lankia et al. (2014) that found people with for volunteering preferences cannot be proven from this result. A similar result to the current ones was also found from Gibson et al. (2015), who revealed a similar allocation of willingness.

8.3.2 Other Variables

The socioeconomic variables gender, age and education are common in WTP examinations (e.g., Howley et al., 2012; Kaltenborn & Bjerke, 2002; Kämmerer et al., 1996; Karkow & Gronemann, 2005; Kim et al., 2015, Mitchell & Carson, 1989; Molina et al., 2016; Morrison & Dowell, 2013; Schmitz et al., 2003) and also for WTV analysis (e.g., Measham & Barnett, 2008; Menchik & Weisbrod, 1987; Wilson, 2000). An overview of the control variables is shown in Tab. 8-5.

The questionnaire also asked about the duration of being in the region (Tab. 8-1). On average, every second respondent is a newcomer (Tab. 8-5). The rate of newcomers is highest for LNWM while LPLD have the highest rates for returnees and locals. STB lies in between, but has the lowest rate for returnees.

A question with different statements about the nature is used for the variable bond to nature (Tab. 8-1). These statements were based on a study about the awareness of nature in Germany (BMUB, 2014). Because it should be only one variable indicating the relationship to nature, the statements were transformed into one variable. Tab. 8-5 shows that almost every second participant answered that he or she would have a bond to nature. The bond is highest described by the respondents of LPLD (57.4 %) and lowest by the participants of STB (50.1 %). Because de Groot and van den Born (2003) found that the relationship to nature might be dependable on education and age, it was tested. The correlation coefficient showed a significant value for age (0.2) but no significant value for higher education (0.05). Both coefficients are low and could therefore be used.

Bond to region was created similar to bond of nature (Tab. 8-1) and was used because of the identity that could influence the value of a landscape (Morrison & Dowell, 2013). The statements itself were developed referring to Lalli (1992), Weichhart et al. (2006) and Soini et al. (2012). Almost 50 % of the sample has a bond to region – and in LPLD 65 % of the participants feel a bond to the region (Tab. 8-5). Because Kühne (2011) described that identity to a region might be dependable on time, there might be a correlation between the variables of the residents and bond to region. It was checked and it could be found a low correlation (-0.33). The correlation is higher for newcomers than for returnees. Regarding the frequency, it could be found that more locals and returnees have a bond to region than newcomers.

There was a question about the living environment of each respondent (Tab. 8-1). It can be seen that almost every fourth respondent of LNWM lives in rural structures (Tab. 8-5). In the region LPLD are living more than 66% in rural structures while in STB more than half of the respondents are living in rural structures.

The question about the house where the respondent lives is used as a surrogate if a respondent has a garden (Tab. 8-1). Usually, if a person lives in a house, there is a garden. It can be seen, that even if in LNWM only every fourth respondent lives in rural structures, almost three-quarters of the respondents are living in a detached house (single family), semi-detached house, or row house (Tab. 8-5). In STB the rate of respondents who are living in these houses are lower.

There was a question about the occupational use of the landscape. It was transformed into a dummy and could see that the rate of people working for or in the landscape is relatively low (Tab. 8-1, Tab. 8-5). The highest rate of occupational use can be found in LPLD. For the other regions the rate is lower.

8.3.3 Individual Characteristics with Importance for the Contribution Form

To analyze what characteristics are important for the kind of support, the work assessed each of the control variables on the dependent variable. Only the highest percentages of each variable in the manifestations of the dependent variable are described; others can be seen in Tab. 8-6.

Tab. 8-6: Individual characteristics and support of the landscapes (in percent)

	No involvement				Only volunteering				Only financial				Both forms of involvement			
	All	LNWM	LPLD	STB	All	LNWM	LPLD	STB	All	LNWM	LPLD	STB	All	LNWM	LPLD	STB
Male	25,17	25,58	23,53	27,61	25,40	27,91	26,24	22,39	17,01	16,28	16,29	18,66	**32,43**	**30,23**	**33,94**	**31,34**
Female	29,44	22,78	**33,67**	27,54	19,55	18,99	20,60	18,56	20,45	21,52	18,59	22,16	**30,56**	**36,71**	27,14	**31,74**
Younger than 25 years	20,69	22,22	21,43	19,05	8,62	0,00	7,14	14,29	29,31	0,00	**50,00**	14,29	**41,38**	**77,78**	21,43	**52,38**
Between 25 and 65 years	25,26	21,05	26,35	26,20	23,49	22,56	26,67	19,65	17,43	21,05	13,02	21,40	**33,83**	**35,34**	**33,97**	**32,75**
Older than 65 years	**38,86**	**39,29**	**37,93**	**40,00**	22,86	35,71	17,24	25,00	20,57	14,29	25,29	16,67	17,71	10,71	19,54	18,33
No children	19,67	14,81	22,33	19,51	20,50	22,22	24,27	14,63	19,25	18,52	19,42	19,51	**40,59**	**44,44**	**33,98**	**46,34**
One child	26,80	22,50	**28,28**	27,27	24,74	30,00	**28,28**	14,55	19,59	17,50	16,16	27,27	**28,87**	**30,00**	27,27	**30,91**
Two children	**33,69**	**33,96**	**32,39**	**35,63**	23,40	28,30	19,72	26,44	19,15	13,21	20,42	20,69	23,76	24,53	27,46	17,24
More than two children	28,22	25,13	28,36	27,66	21,47	21,93	25,37	20,48	16,56	18,18	13,43	19,95	**33,74**	**34,76**	**32,84**	**31,91**
Middle school and lower education	**34,46**	32,05	**33,19**	**37,35**	20,51	19,23	20,96	20,48	16,91	15,38	16,16	18,67	28,12	**33,33**	29,69	23,49
High school and College	20,14	17,39	22,89	18,06	24,49	27,17	26,37	20,14	20,82	21,74	19,90	21,53	**34,55**	**33,70**	**30,85**	**40,28**
Monthly income up to 2,000 euros per household	**35,02**	26,92	**35,14**	**40,26**	16,61	17,31	16,89	15,58	20,94	23,08	21,62	18,18	27,44	**32,69**	26,35	25,97
Monthly income up to 4,000 euros per household	21,22	20,27	19,68	23,65	22,68	22,97	25,00	19,59	18,78	22,97	14,89	21,62	**37,32**	**33,78**	**40,43**	**35,14**
Monthly income more than 4,000 euro per household	23,60	21,21	26,15	22,22	28,57	33,33	**32,31**	22,22	18,01	6,06	21,54	20,63	**29,81**	**39,39**	20,00	**34,92**
Locals	**29,83**	25,64	**28,66**	**33,33**	21,69	23,08	24,20	17,17	19,66	17,95	18,47	22,22	28,81	**33,33**	**28,66**	27,27
Returnee	26,76	16,00	**31,65**	23,68	23,94	28,00	24,05	21,05	15,49	8,00	17,72	15,79	**33,80**	**48,00**	26,58	**39,47**
Newcomer	26,50	24,76	27,37	26,59	22,65	22,86	23,16	21,97	18,80	21,90	16,32	19,65	**32,05**	**30,48**	**33,16**	**31,79**
No bond to nature	**33,18**	**26,32**	**34,69**	**34,59**	20,42	23,68	20,41	18,87	22,04	25,00	21,94	20,75	24,36	25,00	22,96	25,79
Bond to nature	22,55	22,34	23,08	21,85	24,22	23,40	26,07	21,85	15,87	13,83	14,53	19,21	**37,37**	**40,43**	**36,32**	**37,09**
No bond to region	26,15	25,00	30,00	22,67	21,54	23,75	20,63	21,33	18,97	20,00	17,50	20,00	**33,33**	**31,25**	**31,88**	**36,00**
Bond to region	28,65	23,33	27,41	**33,75**	23,08	23,33	25,19	19,38	18,65	17,78	18,15	20,00	**29,62**	**35,56**	**29,26**	26,88
Rural environment	28,06	24,00	**33,97**	25,00	22,78	25,60	21,15	22,06	18,47	16,00	18,59	20,59	**30,70**	**34,40**	26,28	**32,35**
Urban environment	27,18	24,44	25,18	**31,03**	22,11	17,78	24,82	18,97	19,07	26,67	17,52	19,54	**31,64**	**31,11**	**32,48**	30,46
No garden	24,63	23,53	24,75	25,49	24,14	23,53	20,79	31,37	15,76	11,76	20,79	9,80	**35,47**	**41,18**	**33,66**	**33,33**
Garden	28,43	24,37	**29,48**	28,96	21,92	23,53	24,32	18,15	19,66	21,85	17,02	22,01	**29,99**	**30,25**	29,18	**30,89**
No occupational use	28,52	23,84	29,39	30,00	20,96	25,17	21,90	17,41	19,79	19,21	19,02	21,11	**30,73**	**31,79**	**29,68**	**31,48**
Occupational use	22,54	26,32	24,10	17,50	30,28	10,53	30,12	**40,00**	13,38	15,79	13,25	12,50	**33,80**	**47,37**	**32,53**	30,00

Note: The bold values are the highest values of the manifestation of the variable for the sample.
Source: Resident survey, own calculations.

Except LNWM, people with lower education do not support landscape protection, while people with higher education support with volunteer work or money. People with a lower income than 2,000 euros monthly and households do not support landscape protection (again except of LNWM), while people with a monthly income between 2,000 and 4,000 euros are involved in both forms of support. The respondents of LNWM are willing to contribute in form of volunteering and money independently of the income. Respondents in LPLD with a higher household income than 4,000 euros a month are only willing to volunteer, while in all other regions the respondents are willing to contribute both forms.

Locals do not seem to support the landscape highly in the analyzed forms of involvement. The percentages are equal in the region LPLD for no support or both forms of involvement. Again, the respondents of the region LNWM are willing to support the landscapes by volunteering and via financial contributions. With except of the returnees in the region LPLD, all returnees and newcomers are willing to contribute volunteer work and money. Bond to nature shows that in all regions, respondents with a bond are willing to support the landscape protection in form of volunteer work and money, while people without a bond to nature are mostly not willing to contribute. Except the respondents of the region STB, people with no bond to region as well as the people with a bond to nature are willing to contribute in both forms of support for landscape protection. The respondents in STB are mostly not willing to contribute if they have a bond to region. The living environmental surrounding does not seem to be crucial for the form of contribution. Respondents with rural or urban environment are willing to contribute in both forms of involvement. Only respondents living in a rural environment in LPLD or living in an urban environment in STB are mostly not willing to support the landscape protection in any of the examined forms. Respondents with and without an own garden are willing to support the landscape protection in volunteer work and money. Only people with a garden in LPLD are mostly not willing to contribute. The description is similar for the occupational use: all respondents are willing to support landscape protection in both forms, expect the respondents of the region STB. There, people with an occupational use are mostly just willing to contribute volunteer work but no money.

In general, younger people with a higher education, household income higher than 2,000 euro monthly and with a bond to nature have higher potential to contribute volunteer work as well as money for the preservation, protection and design of landscape in the sample. A valid manifestation of the other variables cannot be given, because they differ between the regions.

8.3.4 Methodology

First, a multi-nominal logit model was used to address four different categories of the dependent variable *WTP_WTV*. Using this model, it was possible to address the three different regions. But the Hausman test indicates that the most important assumption of this model is violated, and this is why a nested logit model is used. This model overcomes the difficulty stated by Schiappacasse et al. (2013): to find an econometric model that is suitable for this type of analysis. It uses a decision tree to capture if an individual is willing to contribute at first and at second if the individual is willing to contribute, in which form it would support the regional landscape protection (Fig. A 2). To use the nested logit, it is necessary to expand the data set in the first step so that the individual has more than one alternative that could be chosen. For the calculations, the basis is always the manifestation to invest rather than to give money or do volunteer work in the region.

Within the nested model the individual has to decide which alternative it would choose. Therefore, the following model is formulated:

$$chosen_{j,n} = f(X_n, L_n, N_n, IRI_n)$$

where $chosen_{r,n}$ denotes the chosen alternative j of the individual n. X_n is a matrix of the socioeconomic variables (*gender, age, education*) of the inhabitants in the sample and L_n their living circumstances (*children*, *income*, *living environment*, *garden, occupational use*). N_n includes the concern about the nature (bond to nature). IRI_n indicates if the individual moved to a region (*newcomer*) or were born there and left the region for awhile (*returnee*) as well as the individual has a relationship to the region (*bond to region*). Only one model is estimated, using X_n and N_n for the first level estimation and the other two matrices for the second level equation. Due to this two-step estimation, a regional comparison is not possible. The variables of X_n and N_n were chosen because of some previous results of the literature (see hypotheses two to five).

8.4 Empirical Analyses

The first part of this chapter describes the results of the nested logit estimation. In this section, no contribution is always the base outcome. First, there will be description about the first level, followed by a second level. In the end there will be a description of each probability that the alternative is selected. Following are a description of the results of the hypotheses verbalized in Chapter 8.2.4.

8.4.1 Results of the Empirical Model

All variables used that have an impact on the decision whether an individual would support the landscape or not are significant (Tab. 8-7). While gender is significant on the level of 0.05, the other three variables are highly significant on the level of 0.01. The coefficients of gender and age show a negative impact. This would imply that males are more involved in any contribution form than women. For the age, it could be identified that older people are less willing to contribute for their regional landscapes. Regarding education and bond to nature, the coefficients are showing positive directions. This would imply that with a higher education the willingness to contribute increases. Similar can be said for bond to nature. People with a bond to nature are more willing to contribute than people without any.

Tab. 8-7: Coefficients of the nested logit model

chosen *support equations*			*category equations*			
Yes				*Only volunteering*	*Only financial*	*Both contributions*
Gender	-0,41 (0.177)	**	Children	0,09 (0.387)	0,14 (0.486)	-0,29 (0.457)
Age	-0,03 (0.006)	***	Income	0,00 (0.000)	0,00 (0.000)	0,00 (0.000)
Higher education	0,59 (0.191)	***	Rural Environment	-0,04 (0.830)	0,12 (0.892)	0,27 (0.637)
Bond to nature	0,79 (0.181)	***	Garden	-0,43 (1.049)	1,31 (2.221)	-0,39 (1.081)
			Ocupational use	2,39 (4.332)	-2,82 (5.271)	-0,04 (1.071)
			Returnees	0,90 (1.761)	-1,59 (3.320)	0,68 (1.376)
			Newcomer	0,95 (1.535)	-1,13 (2.605)	0,75 (1.072)
			Bond to region	1,22 (2.108)	-0,56 (1.448)	-0,41 (0.899)
			Constant	-8,18 (18.477)	-5,86 (14.037)	-2,40 (8.254)

Note: Stars indicate significance levels: * p<0.1, ** p<0.05, *** p<0.01
Source: Resident survey, own calculations.

Looking the second level estimation, none of the variables have any significant impact on the form of contribution. Even if the coefficients are not significant, some directions could be assumed. Income has at least for each contribution form a positive impact while for every other coefficient the directions differ. An increase of the number of children increases the WTP or WTV – but not both forms. In this case, an increase of children would be lead to a decrease in the willingness for both forms. All other variables show different direction for the WTP and WTV decision. Most of the variables have a positive impact on the WTV and a negative one on the WTP. People with an occupational use of the landscape tend to be more willing to volunteer in the region. Also newcomers and returnees show a tendency for volunteering as well as people having a bond to the region. A reverse impact of these variables could be assumed for the WTP. Regarding both contribution forms, newcomers and returnees show a tendency to be more involved in both forms. Indeed, the occupational use and bond to region show negative directions for both forms of contribution. Garden and rural environment show negative coefficients for the WTV which would imply that people living in more rural areas or having their own garden are less willing to volunteer while these are more willing to pay for their landscapes. The tendencies differ for both contribution forms. For rural environment, it could be seen that the coefficient is positive and therefore people living in rural structures are more willing to contribute in both forms while people with their own garden are tend to be less willing to contribute both forms.

The Wald test implies that all of the integrated coefficients are able to contribute a statistically significant improvement in the fit of the model. The test statistic for the interdependence of irrelevant alternatives (IIA) assumption also indicates that the assumption of IIA can be confirmed for this model.

Out of the 801 cases, about 26 % of the sample was unwilling to contribute at all, while the rest were willing to contribute. These 74 % were distributed into almost 22 % willing to volunteer, 19 % only willing to give a financial contribution, and almost 34 % willing to contribute volunteer work and a financial contribution.

8.4.2 Results of the Hypotheses

The first hypothesis was about a general contribution and a preference to financial support of landscape protection. This could not be approved because the results of Tab. 8-5 show a high preference for volunteering and financial support. This could also be confirmed in Tab. 8-6. With respect to the contribution of both analyzed forms, the respondents prefer no involvement. Not until then is volunteering pursued. This would imply that people generally willing to pay in the form of money and volunteer work at the same time prefer one of the concepts. This also shows the probability measures of the used statistic model. Therefore, the first hypothesis is rejected.

Looking at the hypothesis that analyzed whether men are more involved in the regional landscape protection than women, the results display that men are more willing to volunteer while women are more willing to pay for the landscapes (Tab. 8-6) which was cannot contribute to the formulated hypothesis.

The statistical analyses found that men are more willing to contribute than women (Tab. 8-7). Because of this significant result, the hypothesis about a higher involvement of men in landscape protection could be proven.

Hypothesis three assumed that older people are less likely to contribute than younger people. This can be proven using the descriptive statistic as well as the estimated nested logit model (Tab. 8-6, Tab. 8-7). Hence, this hypothesis could be confirmed.

Education has a positive impact on the landscape protection was the next hypothesis. This hypothesis could be proven via the estimated statistical model and the descriptive statistics (Tab. 8-6, Tab. 8-7). These show a high rate of people having a high school or college degree on both contribution forms, which could be interpreted as them being more willing to contribute than people with a lower degree (most of the rates show no contribution). All told, this hypothesis could also be validated.

Bond to nature has a positive influence on the decision to contribute. This could be proven using the statistical results (Tab. 8-7). The descriptive analysis shows that if using both forms of contributions in general, that people with a bond to nature are more willing to contribute in both forms while people without such a bond are not willing to contribute for the landscape (Tab. 8-6). The hypothesis can be confirmed.

The sixth hypothesis about the negative impact of children on the WTP could not be proved, because there is an indication of a positive influence for volunteering or financial contribution (Tab. 8-7). Only the impact for both contribution forms might be negative. Tab. 8-6 shows only for two children that these people are less willing to contribute, but in all other cases, the highest percentages are shown for both forms of contribution. This is different from the statistical results. Therefore the hypothesis needs to be rejected.

The seventh hypothesis is about a positive impact of income on the WTP and on the WTV. The statistical analysis shows a positive direction of the coefficient. People with a higher income are more willing to contribute both forms of involvement (Tab. 8-6). One exception is the region LPLD, where people with high income are more willing to volunteer. This might be an indication that the hypothesis could be true, but it could not be proved because there are no statistical significant coefficients.

The next hypothesis refers to the higher valuation assigned by returnees and newcomers for all forms of contribution. The positive impact for these two groups is implied for the WTV and for both forms of contribution, but not for the WTP (Tab. 8-7). The coefficients show a negative direction for the two groups. Tab. 8-6 shows that newcomers and returnees (at least for LNWM and STB) are most willing to contribute payment and volunteer work. These results indicate that the hypothesis needs to be rejected, because it could not be proven that the valuation is higher in each form.

The next hypothesis is about the positive influence of bond to region on all forms of contribution. The results of the statistical model indicate a positive influence for volunteering, but not for the financial contribution or both forms. The results of Tab. 8-6 do not show any differences, except of this case of STB: When a person has a bond to region they do not want to get involved in landscape protection. In all other cases, the person is willing to contribute both ways. The hypothesis might therefore be true for the option of volunteering, but could not be verified with robust results due to missing significances in the coefficients.

The tenth hypothesis assumed that people living in the countryside are more willing to volunteer than to pay, compared with the people living in a city. The statistical results imply the other way around. The coefficients show a positive direction for the WTP and a negative one for the WTV (Tab. 8-7). Similar results are shown in Tab. 8-6. While all people no matter what environmental conditions are willing to contribute both forms, for LPLD people living in the countryside are not willing to get involved into landscape protection and for STB people living in a city do not want to contribute for the landscape protection. This hypothesis could not be proved.

Following is the hypotheses that people having an own garden are less willing to contribute volunteer work than payment contributions. Regarding the results of the statistical analysis, the coefficients imply that the garden has a positive impact on the WTP, but a negative one on the WTV. Moreover, both contribution forms might also are negatively influenced by the garden. In regard to Tab. 8-6, the variable garden does not seem to have an effect, because in all regions, the people are more willing to contribute in both forms no matter if they have their own garden. Because of the diverse results and no significant variables, hypothesis cannot be proven.

The last hypothesis is about the occupational use of the landscape and the impact on all forms of contribution. The statistical results show a tendency for a positive impact on WTV and a negative one on the WTP (Tab. 8-7). Both contribution forms might also be influenced negatively. Tab. 8-6 shows that it does not matter if the respondents use the landscape for their occupation: they always tend to involve both forms of contribution – except in the STB region. These results do not allow a confirmation of this hypothesis.

8.5 Discussion

This part of the analysis discusses the results of the empirical analyses and the generalizability of the current results for other regions. Additionally, there is a discussion of further variables that might have impacts on the forms of volunteering. The discussion ends with addressing methodological issues.

8.5.1 Results and Generalizability

The descriptive analyses showed respondents are more willing to contribute money and volunteer work at the same time than doing only payments. This result is different from the results of Schiappacasse et al. (2013), Hung et al. (2007), Rai and Scarborough (2013, 2014), Echessah et al. (1997) and Lankia et al. (2014), but are in line with Gibson (2015). Except for Lankia et al. (2014), all research was done in developing countries, and so it might be the case that there were differences between the development stages. Lankia et al. (2014) concluded that WTV is more common for their analyzed region in Finland, but they used a privately owned landscape, while this analysis used only public landscapes. It might also be the case that the owner of the landscape is an important factor for the decision of an individual in what form the individual contribute to the protection and preservation of the landscape.

If a person is male, the willingness to contribute is higher than for women. This result is different from findings in the United Kingdom, the United States and Australia (Measham & Barnett, 2008; Menchik & Weisbrod, 1987), but in line with results in Germany (Enste et al., 2012; Simonson et al., 2016). Enste et al. (2012) found that younger women are less involved in volunteer work than same-aged men, but with increasing age the rates move closer to each other. Simonson et al. (2016) found an increasing of the rate of involvement for women so that the differences between men and women are decreasing. It might be the case that women are busier with raising kids and therefore have less time than men to get involved.

The older the person, the lower the willingness to contribute. Tidwell and Brunson (2008) found that people willing to volunteer on average are younger than those who are not willing. Menchik and Weisbrod (1987) found an age limit for the age between 35 and 49 – before that the rate of volunteerism increases and afterwards the WTV decreases. This is in line with German findings of volunteering in general (Enste et al., 2012; Simonson et al., 2016). The older a person is the lower is the WTP a financial contribution for the preservation, protection and design of the landscapes in their region. This is also in line with results of other studies (e.g., Carson et al., 2001; Hackl & Pruckner, 1999; Lankia et al., 2014). Vásquez (2014) found that younger respondents are generally more willing to contribute than older ones. The author explains this development by using a shorter lifespan to benefit from improvements. Additionally to this argument, the health condition and the lower income in retirement could be other reasons while older people have lower level of contributions to the regional development.

High education displays as positive impact factor for the willingness to contribution. Positive influence of the level of education on the willingness to contribute was also found by Schervish and Havens (1997), Howley et al. (2012) and other authors and could also be confirmed for Germany by Simonson et al. (2016) as well as Enste et al. (2012). This might be explainable due to more knowledge about nature and its future development if people fail to invest in the preservation and protection of the landscapes (Howley et al., 2012). It might also be the case that people with higher education are thinking more about the next generation and how this generation will live if the landscapes are not being protected today.

Bond to nature is an important variable for the willingness to contribute. The result of volunteering is in line with Bramston et al. (2011), Bruyere and Rappe (2007) and Miles et al. (1998) who found that the relationship to nature is a high important factor for volunteering. It might be the case that people with a bond to nature have more knowledge about the interaction between nature and humans and try therefore to protect the landscape for next generations.

The amount of children is a positive influence factor for the WTV and WTP. Only the impact of the contribution of both forms at the same time might be negative. This result agrees with the results of Freeman (1997) that people with children are more volunteering than without children. Menchik and Weisbrod (1987) found that if the children are at home, then the parents are more likely to volunteer. It might be the case that the parents are more aware of the surrounding than people without any children.

The impact of income could not be identified. Schervish and Havens (1997) found that households with lower income give higher amounts of money to charity than people with a higher income. German examinations (Ahlheim et al., 2010; Simonson et al., 2016) as well as results in developing countries (Schiappacasse et al., 2013; Wilson, 2000) found a positive impact of the income on volunteering. This result might be explainable by the motives the individual has. It might be the case that they are motivated by a warm-glow and would like to do something good for the environment to compensate for their business environment.

There might be a positive influence of being a newcomer or returnee for the WTV and both forms of contribution identified. These two groups might have a negative impact of the age. Schervish and Havens (1997) concluded that people who have lived longer than ten years in a community are willing to give more than people living there for two years or less, but this could not be proven clearly with the current results. A variable about the exact duration and if newcomers or returnees live in the region permanently was not used – indeed, it was controlled for people with different experiences of landscapes. The results could be explained by different motives of the newcomer to get involved than the locals have. It might be reasonable that these individuals volunteers because of social integration (DiEnno & Thompson, 2013; Wilson, 2000). Additionally, with respect to these motives, it could be the case that newcomers and returnees see the landscape as more special for the region than locals do, since they have seen other landscapes before. There might be a bias in the data because it might be the case that locals wanted to move, but had not the chance to move. It was tested if the landscapes are the reason why the participant came into or moved back to the region. The most often reason why a respondent came into the region was a family related one. It could not be tested if the participated locals wanted to move because the data about this question is not available.

The impact of bond to region shows a positive tendency for the WTV, but negative directions for the WTP and both forms of contribution. It could be stated that people with a bond to region seeing the regional landscapes as more worth protecting than those without a bond to region because they identifying themselves with the region and the landscapes (DiEnno & Thompson, 2013).

The rural environment also showed different trends. A positive tendency for the WTV is found for people living in a city environment, while a positive direction might be identified for the WTP and both forms of contribution for more rural areas. People living in the city are not surrounded of the landscape so it might be the case that they valuing it higher because they see what happen if every landscape is build on. It could also be the case that people living in the city are using the opportunity of volunteering to get into the landscape where they are not go into otherwise (Bruyere & Rappe, 2007).

The results regarding gardening might have a negative impact for WTV and both forms of contribution and a positive one for WTP. This could be explained by the work in the own garden. People having their own garden prefer to work there instead of working in the landscape and therefore they are less willing to volunteer than to pay.

As in the variable discussed before, the impact of the occupational use shows a positive direction for volunteering and negative ones for the WTP and both forms of contribution. It might be the case that people working in or for the landscape in their occupation know where the money is going and therefore prefer to work in the landscape to be sure their support is being used in the right way.

Because of the socioeconomic variables gender, age, educational level and bond to nature display common impacts, it might be possible to transfer the results of this work into other regions. But the variables for the categories did not show any significance, so that these results are not robust. Only the variables for the decision if an individual would support the regional landscape might be transferable.

8.5.2 Further Variables

For instance, the experience with the good and its visits might be important (Morrison & Dowell, 2013; Rai & Scarborough, 2014). Carson et al. (2001) and DiEnno and Thompson (2013) concluded that only experienced goods could be valued from the respondents in an adequate way. These arguments might be true, but even if the respondents do not know the landscape the question rises if they perceive the landscape as something special – thus they could be answering as if the landscapes in question are, in their opinion, worthy of protection. There was no question about if the respondents had visited or experienced the examined landscape before or just know the good in the survey and therefore are not able to integrate such a variable into the sample.

The economic development could be integrated in the analysis. Tselios et al. (2015) analyzed the impact of economic development of a neighborhood on the local social integration (including the local social engagement). The authors found that living in a low economic development neighborhood has negative impacts on actual involvement and could therefore be important to explain differences between the regions. It might be possible that some of the region specific variables are some kind of moderator variables for the economic development. In integration of economic development could also be important because it might be the case that less economical developed regions are less willing to contribute in any form of landscape protection because they see other and for them more important fields for contribution. A variable that shows if a respondent is employed were not included, but a variable of employment with interactions to the landscape might be a good idea for further research. Further analyses could be including the variable of economic development of a region and a variable that shows if a respondent is employed at the moment. This analysis could not integrate the employment status, because of the people not working at the moment due to parental leave or the protection of working mothers. There would be the question to which group they would belong to, since by law says these people are employed, even if they are not actually working.

Evidence suggests that environmental values are similar between different countries, but are still bound by culture and the subjective background of the individual (Schultz & Zelezny, 1999; Yu, 1995). Swanwick et al. (2007) also concluded that culture could be a factor for the landscape valuation. For instance, it might be the case that landscapes are an important factor for the region because of its use of for instance geocaching, the valuation might be higher than for regions where the landscape is just available and not used for a special purpose. Especially concerning volunteering, Ahlheim et al (2010) argued that the willingness to contribute volunteer work might be biased by social norms of a region or country. If it is highly esteemed that an individual works on a voluntary basis for the region, more people might be volunteering. If the culture is characterized by a high reciprocity of individuals (Gouldner, 1960) more people might volunteer because they want to repay their benefits. But this argument is probably more evident for the area of social work than for landscape protection. The influence of culture in the selected regions in the used models could not be tested, because there was no possibility to integrate all aspects of culture into the survey.

Donor´s attitude and motives towards helping others and towards a nonprofit organization might be important for the motives to the willingness to invest money or time into something (Webb, Green & Brashear, 2000). Research published by the German Ministry of Family Affairs, Senior Citizens, Women and Youth examines the motives of German volunteers on a regular basis. Their first work found that almost half of the respondents in their sample had almost purely altruistic motives ("Because I want to do something for others, helping others", "Because I want to repay something") for volunteering (Haumann, 2014). But on average, most of the volunteers have more impure or warm-glowing motives ("Because I am happy working voluntary"). A later examination by Simonson et al. (2016) found that a lot of volunteers work voluntary because of having fun or for social reasons (more than 80 % of the sample). These are warm-glowing motives. At least half of the sample volunteers because of gaining competencies. Lower levels of motives are gaining social esteem (31 %) or getting career competencies (25 %). More generally, the motives are important to understand why somebody is volunteering of contributing money for landscape protection. It could just be analyzed what individual characteristics are crucial for the contribution form, but it was not possible to link the motives and the individual characteristics. Because motivation differs between individuals (Clary, Snyder & Stukas, 1996; Houle et al., 2005), it might be even interesting to couple these with individual characteristics. This might be an interesting question for further research. Such an analysis could help get deeper insight on why a person is willing to pay or volunteer for their regional landscapes.

Roesch-McNally and Rabotyagov (2015) examined differences between a mandatory and voluntary model for payment and found that the willingness to give some money depend on the model of giving. For water quality they concluded that the voluntary model collects higher amounts while for recreation the mandatory model got higher amounts. Harbaugh et al. (2007) concluded that the motive of giving is important for the model of payment: While people with altruistic motives prefer mandatory payments, the warm-glowing motivated people prefer voluntary payments. The payment form of contribution for money was asked. Most of the respondents (almost 63 %) answered they would like to pay for the landscapes as voluntary donations which indicate more warm glowing motives of the respondents. Almost every fourth respondent of the survey of Kämmerer et al. (1996), Hartje et al. (2003) or Job and Knies (2001) also voted for donations. The second and third chosen forms in the current survey were a mandatory form: entrée fees or parking prices (both 31 %). Entry fees were also the second most answer in the research of Hartje et al. (2003). These results suggest that respondents are more likely to pay on a voluntary basis than on a mandatory one for landscape protection and that implies that the respondents are acting more for impure altruistic motives than for altruistic motives.

8.5.3 Methodology Issues

Using the method of WTP to determine a value for a public good is chancy regarding the capacity of the stated values. It could be the case that people answering would like to contribute, but they do not if it is not necessary. Concerning the monetary payment, this was stated, for example, by Ryan and Spash (2011) and was proven by Cummings and Harrison (1994). Regarding the WTV, Tidwell and Brunson (2008) found similar results. Ahlheim et al. (2010) and Echessah et al. (1997) concluded that the WTV depends on the social norms and the living conditions of an individual. Additionally, Schiappacasse et al. (2013) concluded that people do not consider the time constraint and stated more time as realistic for them. Therefore, it might be possible that respondents did not answer truthfully about their willingness to contribute. This could be a problem if volunteering would get a monetary value and would estimate both contributions to determine a welfare gain or loss, but both contributions could be used as indicators for the perception.

Close to the argument about the truth about the stated values is the risk of strategic behavior of the respondents. It could be the case that people think about asking before realizing a price for a good and therefore avoid choosing a concrete value for this good (McFadden 1998) or answer how they think it would be social desirable (Kim et al., 2015). If there are social norms that pretend working voluntarily is positive, the risk of answering in a social desirable way increases (Ahlheim et al., 2010). Another important reason, especially in developing countries, is that inhabitants would like to have more research in their region or area and thus they answer the way they think the researcher would prefer (Echessah et al., 1997).

Vásquez (2014) found that Guatemalan are willing to pay more for municipal services while the respondents are neither WTP nor WTV for community-managed services even if they are impatient with the actual condition of the environmental good. There was no information about who has to collect the contributions and uses it afterwards, but one question asked for the possible instrument to pay. The most common answer was voluntary donations for the preservation, protection and design of the landscape. In this case, everyone could choose by him or herself to whom the payment would go or for whom the individual would like to volunteer. But still there might be the question if the WTP and/or WTV are dependable on the executor and for whom he or she is working. This could be a field of further research, especially in the industrialized countries where this phenomenon was not examined yet.

Free-riding[52] is another risk when looking into valuations of monetary and volunteering payments (Hung et al., 2007; Rai & Scarborough, 2014). It might be the case that people think that the landscape is only important for them and not a resource for future generations (Carson et al., 2001). DiEnno and Thompson (2013) found that volunteers usually working voluntarily because they see nature as a resource and want to conserve it for their children. But by using a survey, free-riding could also be a problem for volunteering (Vondolia et al., 2014). It could lower the amount of respondents who are WTP or WTV and therefore distort the value of the landscape. By using WTP and WTV as indicators, this risk might be lower, but it could still be present.

Both contributions require a hypothetical market. Volunteers could value time differently if the context changes (Álvarez-Farizo, Hanley & Barberán, 2001; Becker, 1965; Dalenberg, Fitzgerald, Schuck & Wicks, 2004; Eom & Larson, 2006). For the financial contribution it might be the case that respondents do not have the information that is needed for valuing a good, the results could not interpreted in the right way (Carson et al., 2001; Mitchell & Carson, 1989). In the survey, current landscape photographs were used to show the context of the volunteer work and to imagine the landscapes without describing them much because the landscapes are part of their region. Additionally, a variable for volunteer work out of different fields of volunteering were created; there might be fewer problems with the context of the volunteer work.

Generally speaking, it could be said that the determination of a value of a public good using the WTP and WTV together could be used for the determination of a landscape. There might be some risks using the methods, but it seems to be more adequate than using only the monetary valuation.

52 Chapter 4.3 explains what free riding means regarding the WTP.

8.6 Conclusion

This paper analyzed how many people are willing to contribute for the protection, preservation or design of their landscapes and what individual characteristics are influencing each form of contribution. The results suggest that the decision of an individual to contribute might be performed on two levels. The first will be to support the regional landscape and the second in what form the individual would contribute.

The analyses found that less than 50 % of the respondents of the analyzed regional samples are willing to invest money to preserve or protect their regional landscape, but about 75 % are willing to contribute labor in terms of volunteer work to support the landscape. Combining these results, one third of the sample would be willing to contribute volunteer work and financial support. The people with the less willingness to give any support are older than 65, mostly women, have a lower education, and or enjoy no relationship with nature.

Regional managers could learn from these results that every individual has similar needs and value the landscape in a different way, even if there are similarities. But beyond this, these results suggest some potential for activating volunteers in the conservation in landscapes in these regions and that at least some residents are still interested in volunteer work – even if they are not volunteering at the moment.

Further research could repeat such an analysis with a bigger sample so that significances on the second level could be scrutinized. Beyond this, the results of this analysis suggest further research regarding the interaction of individual characteristics and the motives for an investment for the region in form of volunteering or financial contribution is warranted. One research along these lines was already done from Carstensen (1995). Another interesting approach could be to see how the economic development of the region could influence the willingness to invest something in the regional landscapes.

Additional Analysis 1: Monetization of Volunteering

The monetary value of a good can be determined by using replacement costs, opportunity costs, and social benefit approaches (Sajardo & Serra, 2011; Salamon et al., 2011). Replacement costs assume the same costs as a paid worker doing the same work as a volunteer while opportunity costs quantify the income for the unpaid work with working extra hours of a paid job (Álvarez-Farizo et al., 2001; Sajardo & Serra, 2011; Salamon et al., 2011). The focus of the social approach is on the economic outcome of the work (Salamon et al., 2011). Some of the common methods of the monetary valuation are Average Wage, Comparable Worth, Independent Sector, Living Wage and Minimum Wage (Anderson & Zimmerer, 2003; Sajardo & Serra, 2011). Average Wage means to use the average of the national or regional income of a person from census data. Comparable Worth supposes equality between a paid worker and a volunteer and treats them as a perfect substitute. Independent Sector calculates the value out of the average hour earnings of people working in production and on non-supervisory level in the private sector. Living Wage uses the federal poverty line to calculate a monetary value of the volunteer work. Minimum Wage uses the federal minimum hourly wage as basis of the monetary value (Anderson & Zimmerer, 2003). There might always be the question which technique might be the best one for transferring time into a monetary value (Gibson et al., 2015; Rai & Scarborough, 2013,). Schiappacasse et al. (2013) used the approach of opportunity costs to value the asked monthly hours of volunteer work while Stone et al. (2008) and Hung et al. (2007) used the average wage rates of the local labor market to determine a monetary value. Gibson et al. (2015) tested the differences between the wage rates on the local labor market and the internal opportunity costs in a rural area in Cambodia and concluded that the local wage rates are similar to the internal opportunity costs. Rai and Scarborough (2013) concluded that rural farmers value their working hours differently from the wage rate.

There is some trouble with the construct for giving volunteer work an economically adequate monetary value (Dalenberg et al., 2004) – especially when adding the costs of labor, the costs of the selection process or for social security to the valuation of volunteer work (Sajardo & Serra, 2011). The context in which the individual spent that time and its position on the labor market should always be considered, because that could change the value of time of the individual (Álvarez-Farizo et al,. 2001; Becker, 1965; Dalenberg et al., 2004). Additionally, other difficulties with using a monetary payment could occur, for instance, given that cash has a sensitive nature in the culture of the country (Hossack & An, 2014).

Finally, there might be a huge difference between the calculated value and the real value of the work for the society, because of different techniques to estimate the monetary value (Anderson & Zimmerer, 2003). But not only the techniques of estimating the monetary value of volunteering could be difficult, also the subjective monetary value of working time for each individual could be varying (Hossack & An, 2014) or time as a whole would be valued differently for each individual (Rai & Scarborough, 2013).

But still no method for calculating a monetary value for volunteer work is dominant and therefore it should be recognized that each method provides different financial results and has their own strength and weaknesses (Anderson & Zimmerer, 2003; Dalenberg et al., 2004).

The discussion about a monetary value of volunteering shows that it might be reasonable to analyze the valuation of a public good by integrating volunteer work into the existing construct of the CVM, especially because hypothetical hours of each individual for volunteer work were used.

Even if the problems about monetizing volunteer work are known, a monetization of the hours with the minimum wage in Germany that is at the moment 8.84 euros is implemented. These values are just created to show the potential of volunteering per month[53]. It can be seen that for instance almost 12 % of the whole sample marked that she or he is willing to volunteer for on average 1.5 hours a month (Tab. 8-8). This would be an additional value of 689.52 euros for the analyzed regions as a whole but only for the respondents of the survey and not as a representative value for the region[54]. If all respondents that marked that they would be willing to volunteer would really working voluntarily, the monthly gain for LNWM would be 6,755.76 euros; for LPLD the gain would be even higher with 19,890.00 euros and for STB it would be 9,763.78 euros, respectively [55]. The monthly potential apiece is about 64.94 euros for LNWM, 61.58 for LPLD, and 67.80 euros for the region STB. To express the monetary amount in terms of working hours, the average for the regions are for LNWM 7.3 hours a month, almost 7 hours per month for LPLD and for the region STB 7.7 hours monthly.

Tab. 8-8: Monetization of the volunteer work for or in the region (in euros)

Hypothethical monthly hours for volunteering in or for the region	All monetary	LNWM monetary	LPLD monetary	STB monetary
On average 1.5 hours	689,52	172,38	503,88	212,16
On average 4 hours	5.268,64	1.272,96	3.783,52	1.768,00
On average 8 hours	9.971,52	2.475,20	6.435,52	3.323,84
On average 15.5 hours	7.673,12	1.507,22	4.658,68	2.603,38
On average 30 hours	6.364,80	1.326,00	4.508,40	1.856,40

Source: Resident survey, own calculations.

53 It was only possible to calculate the potential by using the respondents and therefore, the potential might be even higher than the results displayed here.

54 Because it cannot determine a value for the examined regions as a whole it is not possible transferring this amount on the whole region but stay on the level of the survey.

55 These values are dependent on the respond rates in each region.

Additional Analysis 2: Comparison of Sample and the German Volunteers

To have a look in the distribution of volunteers in the sample and in Germany, this part describes similarities and differences between the sample and the German Civic Engagement Report (CER) of the year 2014.

The overall involvement in the sample is almost the same for the German population (Tab. 8-9). The difference is only 1 %. Regarding the involvement by gender, the sample shows differences between men and women, even if the tendency that more men are volunteering is the same. Regarding the distribution in the same, about 50 % of all participated men are volunteering at the moment, while for the women this amount is only about one third. In the CER, the amount just differs by about 4 % between the genders.

Regarding the age, it was only possible to compare the people that are older than 65 years, because the other age classes do not match. It can be seen that the rate of the group of the people over 65 years is relatively high, especially to Germany (47 % to 34 %). This implies an overrepresentation of this age class.

The educational level shows the same tendency than the CER: People with lower education are less involved than people with a higher education. But, the rates for the lowest level of education and the highest are different and could be indicative of an underrepresentation of these groups regarding involvement. The rate of the involved people with a lower education is higher for the sample than in the CER and implies an overrepresentation. The reverse can be observed for the people with higher education, which indicates an underrepresentation in this case. In the case of the middle education, the rates of the sample and the CER are almost the same.

The areas of volunteering also show some underrepresentation. While in the sample more people are involved in the society, the conservation of the environment/ animal protection, politics, religion or fire service, these rates are lower in the CER. Some of the rates are almost doubled in the sample. Regarding education/ kindergarten/ school, sports/ exercise and art/ culture/ museums, the rates show underrepresented values when compared to the CER.

It is noticeable that most of the sample is involved in volunteering activities up to 8 hours a month. The rate in the CER is also high, but still lower than the rate that the sample shows. Compared to the other two rates of working more than 11 hours a month, the rate for up to 8 hours is even higher. While the sample shows an involvement of 16 % to work more than 11 but less than 20 hours, it shows only lesser rate (12 %) for the monthly volunteer work of more than 20 hours. The CER found that in Germany the rate for working between 11 and 20 hours voluntary is about 24 %, and the rate for more than 20 hours monthly is about 18 %. These results imply that people with an involvement of up to 8 hours are overrepresented in the sample while the other two involvements show an underrepresentation.

Tab. 8-9: Comparison between the sample and the German Civic Engagement Report 2014

	Current volunteer work (Sample)	Civic Engagement Report 2014
Overall	42,6%	43,6%
Gender		
Men	51,7%	45,7%
Women	34,2%	41,5%
Age		
65 and older	47,1%	34,0%
Education		
Lower education#	35,2%	28,3%
Middle education#	41,8%	41,1%
High education#	45,5%	52,3%
Areas of volunteering		
Society	33,6%	15,1%
Conservation of the environment/ animal protection	20,9%	8,6%
Education/ kindergarten/ school	16,5%	22,5%
Sports/ exercises	29,2%	43,7%
Art/ culture/ museums	12,7%	19,2%
Politics	12,7%	6,6%
Religion/ church	16,8%	12,3%
Fire or ambulance service	22,2%	5,1%
Hours for volunteer work		
Up to 8 hours*	72,1%	58,1%
11 to 20 hours*	16,2%	23,8%
More than 20 hours*	11,7%	18,1%
Life satisfaction		
Dissatisfied+	4,1%	26,5%
Indecision+	12,9%	33,5%
Satisfied+	82,9%	46,3%

Note: # Regarding the results of the CER the sample is defined as follows: Lower education= no school graduation and main school, Middle education= middle school, Higher education= high school or college degree. * The CER uses hours per week while the sample uses monthly amount of hours. The transformation complies to: Up to 8 hours= all categories including the 10 hours, 11 to 20 hours= the category is the same, more than 20 hours= remains the same. + For the sample, the categories of life satisfaction are defined as: Dissatisfied= 0 to 3, Indecision= 4 to 6, Satisfied= 7 to 10.

Source: Resident Survey, Simonson et al. (2016).

The tendency of people who are satisfied with their life working more in volunteer activities can be confirmed, but again, the rates are differing from these of the CER. People volunteering and dissatisfied with their lives and people who are undecided concerning their life satisfaction are underrepresented. The CER have results of 27 % for the people dissatisfied and 34 % for the undecided ones, while the sample only displays rate of 4 % for the dissatisfied and 13 % for the undecided people. Regarding the satisfaction, the rate of the sample is about 83 %, which is almost as double as the CER found (46 %).

9 Regions and Their Residents: How Do Individuals Assess Different Aspects of the Region?

In this part of the doctoral thesis, a cluster analysis is done to identify different clusters and their preferences regarding WTP, WTV, life satisfaction and the use of the region.

9.1 The Importance of Regional Features Assessment

Rural regions are suffering because of the demographic change. They lose their residents because the cities seem to be more attractive for them than rural areas (von Reichert et al., 2014). But the residents who are still living in the region might see strengths of the region and these might not be obvious aspects of it. But it might also be the case that the assessment of the region differs by the residents because of individual characteristics or past experiences. Some authors already examined such individual characteristics that are responsible for the support of the region and its landscapes (e.g., Brereton, Clinch & Ferreira, 2008; Gibson et al., 2015; Kim et al., 2015; Lankia et al., 2014). The authors did not go a step further nor include other aspects of the region, but only focused on the perception of specific elements out of the landscape support or life satisfaction. In the end, none of the examination did a cluster analyses to determine types of individuals and then analyzed these clusters regarding the features of the region regarding landscape support, life satisfaction and the use of the region.

Therefore, the aim of this research is to answer two research questions: First, which clusters of residents can be distinguished in the research region, and second, what are the differences between the clusters with regard to landscape support (financially or volunteering), life satisfaction and the use of the region? These research questions are answered using a cluster analysis and further with a descriptive evaluation of the variables of interest.

The added value is that these analyses might have the opportunity for the regional management to build target groups to strengthen the landscape support and the attractiveness of the region itself. It might be the case that the analyses show aspects that are not considered before. The results could also be used for the identification of a target groups to get a more effective marketing strategy.

J. Matloch, *The Assessment of German Cultural Landscapes*, RaumFragen: Stadt – Region – Landschaft, https://doi.org/10.1007/978-3-658-21416-6_9

If the target group is addressed first, it might be possible to save some money that could then be used for other projects in or for the region.

This introduction is followed with the theoretical background that consists of some theoretical backgrounds of life satisfaction and the use of the region. The next section is about the description of the manifestation of the variables for the cluster. The following part describes the results of the cluster analysis. Section four describes the variables for the analyses of the clusters and states some hypotheses regarding the variables of interest. The results of these variables are shown in the following part. Subsequently, a discussion about the results is done. In the end, the paper summarizes the results in the conclusion.

9.2 Theoretical Constructs and Measurements

This part describes some theoretical constructs and measurements of the variables under examination. The approach of the WTP is described in Chapter 4, while the concept and the measurement of the WTV are presented in Chapter 8.2.2. A short description of the life satisfaction is given, before the chapter ends with theoretical constructs of the use of the region.

9.2.1 Life Satisfaction

The highest goal of an individual is to be happy in life (Enste & Ewers, 2014; Frey, 2008; Frey & Stutzer, 2002; Harding, 1985) and therefore research developed an index to measure the happiness of people. Life satisfaction or happiness is usually used as an alternative prosperity indicator instead of using the gross domestic product. This indicator leads to the fact that the individual needs to value his or her life regarding his or her own life conditions. The economic output of a society cannot accomplish these soft aspects (Enste & Ewers, 2014; Kahnemann & Krueger, 2006; Veenhoven, 2004; Veenhoven, 2011). Even if in countries with higher incomes the residents might be happier, Easterlin (1995), for instance, describes that an increase of income for all people does not increase the life satisfaction of all people. This is because of individual norms that might be important for the value of life satisfaction. Even if this approach is a psychological one, economists try to examine the economic determinants of this approach and use them for political decisions or to predict the impacts of institutional changes (Frey, 2008; Frey & Stutzer, 2002; Welsch, 2009).

There are again two approaches for measuring life satisfaction: the objective and the subjective approach. While the objective approach asks about the measurable fact as for instance income, while the subjective one gather the softer facts like perceptions (Veenhoven, 2004). But, the subjective approach has also some measurable indicators in it. Furthermore, the literature suggests two perspectives of life satisfaction: the hedonic and the eudaimonic one (Ryan & Deci, 2001). While the hedonic perspective looks deeper in the expectations of the individual, the eudaimonic one is about the degree of self-fulfillment and is therefore more orientated towards the long run. The hedonic perspective allows a comparison between different conditions, while the eudaimonic one is more orientated concerning the external influences (Frey, 2008; Kahnemann, Diener & Scharz, 1999; Veenhoven, 1996; Veenhoven, 2011; Waterman, 1993). Nettle (2005) distinguishes further three kinds of measurement: the author called the monetary emotions about joy and pleasure "happiness", the overall conditions "life satisfaction" and the self-fulfillment "the good life". The following analyses use the term life satisfaction, happiness and well-being synonymously for the overall conditions.

9.2.2 Use of the Region

There are two theories to predict behavior: Theory of reasoned actions (Ajzen & Fishbein, 1980; Fishbein & Ajzen, 1975) and theory of planned behavior (Ajzen, 1985). Both theories predict volitional behavior of individuals and its determinants, but the theory of reasoned action is based on the theory of planned behavior (Madden, Scholder Ellen & Ajzen, 1992). While the theory of reasoned behavior is based on the person´s intention as a function of the individual attitudes and norms (Ajzen, 1985; Ajzen & Fishbein, 1980; Fishbein & Ajzen, 1975; Madden et al., 1992), the theory of planned behavior includes unforeseeable instances, so that only the intention to do something could be predicted but not the performance of the act itself (Ajzen, 1985; Madden et al., 1992). These theories are used in different areas of psychology research. For instance, Sniehotta, Presseau and Araújo-Soares (2014) wrote about its application in health psychology, Hausenblas, Carron and Mack (1997) used the theories for sportive activities and Southey (2011) applied it in the context of business. Because of this variety it could therefore also be applied for explain differences in the use of the region.

A more economic orientated theory is the allocation of time by Becker (1965). The author argues that each individual has time left when he or she comes home from work so that he or she could spend for the consumptions of goods. The theory is based on the assumption that an increase in earnings would lead to more time spend in leisure activities (Becker, 1965; Gronau, 1977). Gronau (1977) extended the theory with the fact that it could be some work at home like child-care, which reduces the leisure time of an individual. These theories are hard to apply on the use of the region in this examination, because it only gives a hint how much time an individual would have to spend in the region rather than how often it would use it.

9.3 Sample of the Cluster Analysis

This chapter is about the data used to answer the research questions. The first section presents the used variables for the cluster analyses and its methodology is described.

9.3.1 Description of the Variables

The socioeconomic variables gender, age and education are common in all of the examinations of the variables of interest later on (Agahi & Parker, 2005; Howley et al., 2012; Measham & Barnett, 2008; Morrison & Dowell, 2013; Palmore & Luikart, 1972; Sørensen, 2014; Wilson, 2000). Because of this almost being standard, the cluster analysis also uses these variables (Tab. 9-1).

The sample consists of more men than women (Tab. 9-2). The biggest group of respondents is 45 to 55 years old, followed by the group from 55 to 64 years of age, while the group of younger respondents is only about 6 % of the sample (Tab. 9-2). This shows again that the older ages are overrepresented. The education variable was transformed into a binary one that shows if a respondent passed the high school or a higher education institution (Tab. 9-1). Almost half of the sample has a higher education level (Tab. 9-2).

Tab. 9-1: Overview about the questions and variables used

Question in the survey	Possible answers	Transformation of the variable	Name of the variable in the data set
What is your gender?	female male	with 0 = male and 1 = female	gender
How old you are?	0 - 17 years 18 - 24 years 25 - 34 years 35 - 44 years 45 - 54 years 55 - 64 years 65 - 74 years older than 75 years	manifestations of the categorizes are displayed by the mean of each class	age
How many children do you have?	open answers	amount of children	children
What is your highest school graduation?	No school graduation Main school Middle school High school College	transformed into a variable that shows if a respondents have higher education with 1 = yes (High school, College) and 0 = no (no school graduation, main school, middle school)	higher education
To what extent do you work?	Fulltime part time protection of working mothers job-seeking pension because of other reasons not working	transformed into a a binary variable that shows if a respondent is working at the moment with 0 = not working and 1 = working	employed
What is your your net income for all members of your household (monthly)?	lower than 1,000 euros 1,001-1,500 euros 1,501-2,000 euros 2,001-2,500 euros 2,501-3,000 euros 3,001-4,000 euros 4,001-5,000 euros more than 5,000 euros	manifestations of the categorizes are displayed by the mean of each class	income
Since when do you live in your region?	Since birth Since birth with interruptions I moved to the region	transformed into binary variables with 1 = yes and 0 = no	local returnee newcomer
Please evaluate the following statements concerning your bond to nature: (1) I try to be as often as possible to be in the nature. (2) Nature has only a minor part in my live. (3) I feel a strong bond to the nature and the landscapes of my region. (4) I think human beings are benefit from an in working order nature.	All statements could be rated from -- = radical refusal over 0 = irresolute to ++ = unreserved approval	transformed into a binary variable with 1= yes (everyone who marked statements 1, 3 to 6 with ++ or + and statement 2 with - or --) and 0 = no (otherwise)	bond to nature

Question in the survey	Possible answers	Transformation of the variable	Name of the variable in the data set
(5) I think it is an obligation to preserve and protect the nature. (6) I feel like I am part of the nature.			
Please evaluate the following statements concerning your bond to your region: (1) I feel really like home in my region. (2) With my region, i have a lot of personal memories. (3) My personal future is strongly linked to the region.	All statements could be rated from -- = radical refusal over 0 = irresolute to ++ = unreserved approval	transformed into a binary variable with 1 = yes (everyone who marked every statements with ++ or +) and 0 = no (otherwise)	bond to region
In what kind of environment is your house located?	in a city in the city limits in a village outside of a city or a village	transformed into a binary variable with 1 = living in a village or outside a city or village and 0 = living in a city or in the city limits	rural environment

Source: Resident survey.

Tab. 9-2: Overview about the manifestations of the variables (in percent)

Variables and its manifestations	All	LNWM	LPLD	STB
Gender				
male	52,5	51,8	52,5	44,6
Age				
18-24	6,4	5,3	6,7	6,7
25-34	12,6	16,4	12,7	10,3
35-44	13,9	15,2	13,6	13,5
45-54	25,6	24,0	24,7	27,6
55-64	22,5	22,8	22,2	22,8
65-74	12,1	11,7	12,5	11,9
75 and older	7,0	4,7	7,6	7,4
Children				
No children	27,2	31,6	25,1	27,5
Child	22,0	23,8	23,9	18,2
2 children	32,1	32,1	35,5	28,8
3 children	12,4	7,1	9,7	19,2
4 children	4,5	4,2	4,8	4,3
more than 4 children	1,8	1,2	1,9	2,0
Employed				
Employed	67,9	71,9	65,8	68,6
Higher education				
Higher education	48,3	55,0	46,9	46,5
Income household				
Lower than 1,000 Euro	8,9	6,9	11,4	6,6
1,001-1,500 Euro	12,7	12,5	14,9	9,7
1,501-2,000 Euro	11,0	12,5	10,9	10,3
2,001-2,500 Euro	13,4	10,0	14,9	13,1
2,501-3,000 Euro	16,4	13,8	15,6	19,0
3,001-4,000 Euro	18,5	23,1	16,1	19,3
4,001-5,000 Euro	10,1	11,3	8,2	12,1
more than 5,000 Euro	9,0	10,0	7,9	10,0
Duration Region				
Returnee	15,7	14,6	18,5	15,7
Newcomer	51,4	62,6	44,1	51,4
Bond to Nature				
Bond to Nature	52,8	56,1	54,5	48,7
Bond to Region				
Bond to Region	57,2	58,8	62,4	51,9
Rural Environment				
Rural Environment	53,9	25,7	63,5	56,1

Source: Resident survey.

Additionally to these socioeconomic variables, there was a variable integrated that shows if the respondent has children. It might be possible that children change the view of the world and therefore the perception of the region and its landscapes. The variable children was used continuously throughout the cluster analyses (Tab. 9-1). Most of the respondents have either two children or no children (Tab. 9-2). Only a few respondents have more than three children.

There is a variable included that shows if a respondent is still working (Tab. 9-1). It might be that people who are still working might have a different view of the region because they might not able to use it on a regular basis because of limited time (see Becker, 1965). This binary variable was created out of some answers given by the respondents. Only respondents who marked that they are working at the moment are labeled as employed, even if the law denotes protected mothers or people in parental leave are still employed. In fact, they are not working at that time. Again, the overrepresentation of older people could be seen in this variable (Tab. 9-2). Almost 68 % of the sample is working at the moment.

The income of the household is a central factor in especially WTP examinations, because people with a low amount of money are less willing to pay something for the landscapes (Howley et al., 2012; Kämmerer et al., 1996; Marangon & Visintin, 2007). Most of the respondents have 2,500 income and more, but almost 45 % of the sample has lower income of the household (Tab. 9-2).

Because the view of the region might differ between people who have left the region for duration of time or moved into the region, variables for newcomers and returnees were included in the analyses. These variables are binary ones and could be interpreted later on in comparison to the locals (Tab. 9-1). Most of the sample consists of newcomers (more than 50 %) (Tab. 9-2). Returnees constitute the smallest group of the sample with almost 15 %.

The relationship to nature might also influence the perception of the region and therefore a variable called bond to nature was developed (Tab. 9-1). The statements used in the survey were based on a study about the awareness of nature in Germany (BMUB, 2014). It might be easier to have only one variable indicating if there is a bond or not, a transformation of the statements into one variable was made. Almost half of the sample stated with their answers that they have a bond to nature (Tab. 9-2).

Bond to region was similar created as bond of nature (Tab. 9-1). It might be important because the identity of an individual could influence the value of a regional landscape (Morrison & Dowell 2013). The statements itself were developed referring to Lalli (1992), Weichhart et al. (2006) and Soini et al. (2012). Almost 60 % of the sample has a bond to region (Tab. 9-2).

The view of the regional landscapes might be influenced by the place a person lives in. Therefore, a binary variable was creates for the fact if the respondent is living in a city or in the countryside (Tab. 9-1). More than half of the sample is living in rural areas (Tab. 9-2).

9.3.2 Methodology

Before starting the cluster analysis, the correlations between the used variables were checked. The cluster analysis is used to determine different clusters of individual characteristics in the whole sample. This could offer the possibility to find different types of people living in the regions and might to some extent explain differences in the willingness to invest in the region.

9.4 Results

The results of the correlation coefficients suggest that all variables could be used for the cluster analysis (Tab. 9-3). The highest coefficient could be found between the variables employed and age. Most of the higher correlations just show common relationships like having children and being older.

The displayed dendrogram for the cluster analysis shows that three clusters might be methodologically the best solution (Fig. 9-1). The Gower Dissimilarity Measure was used because it allows mixing dummy variables like gender and effectively metric variables like the age in this case. The three-cluster solution displays one cluster with most of the men; the others are showing more women. One of the female and the male cluster has almost the same age on average and therefore is employed with the same extent. But this does not seem to be adequate to identify differences between groups, because the groups are too similar to each other. A differentiated analysis is not possible if the clusters are as similar as the three-cluster solution suggests.

Tab. 9-3: Correlation coefficients

	Gender	Age	Higher education	Employed	Income	Children	Rural environment	Returnee	Newcomer	Bond to nature	Bond to region
Gender	1.00										
Age	-0.14	1.00									
Higher education	0.01	**-0.17***	1.00								
Employed	-0.02	**-0.53***	**0.17***	1.00							
Income	-0.12	-0.14	**0.32***	**0.31***	1.00						
Children	-0.00	**0.37***	-0.07	-0.11	0.05	1.00					
Rural environment	0.00	0.01	-0.06	0.03	-0.02	0.04	1.00				
Returnee	-0.01	-0.05	0.13	0.06	0.05	-0.01	-0.01	1.00			
Newcomer	0.05	**0.16***	0.13	-0.10	0.04	0.09	-0.01	**-0.44***	1.00		
Bond to nature	0.09	**0.20***	0.02	-0.12	0.01	**0.13***	0.07	0.04	-0.03	1.00	
Bond to region	0.01	0.11	**-0.15***	-0.02	-0.02	0.07	0.01	0.09	**-0.33***	**0.20***	1.00

Note. * Coefficients that are significant on the level of 0.01.
Source: Resident survey, own calculations.

Fig. 9-1: Dendrogram of the cluster analysis for 10 groups

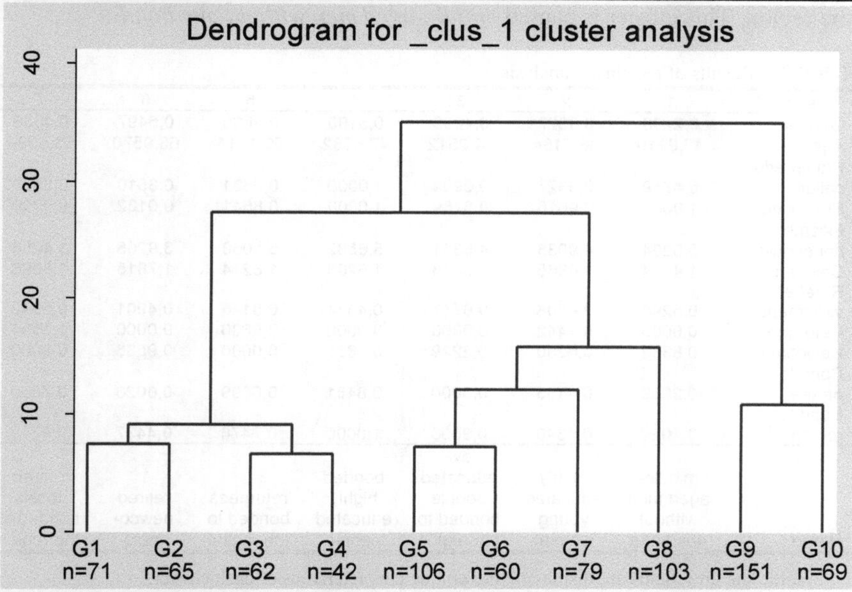

Source: Calculations by STATA. Illustration by STATA.

The solution of five clusters also show similar characteristics (the gender is almost similar distributed than the solution before, the same can be said for the employed) within the clusters and does not make it possible to distinguish between special types of the survey. As argued above, the clusters are too similar to analyze individual characteristics of the sample and identify types for protection of the region. The best result is done by the seven-cluster solution, because in this solution the mixed groups are separated from each other. In this case, there are enough characteristics of the individual to differentiate them in groups and analyze its differences. The creation of more clusters is further not suitable, because the characteristics are the same between the clusters as in the solution with seven clusters and could therefore not show more information than the solution of seven clusters. Further, the amounts of observations decrease with more clusters.

The first cluster contains men in their middle age and with a mid-level education, but with relatively high income because they are still working (Tab. 9-4). These people have on average one child and live in more rural areas. Most of these are newcomer into the region and have neither a bond to nature nor to the region. This cluster is named *middle-aged men without any bond.*

Tab. 9-4: Results of the cluster analysis

Cluster	1	2	3	4	5	6	7
Gender	0,2206	0,9327	0,4699	0,5190	0,4660	0,5497	0,4638
Age	45,0110	38,7164	44,0512	47,6582	50,1214	63,9570	63,6594
Higher education	0,4779	0,9327	0,0904	1,0000	0,5631	0,3510	0,3333
Employed	1,0000	0,9615	0,9759	1,0000	0,8641	0,0132	0,0000
Income household	5,0294	4,9038	4,5361	5,6582	5,1068	3,9205	3,4058
Children	1,4044	1,0865	1,3373	1,6203	1,6214	1,7815	1,8696
Rural environment	0,5294	0,4808	0,6747	0,4684	0,5146	0,4901	0,5942
Returnee	0,0000	0,1442	0,0060	0,0000	0,9806	0,0000	0,1594
Newcomer	0,8382	0,6250	0,2229	0,6835	0,0000	0,9536	0,0000
Bond to nature	0,2059	0,4135	0,5000	0,8481	0,6699	0,6026	0,7536
Bond to region	0,1029	0,1346	0,9096	1,0000	0,7476	0,4437	0,9710
Name	middle-aged men without any bond	highly educated young women	low-educated people bonded to the region	bonded highly educated people	returnees bonded to the region	retired newcomers	retired locals bonded to the region

Note. The grey fields are displaying the unique characteristics of each cluster.
Source: Resident survey, own calculations.

16.8 % of the whole sample is part of the first cluster (Tab. 9-5). Regarding the distribution into the examined regions, the clusters have different shares. While the share is almost the same than for the whole sample for the region LNWM, it is lower for LPLD and higher for STB.

The second cluster consists of young to middle-aged women with a high level of education, still working, but with less income than the cluster before (Tab. 9-4). On average, these women do have a child, living in more urban areas and includes newcomers, returnees and locals. Their bond to nature is a bit higher than the bond to region. The name of this cluster is *highly educated young women.* The cluster incorporates 12.9 % of the whole sample (Tab. 9-5). It is especially well marked for LNWM, while it is less marked for LPLD. The manifestation of the cluster for STB is close to the whole sample.

Tab. 9-5: Distribution of each cluster in each examined region (in percent)

Cluster	Sample	LNWM	LPLD	STB
1	16,83	16,77	14,74	19,78
2	12,87	19,35	9,21	14,29
3	20,54	12,26	24,21	20,15
4	9,78	13,55	8,68	9,16
5	12,75	14,19	13,95	10,26
6	18,69	18,06	19,47	17,95
7	8,54	5,81	9,74	8,42

Source: Resident survey, own calculations.

The third cluster is almost equal distributed between women and men in their middle ages with a low level of education and the lowest income in comparison to all clusters that show that the people are still working (Tab. 9-4). These people are living in more rural areas and could be defined as locals with a high bond to region and a low bond to nature. In the following, this cluster is denoted as *low-educated people bonded to the region.* It is the biggest cluster with almost one fifth of the overall sample (Tab. 9-5). It is less marked in the region of LNWM and well marked for LPLD. Again, the region STB is similar to the whole sample.

The next cluster is equally divided between men and women who have a high level of education and are still working (Tab. 9-4). These people have the highest income compared to all other clusters. On average, they have one to two children and live in more urban areas. The rate of the newcomers is slightly high, but this could indicate that the people in this cluster incorporate newcomers as well as locals with a high bond to nature and the region. It is called *bonded highly educated people.* 9.8 % of the sample is pooled in this cluster (Tab. 9-5). While the cluster is similar marked than the whole sample for LPLD and STB, for the region LNWM this cluster is more present.

The fifth cluster incorporates the returnees of both genders (Tab. 9-4). These are the oldest group compared to all other clusters with people who are still working. The education is between the high and the low level and the income is also in the middle when compared with the other clusters. As the cluster before, people have one to two children and live as well in urban and in rural areas. Their bond to region is higher than the bond of nature but only slightly. The name of it is *returnees bonded to the region.* One-eighth of the sample is part of this cluster (Tab. 9-5). The fifth cluster is less marked in region STB while it is similar marked in LNWM and LPLD.

The newcomer cluster contains slightly more women than men not working anymore and with a relatively low level of education and income (Tab. 9-4). On average, they have two children and live in rural as well as urban areas. Their bond to nature is present, but its level is moderate while the bond to region is lower. This cluster is named *retired newcomers*. Almost 19 % of the sample is incorporated in the sixth cluster (Tab. 9-5). This cluster is almost distributed equally over all regions.

The last cluster incorporates women and men who are not working anymore, have a relatively low education and the lowest income in comparison to the other clusters (Tab. 9-4). These people have the highest amount of children on average and could be classified as locals. They have the highest rate of bond to region over all clusters and a high rate of bond to nature. This cluster is denoted as *retired locals bonded to the region*. The last cluster pooled only 8.5 % of the sample (Tab. 9-5). It is less marked in region LNWM. The other two regions have similar manifestations than the whole sample.

After the clustering process, there was done a Chi test with a level of significance of 0.01. The result of this test indicates that the sample for LPLD and STB and the sample of LPLD and STB could be out of the same population (Tab. 9-6). But the samples for LNWM and LPLD are not similar and might therefore be out of different populations.

Tab. 9-6: Results of the Chi test of the samples

	LNWM: LPLD	LNWM: STB	LPLD:STB
Cluster in general	0,001	**0,134**	**0,470**
Cluster WTP	0,000	0,000	0,000
Cluster WTP Water	0,000	0,000	**0,020**
Cluster WTP City	0,000	0,000	0,000
Cluster WTP Village	0,000	0,000	0,003
Cluster WTV	0,000	0,000	0,000
Cluster Satisfaction low	(.)	(.)	(.)
Cluster Satisfaction middle	0,000	0,000	0,000
Cluster Satisfaction high	**0,023**	0,000	**0,181**
Cluster Use relaxing	**0,626**	**0,260**	**0,158**
Cluster Use sports	**0,021**	**0,079**	0,000
Cluster Use education	0,000	0,000	0,000
Cluster Use culture	0,000	(.)	0,000

Note. LNWM: LPLD = Comparison of the samples of LNWM and LPLD, LNWM :STB = Comparison of the samples of LNWM and STB, LPLD: STB = Comparison of the samples of LPLD and STB, (.) indicates omitted values because of a division of zero, level of significance 0.01.
Source: Resident survey, own calculations.

9.5 Analyses of the Clusters

This part considers the variables that are analyzed in more detail within the determined clusters. The first section describes the variables of interest, followed by some results of the literature and the hypotheses under examination.

9.5.1 *Variables of the Analyses of the Clusters*

The question of the WTP was split into typical landscapes in the region[56]. Because of the comparison between the regions, it could only be used landscapes that are common in each region. These are water, city and village landscape. The photographs of the landscapes covered with water show waterfronts or banks at one place in each region. Around the water there are some trees and grasslands and every photo offer an unrestricted view into the region. The pictures of the cities contain mostly old houses. Two of the pictures showed special buildings like a church or a city hall in the background while the last one highlights a terrace. The last landscape type under research displays a typical village in each region. The pictures show some houses with trees and grasslands around or a pathway. The sky in each pictures is blue or with some white clouds.

Because the interest is about the clusters regarding their characteristics and what they are willing to pay for a regional landscape, a transformed variable shows if an individual is willing to pay for one of the landscapes in general and than if the individual is willing to pay for a special landscape type (Tab. 9-7).

The creation of the variable for volunteer work is more complicated, because volunteering was not prompted in the context of the landscape (Tab. 9-7)[57]. The question of the hypothetical volunteering was chosen, only using the categories that have a direct impact on the regional landscapes. This was matched with the hypothetical hours the individual stated in the questionnaire and with the respondents who are already working on a voluntary basis. These categories were then transformed into one variable that shows if an individual is willing to volunteer in or for the landscape.

The questionnaire asked the respondents to mark how satisfied they are with their life at the moment (Tab. 9-7)[58]. The analyses of each of the manifestations might not be an added value to the analyses, and therefore the variable of satisfaction was transformed into three levels of satisfaction, namely dissatisfaction, undecided and satisfaction. The first category express that the individual is not satisfied with the life as a whole and the last part indicate that the individual is very happy with his or her life. The second category is in between these emotions. This kind of transformation was also used from Simonson et al. (2016).

56 A detailed description of the WTP question is done in Chapter 5.1.
57 The questions regarding the WTV are presented in detail in Chapter 5.3.
58 The description of the answers of the respondents can be found in Chapter 5.4.

Tab. 9-7: Overview of the questions used for the analysis of the cluster

Question in the survey	Possible answers	Transformation of the variable	Name of the use variables/ categories for the analyses
What amount of money are you personally willing to pay monthly and additionally to the already paid taxes for preservation, protection and design of the landscapes that are shown on the following photographes?(water, city and village landscape)	0 euros 1 - 5 euros 6 - 10 euros 11 - 20 euros 21 - 50 euros 51 - 100 euros more than 100 euros	transformed into a binary variable with 1 = is willing to give some money for each landscapes and 0 = is not willing to give any money for the corresponding landscapes	financial_water financial_city financial_village
In future, in what area could you imagine to volunteer most likely (for the first time, strengthened or additional)?	place development politics conservation of the environment animal protection fire or ambulance service sports/ exercises education/kindergarden/school religion/church art/culture/museums monument conservation local history society other	transformed into a binary variable, that displayed if the person is willing to volunteer for the region (place development, conservation of the environment, animal protection, art/culture/museum, monument conservation and local history)	place development conservation of the environment animal protection art/ culture/ museums monument conservation local history
Please mark how satisfied you are with your life at the moment.	0 to 10	transformed into three variables, how much a person is satisfied. Low satisfaction with a cross for 0 to 3, middle satisfaction from 4 to 6, high satisfaction higher than 6 with 1 = in this category and 0 = otherwise	l_satisfaction m_satisfaction h_satisfaction
For what purpose and how often do you use your region?	(almost) daily at least once a week at least once a month less often not at all	transformed into a binary variable with 1 = yes (everyone who marked using the the region daily or at least once a week) and 0 = no (otherwise)	w_relaxing w_sports w_education w_culture

Source: Resident survey.

For the use of the region there is used a question for what purpose the respondents use their region at least weekly (Tab. 9-7)[59]. To have a deeper look into the use, there is a differentiation between relaxing, sports, education or cultural purposes.

9.5.2 *Previous Findings and Hypotheses*

The examinations of the WTP suggest that a higher level of education is associated with a higher WTP (Howley et al., 2012; Verbič & Slabe-Erker, 2009). Rai and Scarborough (2014) did not find any significant coefficient for education, but the authors concluded that education might have a positive tendency on the WTP.

For age, the results regarding the WTP are not as clear as for education: for instance, Campbell (2007) found a positive impact, while most authors (Asah et al., 2014; Carson et al., 2001; Hackl & Pruckner, 1999) confirmed a negative impact associated with age. However, Tuan (1990), Schiappacasse et al. (2013) and Vásquez (2014) concluded that with an increasing age the WTP decreases, because people with a limited time horizon are less willing to contribute in general.

The living environment might also have an impact on the WTP – even if results are not consistent: Kämmerer et al. (1999) and Kim et al. (2015) found that people live in the city assess their close urban environment higher, while Howley et al. (2012) and Purcell (1992) argued that people living in the city value the rural landscapes higher than built-up environments.

Because of these findings, there will be the assumption that older, less-educated people living in more rural structures clusters show lower WTP than younger, highly educated people with the living environment in more urban areas. Regarding the determined clusters, this could be found in Cluster 2 (*highly educated young women*) and for cluster 7 (*retired locals bonded to the region*). Cluster 2 shows mostly young women with higher education and the living environment in the city. Cluster 7 is one of the clusters with the oldest members, is less educated and lives in more rural areas. Because of the results of the previous research, it is assumed that especially cluster 2 shows a higher WTP than cluster 7.

The literatures about the WTV show some results regarding the age. Especially middle-aged individuals are willing to volunteer (e.g., Asah et al., 2014; Lankia et al., 2014; Menchik & Weisbrod, 1987; Wilson, 2000). Older people are getting lesser involved in volunteer work than younger people (Carstensen, 1995; Tidwell & Brunson, 2008; Tselios et al., 2015). In Germany, the rate of

59 Detailed descriptions are made in Chapter 5.5.

volunteer work in older ages is increasing, but compared to younger ages its rate is lower (Enste et al., 2012; Simonson et al., 2016).

As in the examinations for WTP, it was also found a positive impact of a higher educational level on the voluntary involvement (Ahlheim et al., 2010; Menchik & Weisbrod, 1987), but Wilson (2000) argued that this might depend on the work that has to be done as a volunteer. Especially in Germany, Simonson et al. (2016) found that higher education influences the WTV positively.

Regarding the determined clusters there could by an analysis of Cluster 7 (*retired locals bonded to the region*) and Cluster 4 (*bonded highly educated people*) regarding these aspects. Cluster 7 includes older ages and the members are less educated. For Cluster 4, the members are on average middle-aged and show a high level of education. Therefore, Cluster 7 needs to be the cluster with the less WTV, while Cluster 4 has a high WTV.

While Palmore and Luikart (1972) found only a minor influence of the age on life satisfaction, Schwandt (2013) concluded that the impact of age on life satisfaction is more like a U-shape: in younger years the satisfaction increases while in older age, the satisfaction decreases again. In some cases that could be explained by the health condition of the individual (Enste & Ewers, 2014; Kahnemann & Krueger, 2006; Palmore & Luikart, 1972).

People living in more rural areas are usually more satisfied with their lives than people living in more urban settings (Sørensen, 2014). This could be an indirect effect of lesser stress in rural areas (Abraham, Sommerhalder & Abel, 2010), since some hints suggest that the conditions of the environment have an impact on life satisfaction (e.g., Abraham et al., 2010; Brereton et al., 2008; Levinson, 2012; Rehdanz & Maddison, 2008; Welsch, 2006, 2007).

The determined clusters that could be compared in this case are Cluster 6 (*retired newcomers*) and Cluster 3 (*low-educated people bonded to the region*). Cluster 6 includes relatively old members who live in more urban areas. The individuals of Cluster 3 are middle-aged and living in more rural environments. Therefore, the life satisfaction for Cluster 6 is assumed to be lower than the satisfaction for Cluster 3.

Regarding the use of the region, there are some results associated with the leisure participation in general or for special activities. Overall, Agahi and Parker (2005) found in a panel study in Sweden that the participation rates of older people decreases. Wilcox, Castro, King, Hausman and Brownson (2000) examine different participation rates of American women living in urban or rural areas and concluded that for both living environments an increasing age leads to lesser participation. In addition to these results, a previous review by Dishman, Sallis and Orenstein (1985) also identified that in older ages people participate less in leisure activities.

Agahi and Parker (2005) were able to show that with a higher level of education the participation rate increases as well. In a review of different examinations, Dishman et al. (1985) identified that all these examinations had one common finding: lower education is associated with lower participation rates, especially in physical activities. This finding can be confirmed by the examination of Wilcox et al. (2000).

With the cluster determined by this cluster analysis, there can be a comparison of cluster 4 (*bonded highly educated people*) and Cluster 7 (*retired locals bonded to the region*). While the members of Cluster 4 are middle-aged and possess a high degree of education, Cluster 7 integrates retirees with a lower level of education. The assumption would be that Cluster 4 has a higher use of the region than Cluster 7 does.

9.6 Results of the Analyses of the Clusters

The results of the deeper analyses of the landscape support in terms of financial contribution and volunteering, as well as the life satisfaction and the use of nature is shown. Each variable has its own part, starting with the financial contribution.

9.6.1 Financial Payment for the Regional Landscapes

Regarding the percentages of WTP for the regional landscapes, it can be seen that the cluster that is most willing to invest some money in the regional landscapes in general is Cluster 2 (*highly educated young women*) (Tab. 9-8). This does not differ between the overall sample and the regional samples. Only the percentages of how many individuals of the sample are willing to pay differ. The most people of the cluster that are willing to give some money are located in the region of LNWM (73 %), while the lowest percentage of these high rates can be found in LPLD (57 %). Because the cluster is the same in every sample, the characteristics of the clusters are also the same.

The clusters with the lowest percentage for the willingness to give some money for the regions differ between the regional samples. While the sample of the region of STB is in line with the overall sample, the other two regions show different results. In STB the cluster with the lowest percentage of willingness is the Cluster 6 (*retired newcomers*); for LNWM it is the Cluster 7 (*retired locals bonded to the region*) and for LPLD it is the cluster with the returnees with a bond to the region (Cluster 5). The differences in percentages do not differ that much to each other in the regions (difference 8 %). These three clusters are similar in the case of the age (all other clusters are on average younger) and the amount of children (on average they have two children) while all the others characteristics are differing between the clusters.

The picture for the water landscape is different. While Cluster 2 (*highly educated young women*) are still the cluster with the highest percentage of WTP for this landscape type in the region of LNWM and for the overall sample, for the other two regions, other clusters are most willing to contribute money for the water landscape. For the region of LPLD, Cluster 3 (*low- educated people bonded to the region*) shows the highest percentage of willing to contribute while for STB it is Cluster 1 (*middle-aged men without any bond*). The differences between the percentages are a generally little bit higher for the WTP than for the protection of the regional landscapes (highest difference is about 20 %). These three clusters are similar in the characteristics of the age (these clusters are the youngest ones), individuals are still employed, they have on average one child and their bond to the nature is either low than high.

The clusters that are showing to be willing to contribute the lowest percentages for the protection of the water landscape are similar for the overall sample, LNWM and STB. For these samples it is Cluster 6 (*retired newcomers*) that is willing to contribute the lowest percentage. In the region LPLD, Cluster 7 (*retired locals bonded to the region*) displays the lowest percentage. Again, the percentages of the clusters over all regions are close to each other (5 %). These clusters have similar characteristics regarding their employment status (both are retired) and therefore income, the number of children on average (two), age, and their educational level.

The percentage of people willing to pay for the city landscape is highest in all tested samples in the cluster of the highly educated young women (Cluster 2). Regarding the percentages, the difference between the regions is similar as for the water landscape or the regional landscape in general (highest difference 16 %). Because the same cluster is willing to give the highest amount in each region, these people need to be similar in each of the chosen characteristics.

The lowest percentage of willingness to make a financial contribution for the protection of the city landscape is different in each region and in the overall sample. For the overall sample, Cluster 7 (*retired locals bonded to the region*) displays the lowest percentage of WTP. For LNWM, this can be seen in Cluster 1 (*middle-aged men without any bond*). Cluster 3 (*low-educated people bonded to the region*) is the cluster with lowest percentage of people willing to contribute money for the city in the region of LPLD and for STB Cluster 5 (*returnees bonded to the region*) shows the least will. The difference between the clusters is highest between the overall sample and the region of STB (7 %). The clusters have similar characteristics regarding the gender (consists mostly of men), have on average a lower educational level and live in more rural areas.

Tab. 9-8: Results of the willingness to pay of each cluster for the regional landscape (in percent)

Cluster	Willingness to pay for regional				Water landscape				City landscape				Village landscape			
	All	LNWM	LPLD	STB	All	LNWM	LPLD	STB	All	LNWM	LPLD	STB	All	LNWM	LPLD	STB
1 middle-aged men without any bond	55,15	50,00	50,00	62,96	48,53	38,46	42,86	**59,26**	33,82	***19,23***	33,93	40,74	39,71	30,77	35,71	48,15
2 highly educated young women	**66,35**	**73,33**	**57,14**	**69,23**	**56,73**	**63,33**	40,00	66,67	**45,19**	**56,67**	**40,00**	**41,03**	**51,92**	**46,67**	48,57	**58,97**
3 low-educated people bonded to the region	54,22	47,37	55,43	54,55	45,78	47,37	**43,48**	49,09	25,30	21,05	***23,91***	29,09	46,99	31,58	**52,17**	43,64
4 bonded highly educated people	55,70	57,14	51,52	60,00	48,10	57,14	39,39	52,00	30,38	42,86	27,27	24,00	43,04	28,57	48,48	48,00
5 returnees bonded to the region	46,60	59,09	***39,62***	50,00	41,75	54,55	35,85	42,86	32,04	54,55	30,19	***17,86***	35,92	45,45	***33,96***	***32,14***
6 retired newcomers	***41,06***	35,71	43,24	***40,82***	***33,77***	***28,57***	36,49	***32,65***	27,81	28,57	28,38	26,53	***31,79***	***21,43***	35,14	32,65
7 retired locals bonded to the region	47,83	***33,33***	54,05	43,48	36,23	33,33	***32,43***	43,48	***24,64***	22,22	24,32	26,09	39,13	22,22	45,95	34,78
All	49.89	52.05	48.04	51.28	42.25	45.61	37.88	46.47	29.59	35.09	27.94	28.85	38.86	32.75	40.42	40.06

Note: Bold values display the highest percentage of the region and the italic and bold ones the lowest ones.
Source: Resident survey, own calculations.

The village landscapes show similar percentages of the highest willingness to give some money in Cluster 2 (*highly educated young women*) for the overall sample and the regions of LNWM and STB. Only for the region LPL does Cluster 3 (*low-educated people bonded to the region*) displayed a highest percentage. The differences between the samples are about 12 %, while LNWM shows the lowest and STB the highest percentage. The clusters are similar in the characteristics of the employment status and the average amount of one child.

The lowest percentages of the WTP for the village landscape split between Cluster 6 (*retired newcomers*) for the overall sample and the region of LNWM and Cluster 7 (*retired locals bonded to the region*) for the other two regions. This time, the difference between the percentages is relatively high compared with the difference of the lower percentages of the other landscapes (13 %). The characteristics of these clusters are similar in the case of the age, the level of education, the employment status, the income and the number of children.

The hypothesis regarding the WTP assumed that Cluster 2 (*highly educated young women*) has a higher WTP than Cluster 7 (*retired locals bonded to the region*). It can be found in the results that Cluster 2 shows the highest rates on WTP in most of the cases, But Cluster 6 rather than Cluster 7 shows in more cases the lowest rate. In some characteristics, this cluster is similar to Cluster 7. It also has relatively old people integrated and these people are less educated. Regarding the highest WTP, the hypothesis can be proven, but for the lowest one, it has to be rejected.

The Chi test for each of the examined variables shows that only the sample of LPLD and STB for the landscape of water could be out of the same population (Tab. 9-6). All other results do not display similarities of the samples.

9.6.2 Volunteering as Contribution Form for the Region

Having a look into the WTV in general, the most WTV shows Cluster 4 (*bonded highly educated people*) for the overall sample and for the region of LPLD (Tab. 9-9). The clusters of the highest percentage of WTV for the other regions differ. While in the region of LNWM, the highest percentage can be found for Cluster 5 (*returnees bonded to the region*), for STB it is Cluster 2 (*highly educated young women*). The difference between the percentages is about almost 20 %. The characteristics of these clusters are similar to each other regarding the educational level (on average it is a higher education) and the individuals in these clusters are still employed.

The lowest percentage of the WTV can be found in Cluster 7 (*retired locals bonded to the region*) for the overall sample as well as for the regions LPLD and STB. In the region LNWM, Cluster 1 (*middle-aged men without any bond*) is less willing to volunteer in or for the region in general. The highest difference in the percentages is about 5 % between the region LNWM and the overall sample. Both clusters consist mostly of men who are less educated and tend to live in more rural areas.

Regarding the areas of volunteering, these analyses only describe the results for the overall sample[60]. For the areas of place development, conservation of the environment and art/culture/museums, the cluster with the highest percentage of involvement is Cluster 4 (*bonded highly educated people*). The area of animal protection is preferred by the cluster of the highly educated young women (Cluster 3) while the monument conservation is highly chosen by Cluster 1 (*middle-aged men without any bond*). The local history is preferred among the cluster of returnees with a bond to the region (Cluster 5).

The lowest percentages differ between the areas of volunteering. The lowest percentage of the involvement of place development is shown from Cluster 7 (*retired locals bonded to the region*). In the area of the environment conservation, cluster 1 (*middle-aged men without any bond*) shows the lowest rate while the cluster that consists mostly of women (Cluster 2) displays the lowest percentage for monument conservation. The bonded highly educated people (Cluster 4) are less willing to volunteer for the area of animal protection and the low educated people bonded to the region (Cluster 3) shows the lowest percentages for art/culture/museums. The retired newcomers (Cluster 6) seem to have lower interests in the area of local history and display in this area of volunteering the lowest percentage. Because mostly every cluster is named in the highest and lowest areas of volunteering, this might indicate preferences in each cluster over the area of volunteering.

The assumption regarding the WTV is that Cluster 7 (*retired locals bonded to the region*) is less involved in voluntary work than Cluster 4 (*bonded highly educated people*). It can be seen that Cluster 4 shows in some of the areas of volunteering the highest values, but not in all. Cluster 7 displays only for the overall willingness to contribute labor the lowest rates. In the specific areas, this cluster has only the lowest rate in the field of place development. But there is none of the other clusters that show a higher rate of the lowest rate of volunteering. Because the results are not given for each region, the hypothesis needs to be rejected.

---

60 The regional analyses are shown in the appendix Tab. A 10. Besides the area of place development, the results differ extremely between the regions so that a comparison seems to be highly complex and does not seem to be explained easily by our data. There might be other influences included like the offer of the region or else.

Tab. 9-9: Results of the willingness to volunteer in general of each cluster, areas of volunteering for the whole cluster (in percent)

Cluster		Any kind of volunteering for or in the regional landscape				Place development	Conservation of the environment	Animal protection	Art/ culture/ museum	Monument conservation	Local history
		All	LNWM	LPLD	STB	All	All	All	All	All	All
1	middle-aged men without any bond	49,26	***30,77***	57,14	50,00	19,66	***21,37***	21,37	8,55	**8,55**	6,84
2	highly educated young women	65,38	66,67	60,00	**69,23**	11,88	24,75	**32,67**	13,86	***4,95***	6,93
3	low-educated people bonded to the region	51,81	36,84	57,61	47,27	21,58	27,34	27,34	***4,32***	5,76	7,91
4	bonded highly educated people	**74,68**	71,43	**87,88**	60,00	**30,14**	**32,88**	***16,44***	**17,81**	6,85	9,59
5	returnees bonded to the region	62,14	**81,82**	56,60	57,14	21,65	30,93	23,71	9,28	8,25	**11,34**
6	retired newcomers	48,34	64,29	43,24	46,94	16,24	25,64	26,50	12,82	5,98	***5,13***
7	retired locals bonded to the region	***36,23***	44,44	***35,14***	***34,78***	***10,87***	26,09	21,74	6,52	6,52	10,87
All		53,60	57,31	53,58	51,60	18.86	26.92	25.10	10.40	6.89	8.32

Note: Bold values display the highest percentage of the region and the italic and bold ones the lowest ones.
Source: Resident survey, own calculations.

Regarding the result of the Chi test, it could be assume that the three regional samples are all out of different populations (Tab. 9-6).

9.6.3 Life Satisfaction in the Clusters

The highest rates for dissatisfaction display Cluster 5 (*returnees bonded to the region*) for the overall sample and the region of LPLD and Cluster 4 (*bonded highly educated people*) for the region of LNWM and STB (Tab. 9-10). The rate is highest for the region of LNWM and its difference to the lowest of these rates is about 9 %. These results indicate that the people of Cluster 4 in the region of LNWM are more dissatisfied than in the other regions. Regarding the characteristics the individuals of these clusters are still employed, have on average two children and a relatively strong bond to the region.

For the overall sample Cluster 3 (*low-educated people bonded to the region*) is the cluster that is the least dissatisfied. The regional sample displays more than one cluster that does not say it would be dissatisfied with their life. For the region of LNWM, Cluster 7 (*retired locals bonded to the region*) and Cluster 3 (*low-educated people bonded to the region*) show that they are not dissatisfied with their life in the region. In addition to Cluster 7, the region of LPLD also displays that Cluster 2 (*highly educated young women*) are not dissatisfied with their life. Cluster 2 is the less dissatisfied group in the STB region. The difference between the clusters is really low, at about 2 %. Regarding their, characteristics there are no similarities observed between all of these clusters.

For the overall sample and for the region of STB, Cluster 7 (*retired locals bonded to the region*) shows the highest percentage of indecision. While the cluster with the middle-aged men without any bond (Cluster 1) displays as the cluster with the highest percentage in the region of LPLD, the cluster with the highly educated young women (Cluster 2) show the highest rate in the region of LNWM. The difference between the percentages within the highest rates is about 17 %. There are again no similarities between the three clusters concerning the individual characteristics.

Tab. 9-10: Results of the life satisfaction for each cluster in each region (in percent)

Cluster		Dissatisfaction				Undecided				Satisfaction			
		All	LNWM	LPLD	STB	All	LNWM	LPLD	STB	All	LNWM	LPLD	STB
1	middle-aged men without any bond	6,83	11,36	4,82	6,41	18,05	9,09	**26,51**	14,10	75,12	79,55	***68,67***	79,49
2	highly educated young women	2,78	9,09	***0,00***	***0,00***	15,28	**18,18**	16,67	11,54	81,94	***72,73***	83,33	**88,46**
3	low-educated people bonded to the region	***1,90***	***0,00***	***0,00***	5,41	12,38	4,17	13,64	16,22	**85,71**	95,83	86,36	78,38
4	bonded highly educated people	4,90	**16,67**	1,54	**9,68**	16,67	***0,00***	23,08	***6,45***	78,43	83,33	75,38	83,87
5	returnees bonded to the region	**7,77**	9,09	**9,43**	3,57	13,59	9,09	16,98	10,71	78,64	81,82	73,58	85,71
6	retired newcomers	3,38	3,57	2,70	4,35	***11,49***	17,86	***9,46***	10,87	85,14	78,57	**87,84**	84,78
7	retired locals bonded to the region	4,35	***0,00***	5,41	4,35	**21,74**	***0,00***	18,92	**34,78**	***73,91***	**100,00**	75,68	***60,87***
All		4.93	6.43	4.16	5.19	15.13	12.87	17.78	12.66	79.93	80.70	78.06	82.14

Note: Bold values display the highest percentage of the region and the italic and bold ones the lowest ones.
Source: Resident survey, own calculations.

Regarding the lowest rates, there is also a variety of clusters in the samples. While for the overall sample and the regional sample of LPLD Cluster 6 (*retired newcomers*) is the less undecided one, for the region of STB it is the cluster with the bonded highly educated people (Cluster 4). For the region, of LNWM, the analyses identified two clusters (*bonded highly educated people* and *retired locals bonded to the region*) with less indecision about their life satisfaction. The difference between the rates is about 11 %. The similar characteristics of these clusters are the number of on average two children and the tendency to have a bond to the nature.

The satisfaction displays the highest percentages in different clusters. For the overall sample the highest rate of satisfaction is shown by cluster 3 (*low-educated people bonded to the region*). The region of LNWM displays the highest rate for the cluster of the retired locals with a bond to the region (cluster 7) with about 100 %, which means everyone in this cluster is satisfied with his or her life. The other two regions show rates of about 88 % for Cluster 6 (*retired newcomers*) for the region of LPLD and for STB for cluster 2 (*highly educated young women*). The difference between the percentages is about 14 %. No similarities of the characteristics of these four clusters could be identified.

The lowest percentage of satisfaction for the overall sample and the region of STB show Cluster 7 (*retired locals bonded to the region*). While for the region of LNWM, the lowest rate of satisfaction is observed for Cluster 2 (*highly educated young women*), it is displayed for Cluster 1 (*middle-aged men without any bond*) for the region of LPLD. The difference between the percentages is about 13 %. As seen for the undecided group, there are no similarities in the characteristics between these clusters.

The assumption regarding the life satisfaction was that the satisfaction is lower for Cluster 6 (*retired newcomers*) than for Cluster 3 (*low-educated people bonded to the region*). Cluster 3 shows the lowest rates of dissatisfaction and one of the highest rates in the section of satisfaction in the case of the overall result. For Cluster 6, the results are more complex. Some of the lowest rates of indecision are shown in Cluster 6, but in the section on satisfaction, it shows the highest percentage once and lowest rate not even once. With the results, the hypothesis could not be proven and has to be rejected since the results are inconsistent.

The samples were again tested with a Chi test. The results could not be shown for the dissatisfaction because of division by zero (Tab. 8). The result for the undecided indicate that the samples are out of different populations, but for the satisfied samples, at least the regional sample of LNWM and LPLD as well as the sample of LPLD and STB are out of the same populations. Only the regional sample of LNMW and STB are out of different populations.

9.6.4 Use of the Region in the Clusters

The cluster analysis shows that the clusters that are using the region for relaxing purposes are highest for Cluster 4 (*bonded highly educated people*) in the overall sample and in the regional samples of LNWM and LPLD and cluster 2 (*highly educated young women*) for the region of STB (Tab. 13). The difference between the percentages is about 20 %. These clusters share the same characteristics regarding their high education, employment status, the living environment in more urban areas and their rate of newcomers.

The cluster with the lowest rate of using the region for relaxing purposes is displayed in Cluster 1 (*middle-aged men without any bond*) for the region of LNWM and STB and Cluster 5 (*returnees bonded to the region*) for the overall sample and the region of LPLD. The difference between the values is highest between LPLD and STB with about 12.5 %. The characteristics of the clusters that are similar are employment status, household income and their living environment in more rural areas.

The use of the region for sports is more complex because of a broader distribution of the percentages along the clusters. The highest percentages could be found for the overall sample and the region of LPLD for Cluster 2 (*high educated young women*). Cluster 4 (*bonded highly educated people*) is the cluster with the highest rate for the region of STB, while the retired newcomer (Cluster 6) show the highest percentage of using the region for sports in the region of LNWM. The difference between the highest and the lowest percentage within the high percentages is 18 %. These clusters have in common that the living environment is more characterized by urban structures and the rate of newcomers in the clusters is relatively high.

The lowest percentage for the use of the region for sportive activities could be found for Cluster 3 (*low-educated people bonded to the region*) for the overall sample and the region of LNWM and STB. For the region of LPLD, the lowest rate is shown from Cluster 4 (*bonded highly educated people*). The highest difference between these rates is about 12 %. Similar characteristics, shared by these clusters are the age (both clusters are between 44 and 48 years old), these individuals are still employed and have a strong bond to the region.

While the overall sample shows the highest rates of use of the region for educational purposes for Cluster 2 (*highly educated young women*), the region of LPLD displays it for Cluster 1 (*middle-aged men without any bond*). The other two regions have the highest percentages of use for educational purposes for Cluster 4 (*bonded highly educated people*). Between this percentages there could be find the highest difference of 11 %. Similar characteristics could be found in the employment status and a relatively high rate of newcomers.

For the lowest use of the region regarding the educational use shows Cluster 5 (*returnees bonded to the region*) for the overall sample and the regions of LPLD and STB. For the region of LNWM Cluster 1 (*middle-aged men without any bond*) displays the lowest percentage. Surprisingly, Cluster 5 in the region of STB shows a percentage of 0. This would imply that this cluster in the region does not use the region for educational purposes at all. The highest difference between the percentages is about 10 %. The clusters have the employment status, the level of the income of the household and the more rural living environment in common.

The use for cultural purposes is rated highest in Cluster 5 (*returnees bonded to the region*) in the overall sample and the regional sample of LPLD and in Cluster 7 (*retired locals bonded to the region*) for the regions of LNWM and STB. The highest rate is displayed from the LPLD region and the lowest for the region of STB. The highest difference for the use of this purpose is 14 %. These clusters consist mostly of men that have on average two children, live in more rural areas and have relatively high bonds to the region and nature.

The lowest used of the region for cultural purposes is shown for the overall sample and the region of STB for Cluster 4 (*bonded highly educated people*). STB shows another cluster that has the lowest percentage of use for cultural purposes is Cluster 5 (*returnees bonded to the region*). Both clusters in this region show a usage percent of zero. Zero as well is displays by the percentage of Cluster 5 in the regional sample of LNWM. The lowest percentage that is not a zero for the LPLD region is shown in Cluster 1 (*middle-aged men without any bond*). The difference between the lowest and the highest percentage of the lowest rates is about 3 %. The characteristics that are similar in these clusters are the employment status and the age ranges between 45 and 50 years old.

The hypothesis was that Cluster 4 (*bonded highly educated people*) has a higher use of the region than Cluster 7 (*retired locals bonded to the region*). Actually, Cluster 4 shows on average relatively often the highest percentage of the use of the region. But there are a few exceptions: in the case of use of sports in the region of LPLD this cluster shows the lowest percentage. The lowest percentage is also displayed by the use for cultural purposes in the overall sample and the region of STB. However, Cluster 7 displays twice the highest percentage of use regarding the cultural purpose and not once the lowest percentage. These results might indicate that the different types of use are important for the frequency of use by the clusters and the hypothesis needs to be rejected.

Tab. 9-11: Results of the use of the region for each cluster in each region (in percent)

Cluster		Use for relaxing				Use for sports				Use for educational purposes				Use for cultural purposes			
		All	LNWM	LPLD	STB	All	LNWM	LPLD	STB	All	LNWM	LPLD	STB	All	LNWM	LPLD	STB
1	middle-aged men without any bond	59,48	***56,67***	58,70	***62,50***	59,48	60,00	69,57	47,50	15,52	***10,00***	**28,26**	5,00	3,45	6,67	***2,17***	2,50
2	highly educated young women	81,13	75,00	76,47	**90,00**	**67,92**	56,25	**82,35**	65,00	**20,75**	25,00	17,65	20,00	9,43	6,25	17,65	5,00
3	low-educated people bonded to the region	73,53	73,33	72,73	74,42	***49,02***	***40,00***	54,55	***46,51***	11,76	13,33	15,91	6,98	7,84	6,67	9,09	6,98
4	bonded highly educated people	**82,29**	**84,21**	**79,17**	86,21	61,46	57,89	***52,08***	**79,31**	18,75	**31,58**	12,50	**20,69**	***3,13***	5,26	4,17	***0,00***
5	returnees bonded to the region	***56,45***	57,14	***50,00***	68,42	54,84	57,14	52,78	57,89	***6,45***	14,29	***8,33***	***0,00***	**12,90**	***0,00***	**22,22**	***0,00***
6	retired newcomers	76,71	67,86	76,39	82,61	67,81	**64,29**	69,44	67,39	17,12	17,86	19,44	13,04	8,22	3,57	11,11	6,52
7	retired locals bonded to the region	78,17	85,29	73,27	82,26	58,38	58,82	57,43	59,68	16,75	14,71	15,84	19,35	9,14	**14,71**	7,92	**8,06**
All		73.68	72.73	72.22	76.27	59.50	58.79	60.87	57.97	15.68	16.97	16.91	13.22	7.44	7.88	8.94	5.08

Note. Bold values display the highest percentage of the region and the italic and bold ones the lowest ones.
Source: Resident survey, own calculations.

The results of the Chi test imply that all samples for the use of the region for relaxing reasons are out of the same population (Tab. 9-6). Regarding the sportive activities in the region, only the sample of LNWM and LPLD and the regional samples of LNWM and STB are out of the same population. The samples for LPLD and STB stem from different populations in this case. For the educational purpose, all samples are from different populations. For the aspects of cultural use of the region, it could only be stated that the samples of LNWM and LPLD and these of LPLD and STB are out of different populations. The value for the samples of LNWM and STB could not be calculated because of division by zero.

9.7 Discussion

The following part discusses the results of the variables regarding the analyses of the clusters. The variables are discussed each in the same order than in the chapters before.

The results of the Chi test indicate that the regional samples are in most cases stem from different populations. It might be the case that in these instances the regional aspects are important additionally to the individual characteristics that the analyses used. The questionnaire did not asked about different regional aspects or conditions, so that this examination could not consider these in the analyses.

9.7.1 *Financial Contribution*

One of the key results of the analyses of the financial contribution was that Cluster 2 (*highly educated young women*) are the most willing to pay money for the regional landscapes in general as well as for special landscapes, with few exceptions. It might be the case that young women having one child and are employed value the landscapes higher because they spend a lot of time outside with their kid(s). Additionally, they might think about the future and want to maintain the environmental conditions for their child(ren) and their grandchild(ren) later on. These individuals show the highest educational level of the clusters, which could be responsible for the high WTP, since this cluster might know the most about the environment, protecting it and embracing its value for future generations. Surprisingly, this cluster does not have a strong relationship neither to nature nor to the region.

For the water landscapes, Clusters 1 (*middle-aged men without any bond*) and 3 (*low-educated people bonded to the region*) as well as Cluster 2 (*highly educated young women*) have the highest percentage of willingness to contribute a financial amount. And Cluster 3 (*low-educated people bonded to the re-*

gion) shows in addition to Cluster 2 (*highly educated young women*) the highest percentage for the village landscape. It seems like these clusters show similar aspects as well, so that it could be explained why these are most willing to contribute some money. Tuan (1990) argued that people with a limited time horizon are less willing to invest something in their environment. This argument could be applied for the explanation, because the three clusters are the three youngest clusters, which incidentally have the furthest time horizon. Another argument is that these individual are, on average, employed and thus have more income than other clusters that they can invest in landscape protection.

In most of the cases, the clusters with the retirees (Clusters 6 and 7) are less willing to make a financial contribution. This could be based on their lower income resulting from retirement. But it could also be – as Tuan (1990) argued – based on a limited time horizon for these people.

In some cases, Cluster 5 (*returnees bonded to the region*) shows a lower willingness to invest some money in the regional landscapes. The argument of the time horizon could also be applied to the explanation of this cluster, because together with the other two clusters the oldest individuals are included. In addition, all these clusters have more than one child, but this explanation might not be useful because of their ages – the children might have already left the parent´s house. Even if there are other explanations about the characteristics might be true, these similarities of characteristics might responsible for the low WTP for the regional landscapes.

The city landscape is the exception regarding the lowest rates of WTP. In this case, Cluster 1 (*middle-aged men without any bond*), 3 (*low-educated people bonded to the region*), 5 (*returnees bonded to the region*) and Cluster 7 (*retired locals bonded to the region*) show the lowest percentages. On the face of this, these clusters consist mostly of men with lower education. It might be the case that in general men with a lower level of education could have the attitude that city landscapes are financed by the government or they might not have thought about how such a landscape could be maintained if the costs are higher than the money the government invests. Another explanation might be based on their living environment. On average, the individuals of these clusters live in more rural areas. It might be the case that the city landscape is not more valuable to them than another landscape because they think they have everything they need in their rural area. Maybe these individual are less willing to contribute some money because they think the residents of the city are responsible for that.

The differences regarding the percentages could be due to difference in the regions. In most cases LNWM is the region with the highest percentage of the willingness to give some money. This region is the most urban area of the chosen ones. It is mostly followed by the region of STB, which is in between the other ones in the comparison and LPLD shows the lowest percentages. Some exceptions are found especially for the village landscape that has the highest percentage of WTP in the region of STB and the lowest percentages for the region of LNWM. This might be due to different structures of the region. While LNWM is mostly characterized by the historic city of L, the other two regions are mostly characterized from their villages. Especially STB has some small villages that are characterized by special building that were used for farming and that are still maintained. In one part of the region of LPLD is this also the case because of the circular villages (*Rundlingsdörfer*), but because of the consideration of this region with the other one, it might be the case that the willingness is not displayed as that strong anymore.

The results could prove the highest percentage of Cluster 2 (*highly educated young women*), but not the results for the lowest rate. There is a mismatch between the results and the hypothesis because Cluster 6 (*retired newcomers*) shows the lowest rate more often than Cluster 7 (*retired locals bonded to the region*). The mismatch with Cluster 7 might be over the living environment. The individuals of Cluster 6 live in more urban structures than Cluster 7 do. This might account for the fact that people of Cluster 6 are less willing to contribute money than the members of Cluster 7.

9.7.2 *Volunteering*

The cluster with the highest involvement in the areas of volunteering defined as in or for the region is Cluster 4 (*bonded highly educated people*). This result is not surprising, because Simonson et al. (2016) found that involvement increases with higher education and decreases with older age. Because this cluster included middle-aged people with the highest education, this result is robust. The same explanations might be apply to the cluster of the young and highly-educated women (Cluster 2), because the individuals in this cluster also have a higher education and on average they are younger than the individuals in Cluster 4.

Simonson et al. (2016) also found that men are usually more involved in volunteering than women, but in these analyses there were two areas in that men are more involved than women. These are monument conservation and place development (Tab. 9-8). Only in the area of monument conservation the cluster consists of men (Cluster 1) shows the highest percentage of involvement. It might be the case that men are more talented regarding physical labor and work on buildings than women while women are more involved in creative areas.

Returnees are more involved in the area of local history, which might be explained by the reason of returning. If it is the case that these people are returning because of experiences of the past, it might be the case that they are also more interested in the past of the region. In some cases this could also include the family history if the family has been connected to the region for years. Another explanation might be that these people start to think about the region and its history when they are outside the region because they miss the features of the region. If they then return, they are still interested in the history and the deeper insights of the region. Cluster 6 (*retired newcomers*) in return shows the lowest percentage for the area of local history, which could be explained using past experiences as well: They might not have any connection to the region when they moved into it, so that they might have higher interests on the future than in the past.

It seems to be in line with other analyses, that in general especially older people are not as involved to the same extent as younger people (Enste et al., 2012; Simonson et al., 2016). The authors found that people older than 65 are not involved to the same extent as younger people. This might be a reason why the clusters with the retired locals (Cluster 7) do not display any of the highest percentages for any area of volunteering. One exception of this explanation is Cluster 1 (*middle-aged men without any bond*) for a general involvement in the region of LNWM. This cluster displays the lowest percentage for working voluntary in or for the region. This cluster is similar to the clusters with the retirees regarding the level of education, which was also found to have lower involvements in volunteering (Simonson et al., 2016). The gender might not be adequate in explaining these differences because Simonson et al. (2016) concluded that usually more men are volunteering than women. Another characteristic that is similar is that these individuals determined to one of the clusters where the members live in more rural areas. In some case people are free-riding, pretending to work on a voluntary basis but in reality not doing so.

The conservation of the environment shows the lowest rate of WTV for Cluster 1 (*middle-aged men without any bond*). The explanation that is given above about the lower education might be also applied for this case. It might be the case that individual with lower levels of education might not know that volunteering is an adequate method to protect the environment. The low-educated people with a bond to the region (Cluster 3) are with the lowest percentage involved in the area of art/culture/museums. Again the lower education might be the explanation why this cluster is not that much involved. In a common sense, lower educated people are less interested in cultural events than people with higher education.

The area of animal protection is rated lowest among the cluster of the bonded highly educated people (Cluster 4). In this case it might be possible that the individuals of this cluster think that others will work in this area and therefore they do not have to. But the result is confusing because the cluster of highly educated women (Cluster 2) displays the highest percentage in this area of volunteering. Therefore, the level of education might not be the adequate towards explaining the involvement in animal protection, but the explanation might be based on the bonds to nature and region that Cluster 4 has and Cluster 2 does not.

The variable for the any kind of volunteering uses more individual characteristics asked in the questionnaire then the variables for the areas of volunteering because the variable of volunteering in general was matched with the hypothetical hours to work voluntary and the fact to have currently volunteer work. This was not possible for the areas and therefore there might be differences between involvements in general and in specific areas.

The results regarding the hypothesis for WTV indicate a rejection of it. Cluster 7 (*retired locals bonded to the region*) is not in each region and is the cluster with the less involvement, while Cluster 4 (*bonded highly educated people*) does not display the highest rates for overall involvement. The explanation why other clusters show also the lowest or highest rates could not only be determined by the individual characteristics. The clusters are too different to each other regarding the individual characteristics, which would imply that regional aspects might be important for the involvement additionally to the characteristics.

9.7.3 Life Satisfaction

The most dissatisfied clusters are Cluster 5 (*returnees bonded to the region*) and 4 (*bonded highly educated people*). It might be the case that these clusters are dissatisfied with their life situation because they expected more out of the life's conditions. Returnees might have a better image of the life and the employment situation in the region than really exists. For the highly educated cluster, it might be the case that they have realized that they have limited opportunities regarding employment opportunities. Both clusters are still employed, which could be one explanation for their dissatisfaction. It is questionable if these clusters are most dissatisfied because of a strong bond to the region. Indeed, both clusters are characterized from a bond to the region, but there is no explanation why this could have a negative impact on the life satisfaction.

The lowest rate of undecided people is found in clusters 4 (*highly educated people*), 6 (*retired newcomers*) and 7 (*retired locals bonded to the region*). All of the clusters have on average two children, but this could not be the explanation, because on average the children of the individuals of clusters 6 and 7 have already left their parents′ house, while in Cluster 4 they might still live with their parents. Having a bond to nature could be a characteristic that could explain the low rate of indecision. It might be the case that these clusters have exact impressions about their environment and therefore tend to be either dissatisfied or satisfied.

The hypothesis regarding the life satisfaction needs to be rejected because the results of the assumed Cluster 7 (*retired locals bonded to the region*) and Cluster 4 (*bonded highly educated people*) are different between the regions and the degree of satisfaction. Because not only one cluster shows the highest or lowest rate of satisfaction, it is hard to find an explanation for these phenomena. It might be the case that the cluster analysis did not integrate all the individual variables that are used to explain differences in satisfaction. Additionally, there might be some regional conditions important or that impact one's satisfaction with life.

9.7.4 Use of the Region

On average, Cluster 4 (*bonded highly educated people*) as well as Cluster 2 (*highly educated young women*) are the clusters that uses the region for the examined purposes – except for cultural purposes – most often. This could be explained by their higher education, because it might be the case that the individual get a different feeling of the importance of the nature with increasing education and how it should be used by the residents, even because they are living in more urban areas and might not have that much nature as in the more rural areas. It might be the case that they are more interested in using this, given that the city offers fewer opportunities to do so.

The lowest rate of using the region for relaxing as well as for the educational purpose is showed from Cluster 1 *(middle-aged men without any bond*) and Cluster 5 (*returnees bonded to the region*). These clusters are living in more rural areas. Reversing to the explanation to the clusters before, it might be that these clusters using their houses and gardens for relaxing instead of other parts of the region and thus on average use the region lesser for relaxing activities.

In addition to Cluster 2 (*highly educated young women*) and Cluster 4 (*bonded highly educated people*), Cluster 6 (*retired newcomers*) shows the highest rate of using the region for sports activities. The individuals of Cluster 6 live in more urban settings like the other two clusters, so that the explanation above could also be true. All clusters show a relatively high rate of newcomers, which could be another explanation for the high percentages for sports in the region. It might be the case that newcomers are doing more sports outside or in general than returnees or locals do. Therefore, the rate of using is higher for these clusters than for others. One exception might be pertaining in the region of LPLD, where Cluster 4 displays the lowest percentage of use.

Cluster 3 (*low-educated people bonded to the region*) is the cluster with the lowest rates of using the region for sports. With the Cluster 4 (*bonded highly educated people*) is common that the individuals have a bond to the region, but it seems to be hard to find an explanation regarding the bond to region – especially because of the low region use. However, it seems like that this characteristic are the only one that could explain this low rate. The individuals of the other clusters are also still employed and between 44 and 48 years of age, but these clusters do not show such low rates.

For the use of the educational purpose, in addition to Cluster 2 (*highly educated young women*) and Cluster 4 (*bonded highly educated people*), Cluster 1 (*middle-aged men without any bond*) display high percentages. All these clusters show a relatively high rate of newcomers. It might be the case that newcomers are more interested in educational aspects of the region than the people who are living there since birth or who left the region for a while. This could account for the high rate among these clusters. Conversely, Cluster 1 displays the lowest rate for the use in the region of LNWM. It seems like that the people in this cluster of the region LNWM are different from the other regional sample. This could also be shown from the results of the Chi test (Tab. 9-6).

Cluster 5 (*returnees bonded to the region*) and Cluster 7 (*retired locals bonded to the region*) displayed the highest use of the region for cultural purposes. These clusters live in more rural areas. It might be the case that in this area the frequency of cultural events is higher than in urban areas because the questionnaire suggested some events like block parties. These are usually more common for rural areas than for urban areas. Additionally, these clusters show a bond to nature and to the region. Eventually, these people are more interested in cultural events because of networking and other purposes concerning their region. But Cluster 5 shows for the region of STB and LNWM the lowest percentage of the use. Regarding the cultural events, there might be regional differences and therefore the different rates for the same cluster could be explained. At least the Chi tests confirmed the assumption of differences of the sample for LPLD and LNWM as well as STB (Tab. 9-6).

For the region of LPLD, the cluster with the middle-aged men without any bond (Cluster 1) shows the lowest rate of the use for cultural purposes. In addition to Cluster 5, STB shows the lowest rates for Cluster 4 (*bonded highly educated people*). This cluster is also displayed by the overall sample. These clusters have the age and the employment status in common, but in these characteristics no explanation could be found since other clusters have similar ones.

The hypothesis regarding the use of the region needs to be rejected because the areas of use seem to be more differentiated in their evaluation. Clusters 4 (*bonded highly educated people*) and 7 (*retired locals bonded to the region*) show in some cases the assumed coefficients, but in others they did not. This would imply that the purposes of use are different and therefore different characteristics are responsible for the use. Additionally, the regional conditions could be another explanation for these differences.

9.7.5 Further Discussion

In some cases there might be an interaction between the variables that are analyzed with the results of the cluster analysis. For instance, Simonson et al. (2016) found that German residents are more often involved if they are satisfied with their life. It could be that more satisfied people are feeling like they would give some of their positive emotions back to the community, the environment or to other aspects of their region. Therefore, it might also be the case that the variables are important to explain some of the variances of each other and would be part of the explained variables. This interaction was not tested within this analysis, but it might be an interesting approach.

Frey (2008) found that happy people are usually more optimistic about their lives and therefore tend to be more successful than people with lower satisfaction. This could have impacts on all the other variables analyzed in this examination. As seen in the example of volunteering, life satisfaction could also influence the WTP, because if someone is more successful, the earnings are increasing and therefore, the individual would have more money to spend on recreational aspects (in this case the region).

Regarding life satisfaction it might be a good idea to go deeper into some aspects. As described by the theoretical part, it could be analyzed as one indicator, but also in many different indicators. Therefore, it might be that some individual are more satisfied with single aspects than with others, but the aspects the individual is dissatisfied with have more weight and therefore the individual is less satisfied than another individual.

The same cases could be applied to the other research fields. It could already been seen that especially for the volunteering part and the use of the region, there will not be the one person who is more likely to volunteer or to use the region. With a deeper look into the single constructs, the possibility to analyze some variables in more details might be given.

The sample of the regional analyses seems to be really small and in the most cases, the regional samples are not out of one population. Therefore, it might be important to include some regional specific data to get a more detailed analysis on the regional scale or to integrate more individuals to get a more generalized view of these variables. The questionnaire does not offer this possibility.

9.8 Conclusion

This paper determined different clusters of individual characteristics to define different types of individuals and differentiated this to the research region. These identified clusters are then used to examine differences in the willingness to support the regional landscapes (financially or volunteer work), the life satisfaction and the use of the region.

One of the central results of the WTP is that highly educated young women (Cluster 2) are usually the most individuals willing to contribute some money for the protection of the landscapes. The least willing to pay clusters are the retired ones (*retired newcomers*, *retired locals bonded to the region*) and in some cases the cluster with the retired returnees with a bond to the region.

The clusters with a higher level of education are more involved in volunteer activities for and in the region in general than clusters with lower educational levels (Clusters 2, 4, 5). The lowest rates of the WTV are displayed from clusters with individuals that are retired locals (Cluster 7).

The only pattern that could be identified for the life satisfaction is that Cluster 4 (*bonded high educated people*) and cluster 5 (*returnees bonded to the region*) are more dissatisfied with their current situation of life than other clusters. Regarding the lowest dissatisfaction the cluster 3 (*low educated people bonded to the region*) and cluster 2 (*high educated young women*) are into the focus. For the undecided and the satisfied clusters could not be observed such a pattern, neither for the highest nor the lowest ratings.

The clusters with bonded highly educated people (Cluster 4) or highly educated women (Cluster 2) use the region more often than other clusters. For cultural purposes, the cluster with the returnees (Cluster 5) and the retired locals (Cluster 7) use the region more frequently than other clusters do. The lowest use of the region can be observed for Cluster 5 – except for sports activities. In this case, the low-educated cluster with a bond to the region (cluster 3) is using the region less.

As displayed in these results, a higher education might lead to a higher WTP for the landscapes as well as an increased use of the region compared with lower levels of education. This would indicate that the regional management should integrate these people in particular if the interest is somewhere around the monetary support or the use of the region. Surprisingly, people with less education are more involved in more specific areas of volunteering that are defined for this analysis than the highly educated ones, but this could even be used from the regional management to improve the attraction to the region.

The dependencies of the variables that are examined within these clusters could be a topic of further research. It might be interesting to have a look if, for instance, people with a higher WTP for regional landscapes are also more satisfied with their life (as Simonson et al., 2016 indicated for volunteering) in the region or automatically use the region more often than other people. Additionally, the regional characteristics that might be able explain differences in manifestations of the examined variables could also be a topic of interest research, since if the region knows the individual's characteristics, it has more starting points for methods of improving its attractiveness.

Additional Analysis: Explorative Analyses of the Clusters Using Logit Regression Models for Willingness to Pay and Willingness to Volunteer

This section uses the determined clusters to estimate the influences of each variable on all clusters regarding the WTP and WTV. First, some hypotheses about the impacts of the variables within the clusters are formulated. In the first place, the descriptive analysis of Chapter 9.4 is used to look into the hypotheses. Following is a short description of the methodology of this part. Subsequently, the results of the estimated models for the volunteer work and the financial contribution are presented. A discussion about the results completes this excursion.
Hypotheses

The aim of these analyses is to use the results of the cluster analysis to analyze variables that have controversial findings in previous literature. It might be the case that some hypotheses can be answered by the descriptive analyses of the clusters; others might need the statistical models. First there will be a description of some findings and subsequently some assumptions of the variables in its interaction with each other are formulated.

The discussion about the literature regarding the impact of gender is not distinct. The international literature suggests that men are more often willing to contribute money than volunteer work compared with women (Rai & Scarborough, 2014) and women on average are more often volunteering than men (Measham & Barnett, 2008). In European countries and as part of it, Germany, the results show that more males volunteer than females (Wilson 2000, Simonson et al. 2016).

For the financial contribution and the contribution of volunteering, the results are pretty clear. A negative influence brought by age was found (e.g., Carson et al., 2001; Hackl & Pruckner, 1999 for the WTP; e.g., Lankia et al., 2014;

Wilson, 2000 for the WTV). Asah et al. (2014), Tidwell and Brunson (2008), and Tselios et al. (2015) found that younger people are more willing to contribute volunteer work than make monetary payments than older people. In general, older people seem less interested to contribute for environmental issues (Schiappacasse et al., 2013; Vásquez, 2014) because of different time horizons (Carstensen, 1995).

The first assumption involves the variables gender and age. If the result of Wilson (2000) is true, in younger ages, men are more willing to contribute payments and volunteer work than women, but the rates are getting closer with getting older. Therefore, the cluster with relatively young men needs to have higher rates for both contribution forms than this for women. Statistically, in younger ages the coefficient for women might be lower than with the increase of age.

There were no results identified regarding the bond to region and the willingness to contribute, but it was found that the identification with the region is seen as important factor for the behavior of giving in general (Morrison & Dowell, 2013; Schervish and Havens, 1997). Because of the desire of newcomers to get a feeling of belonging, these are more often voluntarily working than locals or returnees (e.g., DiEnno & Thompson, 2013; Wilson, 2000). Differences between returnees or locals in the literature neither in WTP nor in WTV researches could not be identified yet.

Because of the interaction of bond to region and the time an individual lives in the region, the assumption is that WTP or WTV differ between the groups of newcomers, returnees and locals. The coefficient of bond to region needs to be lower for the group of newcomers than for both other groups in the field of volunteering. This is because newcomers usually do not have a strong relationship to the region and thus it might not be that important for them in valuing the landscapes.

In examinations of WTP and WTV, it was found that education is positively associated (Ahlheim et al., 2010; Howley et al., 2012; Verbič & Slabe-Erker, 2009). This result was also found in Germany (Simonson et al., 2016). Some exceptions are Rai and Scarborough (2014), Karkow and Gronemann (2005), and Stone et al. (2008) who did not find any significant values of the educational level neither for WTP nor WTV.

The relationship to nature is the most important variable for valuing a landscape monetary (Verbič & Slabe-Erker, 2009). This was also proven from research on the motivation for volunteering (e.g., Asah & Blahna, 2012; Bramston et al., 2011; Bruyere & Rappe, 2007).

Because bond to nature seems to have an impact on the WTV and WTP for the landscape, the assumption is that the shape of the relationship to nature is dependable on the education an individual has obtained. For the statistics that would mean that the coefficient for bond to nature of highly educated clusters might be lower than for low-educated ones because its influence is lower for highly educated clusters.

For the amount of children was no literature found to examine the impact of them to neither WTP nor WTV. Because children were not used in analyses before, there is the assumption that people having children are more concerned about the environment and the future use of the nature. This would imply that bond to nature might be more distinct if people having children and therefore work or pay for the landscape protection. More concrete, this leads to the fact that the coefficient of bond to nature is higher for people having only one child than for people having more than one. With one child, it might be the case that the preservation of the environment might even be higher than for people having no child. The effect of the importance might be decreasing with an increasing number of children.

Descriptive Results Concerning the Hypotheses

The first hypothesis stated that men are more willing to pay or volunteer for the landscape protection in younger years and the rates of men and women come closer to each other with older ages. With the descriptive results, this could not be ascertained (Tab. 9-8). For the overall result and the regional samples, Cluster 2 (*highly educated young women*) showed the highest rate of willing to contribute some money for the landscape protection (every type of landscape). The cluster that consists mostly of men (Cluster 1) shows only once the highest rate for the water landscape in the region of STB. For volunteering, it can be said that Cluster 1 (*middle-aged men without any bond*) showed the highest rate of getting involved in volunteer work in the area of monument conservation for the overall result (Tab. A 10).

The willingness differences between the groups of newcomers, returnees and locals regarding their bond to region were covered in the second hypothesis. For the water landscape, it could be found that the highest rate of not willing to pay any money for the landscape protection is displayed from Cluster 6 (*retired newcomer*) (Tab. 9-8). The same can be said for the overall result and for the village landscape. For the city and the village landscape the returnees show the lowest rate in the region of STB in both cases and for LPLD only for the village one. For the city landscape, the highest rate of these cluster are displayed in Cluster 2 (*highly educated young women*) (one exception is the region of LPLD, with the highest willingness of Cluster 3 (*low educated people bonded to the region*) and the lowest rate in Cluster 5 (*returnees bonded to the region*) and again Cluster 6. For volunteering, the newcomers (Cluster 6) are more involved in the areas of animal protection and art, culture and museums than the locals or returnees (Tab. A 10). In general, the returnees are more willing to volunteer for the landscape protection and the locals less so. The group of the newcomers is in between. This can be almost confirmed by the regional analysis as well, except of the region of LPLD where the newcomers are highly involved in the conservation of the environment, animal protection and the art, culture and museums. Since it was not possible to identify clusters for returnees and locals without a bond to region or newcomers with a bond to region, the hypothesis cannot be clearly confirmed.

The third hypothesis assumed that the manifestation of bond to nature depends on the level of education. Clusters 3 and 4 (*low-educated people bonded to the region*, *bonded highly educated people*) can be compared, because these clusters are different in their education and the manifestation of bond to nature (Tab. 9-4). It can be seen that Cluster 3 (*low-educated people bonded to the region*) are willing to pay higher rates for all close-nature landscapes in the region of LPLD than the members of Cluster 4 (*bonded highly educated people*). For volunteering, it can be seen that in most of the areas Cluster 4 (*bonded highly educated people*) have a higher rate of involvement in voluntary work – except in the area of animal protection and local history (). In this area, Cluster 3 (*low-educated people bonded to the region*) has higher rates. Overall the members of Cluster 4 (*bonded highly educated people*) are more involved in volunteering for and in the landscapes (Tab. 9-9). However, because Cluster 3 does not show the same amount of having a bond to nature, the comparison of these two clusters might not be adequate to get an answer for the hypothesis.

Children have a positive impact on bond to nature and therefore people with only one child volunteer more often or pay more money than people with more than one child. The first three clusters have on average less than two children and the others average more than one child (Tab. 9-4). Because the clusters with more than one child have all a higher rate for bond to nature, this hypothesis cannot be examined as it is. It can only be examined if people with more children are more willing to pay or volunteer than these with only one child. On average, there is no pattern for this assumption, excluding a few exceptions: Cluster 1 to 3 (*middle-aged men without any bond*, *highly educated young women* and *low-educated people bonded to the region*) show higher rates for payment amounts for each landscape (Tab. 9-8). For volunteering, only for LPLD and animal protection the first three clusters are more involved than the others (Tab. A 10).

For volunteering in the whole sample no differences could be identified. Because of these confusing results and the lost connection to bond to nature, this hypothesis cannot be validated.

Methodology

Since the results of the descriptive part could not clearly answer the hypotheses a logit estimation was done to explore the impact of the used variables in the hypotheses. The samples in each cluster are relatively small, therefore these results need to be validated by further research and could only be used and interpreted carefully.

Subsequently, for each identified cluster a logit model is estimated to show differences between the WTP or WTV for landscape protection in the clusters of characteristics. The following model is formulated:

$$WTP_n = f(X_n, E_n, N_n, B_n)$$
or
$$WTV_n = f(X_n, E_n, N_n, B_n)$$

where WTP_n respectively WTV_n denotes the willingness to invest payment (P) or volunteering (V) in the region of person n. X_n is a matrix of the socioeconomic variables (*gender, age, education, children*) of the inhabitants in the sample. E_n displays a matrix of the professional circumstances (*income, employed*) while N_n decribes the other circumstances (*rural environment, newcomer, returnee*). B_n includes the concern about the nature and the region (*relationship between nature and person, relationship between region and person*).

For each cluster a separate model for the WTP and WTV was estimated to see the differences between the clusters. Afterwards the marginal effects are calculated. First, the results for the financial contribution are described and second, the results for volunteering as a form of contribution are shown. In the end of this section the results of the models are compared.

Results of the Logit Models for the Payment

The used variables are not important for each cluster (Tab. 9-12)[61]. Clusters 5 and 7 (*returnees bonded to* the *region, retired locals bonded to the region*) have no variables that show a significant influence, but clusters 1, 2, 3, 4 and 6 show significant influences in particular. For the first cluster (*middle-aged men without any bond*) the only significant variables is the age. It displays negative, but its effect is low[62]. This means older men in this cluster are lesser willing to contribute money than younger men. The second cluster (*highly educated young*

61 The coefficients and the pseudo R^2 are displayed in Tab. A 11.

62 Following, the marginal effect of 0.000 to 0.05 is described as low effect, an effect between 0.051 and 0.2 is presented as a medium high effect and the higher effects are used as high effect.

women) and the sixth cluster (*retired newcomers*) show significant impacts: While the age is again negatively and with a low effect, the coefficient of bond to nature has a positive impact and a high effect. Younger women (for Cluster 2) or newcomers in general (for Cluster 6) with a bond to nature are more willing to pay for landscape protection than women (for Cluster 2) or locals (for Cluster 6) without a bond to nature. The third cluster (*low-educated people bonded to the region*) also has two significant variables: higher education and bond to nature. Both variables have a positive influence and a great effect on this contribution form, although the effect of the education variable is even higher. People in this cluster are more willing to give some money for the landscape protection if they have a higher education degree. Also members of this cluster with a bond to nature are more willing to contribute money for landscape protection. For the fourth cluster (*bonded highly educated people*) the highest amount of variables displays significant. While gender has a negative influence of the willingness to contribute and a high effect, the effects of the rural environment and bond to nature are even higher and also significant. Because gender has a negative impact, it can be stated that men are more willing to contribute money for the protection of the regional landscape. Members of this cluster have a higher valuation of the landscape in terms of money if they are living in a rural environment or have a bond to nature.

Regarding a comparison of these variables, it can be said that the variable bond to nature is important for the financial contribution of four out of the seven clusters (*highly educated young women*, *low-educated people bonded to the region*, *bonded highly educated people*, *retired newcomers*). The effect of this variable is highest for the bonded highly educated people (Cluster 4) and lowest for the retired newcomers (Cluster 6), but its effect is still high and positive in each of the clusters. A different picture is drawn with age: its effect in three of the clusters (*middle-aged men without any bond*, *highly -educated young women*, *retired newcomers*) is really low and negative. This implies that the variable age has a minor impact on the valuation of landscape protection in terms of money. The variables gender, higher education and rural environment are just of importance in special cluster.

Tab. 9-12: Marginal effects of the logit model for financial contribution (margins, standard errors in parentheses)

	Cluster 1	Cluster 2	Cluster 3	Cluster 4	Cluster 5	Cluster 6	Cluster 7
	Middle-aged men without any bond	Highly educated young women	Low-educated people bonded to the region	Bonded highly educated people	Returnees bonded to the region	Retired newcomers	Retired locals bonded to the region
Gender	-0,045	-0,033	-0,034	**-0,286** *	0,005	0,042	0,111
	(0.132)	(0.220)	(0.095)	(0.164)	(0.120)	(0.092)	(0.138)
Age	**-0,010** **	**-0,016** ***	-0,002	-0,011	-0,003	**-0,011** ***	-0,004
	(0.004)	(0.006)	(0.004)	(0.007)	(0.005)	(0.003)	(0.004)
Higher Education	0,087	0,000	**0,438** **	0,000	0,199	0,030	0,048
	(0.111)	(omitted)	(0.183)	(omitted)	(0.132)	(0.093)	(0.167)
Employed	0,000	0,263	-0,052	0,000	-0,144	0,000	0,000
	(omitted)	(0.283)	(0.341)	(omitted)	(0.190)	(omitted)	(omitted)
Income	0,007	-0,036	-0,009	-0,061	-0,026	0,034	-0,036
	(0.024)	(0.027)	(0.023)	(0.043)	(0.030)	(0.023)	(0.045)
Children	0,043	0,071	-0,057	-0,051	-0,015	0,030	-0,020
	(0.040)	(0.056)	(0.041)	(0.045)	(0.052)	(0.034)	(0.057)
Rural Environment	-0,018	0,000	-0,063	**0,406** ***	-0,028	0,040	-0,033
	(0.096)	(0.118)	(0.093)	(0.142)	(0.110)	(0.089)	(0.136)
Returnee	0,000	0,064	0,000	0,000	0,000	0,000	0,208
	(omitted)	(0.180)	(omitted)	(omitted)	(omitted)	(omitted)	(0.229)
Incomer	0,063	0,121	-0,086	0,152	0,000	0,431	0,000
	(0.129)	(0.147)	(0.110)	(0.154)	(omitted)	(0.275)	(omitted)
Bond to Nature	-0,001	**0,261** **	**0,268** ***	**0,384** *	-0,059	**0,229** **	0,004
	(0.118)	(0.128)	(0.093)	(0.221)	(0.122)	(0.093)	(0.162)
Bond to Region	-0,072	0,028	-0,073	0,000	-0,048	0,017	-0,005
	(0.1589)	(0.185)	(0.163)	(omitted)	(0.122)	(0.089)	(0.420)
N	136.00	97.00	165.00	79.00	101.00	149.00	69.00

Note: Stars indicate significance levels: * p<0.1, ** p<0.05, *** p<0.01. N displays the number of cases.
Source: Resident survey, own calculations.

Results of the Logit Models for Volunteering

For the contribution form of volunteering, in every cluster could be at least one variable identified with a significant effect (Tab. 9-13)[63]. The first two clusters (*middle-aged men without any bond, highly educated young women*) show bond to nature as significant variable. In both clusters, its effect is positive and high, even if the effect is highest in the first cluster. This would imply that middle-aged men (for Cluster 1) as well as young women (for Cluster 2) with a bond to nature are more willing to contribute volunteering as form of protection for the landscapes than these people without any bond to nature. In Cluster 4 (*bonded highly educated people*), the newcomers' variable is significant. Its impact is positive and high and would mean that members of this cluster who moved into the region are more willing to contribute labor on a voluntary basis than locals.

63 Tab. A 12 shows the coefficients and the pseudo R^2 of the estimation.

Bond to region has a positive high effect on the WTV for the members of Cluster 5 (*returnees bonded to the region*). This indicates that people having a bond to the region are more willing to contribute volunteer work than people without a bond. For the third cluster (*low-educated people bonded to the region*) bond to nature as well as gender displays significant. While the effect of gender is negative, it is positive for bond to nature. The effect for gender for Cluster 3 and bond to nature for Cluster 6 is medium-high, while the effect is high for bond to nature in Cluster 3 and gender in Cluster 6. The negative impact of gender would imply that members of this cluster would be more willing to volunteer if it is male. These members would also be more willing to volunteer if they have a bond to nature. The seventh cluster (*retired locals bonded to the region*) displays gender and age as significant coefficients. Both, gender and age have negative impacts, but differ in their strength. While the variable gender has a high effect, it is low for the variable age. This would imply that men are higher willing to contribute volunteering than women incorporated in this cluster.

As in the analyses of the willingness to contribute money, bond to nature seems to be an important factor for the willingness to contribute labor as well. Again, four out of seven clusters (*middle-aged men without any bond*, *highly educated young women*, *low-educated people bonded to the region*, *retired newcomers*) display this variable as significant, but the effects differ. While the effect of clusters 1 to 3 is high, it is medium-high for Cluster 6. The second often variable that could identify was the variable gender. It is significant for at least three out of seven clusters (*low-educated people bonded to the region*, *retired newcomers*, *retired locals bonded to the region*). The effect is lowest for Cluster 3 with a medium-high effect. For both other clusters, the effect is high. These variables have a negative impact on the WTV. Age also displays as significant variable in more than one cluster (*retired newcomers*, *retired locals bonded to the region*). Because the effect of age is relatively low, it could be neglected, even if it implies that younger people are more willing to contribute labor than older ones. The other variables (newcomer, bond to region) are just special for only one cluster (bonded high educated people for the variable newcomer and returnees bonded to the region in the case of bond to region).

Tab. 9-13: Marginal effects of the logit model for volunteering as form of contribution (margins, standard errors in parentheses)

	Cluster 1	Cluster 2	Cluster 3	Cluster 4	Cluster 5	Cluster 6	Cluster 7
	Middle-aged men without any bond	Highly educated young women	Low-educated people bonded to the region	Bonded highly educated people	Returnees bonded to the region	Retired newcomers	Retired locals bonded to the region
Gender	0,214	-0,146	**-0,183** *	-0,184	-0,174	**-0,264** ***	**-0,353** **
	(0.137)	(0.221)	(0.095)	(0.115)	(0.121)	(0.095)	(0.154)
Age	0,002	-0,003	0,001	-0,004	-0,002	**-0,007** **	**-0,009** **
	(0.004)	(0.006)	(0.004)	(0.005)	(0.005)	(0.003)	(0.004)
Higher Education	0,090	0,000	-0,028	0,000	-0,060	0,022	-0,079
	(0.117)	(omitted)	(0.161)	(omitted)	(0.130)	(0.093)	(0.166)
Employed	0,000	0,277	0,020	0,000	0,062	0,000	0,000
	(omitted)	(0.286)	(0.328)	(omitted)	(0.184)	(omitted)	(omitted)
Income	0,035	-0,001	0,009	-0,052	0,000	0,008	0,035
	(0.025)	(0.028)	(0.023)	(0.032)	(0.029)	(0.023)	(0.045)
Children	-0,046	-0,044	-0,010	-0,016	-0,040	-0,013	-0,036
	(0.043)	(0.056)	(0.041)	(0.031)	(0.050)	(0.033)	(0.058)
Rural Environment	-0,129	0,053	0,034	0,140	0,020	-0,040	-0,037
	(0.101)	(0.121)	(0.094)	(0.098)	(0.107)	(0.090)	(0.137)
Returnee	0,000	-0,253	0,000	0,000	-0,093	0,000	-0,019
	(omitted)	(0.180)	(omitted)	(omitted)	(0.373)	(omitted)	(0.207)
Incomer	0,101	-0,089	-0,079	**0,249** **	0,000	-0,112	0,000
	(0.132)	(0.155)	(0.110)	(0.109)	(omitted)	(0.212)	(omitted)
Bond to Nature	**0,476** ***	**0,271** **	**0,219** **	0,125	0,146	**0,180** **	-0,148
	(0.141)	(0.126)	(0.092)	(0.156)	(0.121)	(0.091)	(0.152)
Bond to Region	0,041	-0,229	0,051	0,000	**0,212** *	-0,051	0,000
	(0.168)	(0.186)	(0.155)	(omitted)	(0.117)	(0.090)	(omitted)
N	136.00	97.00	165.00	79.00	101.00	149.00	69.00

Note: Stars indicate significance levels: * p<0.1, ** p<0.05, *** p<0.01. N displays the number of cases.
Source: Resident survey, own calculations.

Results of Both Contributions in Comparison

Two variables have similar effects to each form of contribution: age for Cluster 6 (*retired newcomer*) and bond to nature for clusters 2, 3 and 6 (*highly educated young women, low-educated people bonded to the region, retired newcomers*). The effect for age is almost the same while it differs for bond to nature between the clusters. The effect of bond to nature only differs for Cluster 6 between medium-high and high, but for the others the effect is constant between the contribution forms. Both variables are also significant in different clusters. Age also shows significant coefficients in clusters 1 and 2 (*middle-aged men without any bond, highly educated young women*) for the financial contribution and in the contribution of volunteer work for Cluster 7 (*retired locals bonded to the region*). Similarly, this can be said about bond to nature: it is also significant for Cluster 4 (*bonded highly educated people*) in the case of a financial contribution and Cluster 1 for the volunteering as form of contribution.

All other significant effects differ between the forms of contribution. This would imply that the used variables are not all important for each form of contribution. For instance, a higher education level and the rural environment display significant coefficients only in the financial contribution and only once each variable (Cluster 3: higher education, Cluster 4: rural environment), while the variables newcomers (Cluster 4) and bond to region (Cluster 5) are only significant in the models of volunteering. Gender seems to be important for both forms, but for different clusters (Cluster 4: financial contribution; clusters 3, 6, 7: volunteering).

Results of the Hypotheses

The first assumption was that younger women are less likely to contribute than men and that the rates are getting closer with increasing age. The members of the first five clusters are on average between 39 and 50 years old (Tab. 9-4). It can be observed that the clusters 1 to 4 (*middle-aged men without any bond, highly educated young women, low-educated people bonded to the region, bonded highly educated people*) show a negative direction (men are more willing to pay than women), even if the effect is only significant for Cluster 4 (Tab. 9-12). The sign of the effect changes for Cluster 5 to 7 (*returnees bonded to the region, retired newcomer, retired locals bonded to the region*) where the members are between 50 and 64 years old (Tab. 9-4). In these clusters, the effects again are not significant. Because of these results, the hypotheses could be true for the financial contribution, but the result might not be robust. The effects of gender for the models for WTV change after Cluster 1 (*middle-aged men without any bond*) (Tab. 9-13). In this cluster the women are more willing to contribute than men, but again, these coefficients are not significant. In all other clusters, the effects are negative and imply that men are more willing to contribute volunteer work than women with increasing age. Three out of these coefficients are significant (clusters 3, 6, 7). The hypothesis could not be verified with reliable results.

That the influence of bond to region is lower for newcomer than for locals and returnees can be analyzed with the last three clusters (*returnees bonded to the region, retired newcomers, retired locals bonded to the region*). For the fifth cluster, 75 % of the returnees marked that they have a bond to region and for the locals (Cluster 7) almost everyone (97 %) stated they would have a bond to region (Tab. 9-4). For the newcomers, only 44 % checked off that they have a bond to region. The directions of the coefficients and effects for the financial contribution are different for newcomers and returnees/locals (Tab. 9-12). The effects of bond to region are not significant for the WTP and could therefore not be used to prove it with robust results, even if the directions of the effects are reversed than the hypotheses assumed. For the WTV, the results are as assumed (Tab. 9-13). For Cluster 5 (*returnees bonded to the region*), bond to region is higher than for newcomers and is significant. For locals, this variable was omitted for the estimation. Again, the hypothesis could neither be proven nor rejected.

For the next hypotheses, argues that higher educated people have a lower coefficient of bond to nature than lower educated people. Clusters 2, 3, 4, 6 and 7 (*highly educated young women, low-educated people bonded to the region, bonded highly educated people, retired newcomers, retired locals bonded to the region*) are important to compare (Tab. 9-4), because clusters 2 and 4 pooled almost only highly educated respondents while the other three clusters included respondents with a lower educational level. It can be seen that for the contribution of monetary payment, the effects of clusters 2, 3, 4 and 6 are significant and high (Tab. 9-12). The effect for Cluster 7 is not significant, but it is also positive. The highest effect can be seen for Cluster 4, followed by Cluster 2, Cluster 3 and Cluster 6. These results cannot prove the hypothesis that bond to nature is more important for monetary contribution for lower educated people than for higher educated ones. For the contribution form of volunteering, the picture is similar (Tab. 9-13). The highest significant effect is identified in Cluster 2, followed by Cluster 3 and Cluster 6. Clusters 7 and 4 do not have any significant effect on the WTV. As seen for the monetary contribution, the highly educated Cluster 2 has the highest effect (in regard to the compared clusters) and therefore for the contribution in terms of volunteering the hypothesis needs to be rejected, while it can be proven for the financial contribution.

For the fourth hypothesis, that stated that the higher the amount of children, the lesser the importance of bond to nature integrates all clusters. The first three clusters (*middle-aged men without any bond, highly educated young women, low-educated people bonded to the region*) show that they have on average one child (Tab. 9-4). The other four clusters (*bonded highly educated people, returnees bonded to the region, retired newcomers, retired locals bonded to the region*) have on average two children. Therefore, these clusters can be compared to each other. Bond to nature was rated higher by those clusters having on average two children instead of one (Tab. 9-4). Regarding the statistical models, there could not be found any pattern in the WTP (Tab. 9-12). The highest significant effect is located for the model of Cluster 4 (*bonded highly educated people*), but the lowest significant effect of bond to nature is also in one of the cluster with more than one child (Cluster 6). For the contribution form of volunteering the picture is clearer (Tab. 9-13). The highest effect could be found for Cluster 1 (*middle-aged men without any bond*) and the following ones that are significant are also in the clusters with one child. For Cluster 6 (*retired newcomers*), the coefficient is significant with the lowest effect of the compared clusters. This could indicate that people having more than one child might have a higher bond to nature in general and hence the variable does not have a big influence on the volunteering contribution. This could only be said for volunteering and not for the financial contribution and therefore the hypotheses could carefully be proven for volunteering, but needs to be rejected for the WTP.

Discussion about the Results for the Financial Contribution

In younger years, less women are willing to give some money for landscape support, but after the age of around 50 the rate increases –even if only one coefficient's effect is significant for the age of 48 on average. The descriptive results could not be proven: This result in advanced because of the heterogeneity of the clusters and the manifestations of other variables. With the assumption that with a bigger sample the results could be validated, this could be at least one explanation for the switch found in the estimation. Because of lesser income gains by women during the earlier years in professional life (Mincer & Polachek, 1974), it might be the case that once they reach 50 a woman's average wages are high enough to give some additional amount of money for environmental concerns. In their younger days they might be spending the money on other things that are more important to them.

The hypothesis of lesser importance of bond to region for newcomer than for locals or returnees was the second one. Regarding the WTP, no significant results could be found. In the case of the WTV, only the returnees can be compared to the locals. Returnees had a higher coefficient than newcomers; for locals, this variable was omitted. The difficulty here is that the clusters are not heterogeneous in their other manifestations. For instance, the clusters of the locals and newcomers integrated almost all retired people into these groups while the cluster for the returnees has employed members. This could be lead to different rates of WTP, except the fact that the background with the region is different. For newcomers, bond to region seems to have a positive impact on the WTP while for the returnees and locals bond to region has a negative influence, but none of the coefficients are significant. If the results would be confirmed using a bigger sample, it might be the case that the importance of bond to region is higher for newcomers than for returnees and locals. It might be the case that returnees and locals already have a high bond to region (could be seen in Tab. 9-4) this variable does not have a high influence on the WTP. But for newcomers, this variable has a higher impact on the willingness to contribute money because they do not identify themselves with the region and the landscape and are therefore less willing to pay something for the landscape protection.

During the descriptive analysis it could be found that low-educated people are more willing to pay for landscapes in the region of LPLD. The compared clusters are differing in the living surrounding (Cluster 3 lives mostly in rural areas, Cluster 4 in mostly urban ones), the newcomer variable or in the manifestation of the bond to nature. These differences might be to some extent responsible for the differences between the clusters. In the statistical analysis, it could be found that the variable bond to nature is highest for highly educated clusters. This would refute the hypothesis if the used sample were bigger. The relationship to nature is more important for the perception of people having a high level of education than having a lower level. This hypothesis was based on the assumption that with higher education nature is better perceived, too. But with the rejection of this hypothesis one would assume that people with lower education are more perceptive of nature and thus possess a higher bond. This bond could grow because when the low-educated people were children they might have played outside more than people with a higher education, since these children had to study a lot inside instead of playing outside. The results of Verbič and Slabe-Erker (2009) could still be true, because the authors do not assume a relationship between education and bond to nature but concluded that the education variable by itself is important for the WTP.

The hypothesis about the relationship between bond to nature and children could not be examined with the descriptive data because the clusters with more than one child have all a high degree of bond to nature. Further, the amount of money that one is willing to pay does not differ between the clusters that much. But the estimations show some results: the highest coefficient of bond to nature for the financial contribution was found by one of the clusters with more than one child, and followed by two clusters with only one child on average. This would lead to the conclusion, given that the results are confirmed by bigger samples, that the assumed hypothesis is not right, because there is more than one cluster with only one child or with more than one child. It might be the case that parents with more than one child have the children play inside the apartment or house more often, as the limited space makes it easier to supervise multiple children simultaneously. Outside children have the opportunity to get in touch with dangerous plants or other things without the parent knowing because they are busy supervising more than one child. People with only one child might be more focused on the one child so that they are more time outside, for instance, in playgrounds. But it could also be the other way around. This could be an explanation why the bond to nature is more important for people having more than one child.

Discussion of the Results for the Contribution of Volunteer Work

The descriptive analysis was not meaningful enough to get information about the interaction of age and gender for volunteering, but it could be identified that men are more willing to volunteer for monument conservation than women do (Tab. A 10).

It might be the case that men have a higher affinity for buildings than women. The estimation shows only in the first cluster a positive coefficient of gender (women are more willing to volunteering); all other coefficients are negative (men are more willing to volunteer). Except for the findings in Cluster 1, the results are in line with a German analysis that generally found men more willing to volunteer than women (Simonson et al., 2016). Because of higher significant negative coefficients in the older clusters, the assumption of Wilson (2000) that the rates of volunteering are getting closer to each other over the years could not be proven with the used data. Rather, the rates are set wider apart with older ages of the cluster and therefore the opposite of Wilson (2000) is reflected. But this result needs to be verified with a bigger sample in order to obtain robust results.

The results of the involvement of newcomers, returnees and locals show some patterns in its manifestations. Locals are willing to volunteering in higher rates in the areas of animal protection and art, culture and museums (Tab. A 10). Especially for some parts of the region LPLD, this can be explained. One part of the region is more popular for the high degrees of artists, so that a high rate of people working voluntary in this area might be explained by this high degree. The descriptive data indicates that returnees are generally more willing to volunteer than locals. It might be the case that returnees want to be more involved in their surroundings than locals, or that locals are more involved in social areas of volunteering that are not the subject of this analysis. The results for volunteering as form of contribution in the statistical part show a positive significant coefficient for the returnees and an insignificant negative coefficient for the group of newcomers. Given, that the results are proven by bigger samples, it might be the case that newcomers are on average more involved in volunteering to get social contact within their new community (DiEnno & Thompson, 2013; Wilson, 2000) than returnees. If newcomers are already volunteering in or for the region and its landscapes without having a bond to region, the fact to have a bond to region is not that important anymore and could explain the negative influence on the WTV. This could also explain why people who are moved to the region are more involved in volunteering than both other groups even if they have a strong bond to region.

Generally, the descriptive analysis shows that the higher educated clusters are more involved in volunteering than the lower educated cluster, but there are differences in the degree of bond to nature so that this might explain the differences. As in the statistical part of the analysis over monetary contribution, the higher educated clusters have also higher coefficients for bond to nature. This result indicates that with an increasing level of education the importance of bond to nature is increasing and in the end the involvement in volunteering activities for or in the region increases subsequently.

The effects for bond to nature are highest for the clusters with one child on average and the effects for bond to nature are lower for the clusters with a higher amount of children. One explanation for this might be the time. People having one child might have more time for volunteer work than people having more than one.

Further Research

The analyses found some interesting results, but for the most part it was not possible to prove them with robust and reliable results because the sample was too small. It might be a chance for further research to examine some of the hypotheses using bigger samples.

Some interesting aspect to analyze might be the interaction of bond to nature and education. One would assume that people with higher education have learnt more about the environment and its change in school and therefore their bond is even higher than for people with less education. Alternatively, as described before, people with lower education have a higher bond because they played longer outside while the higher educated people have studied for school. This relationship between these variables could be examined with a difference in difference approach by itself or with regression models with different groups (e.g., one group with high education and low bond to nature, one reverse, one with high education and high bond to nature, and the last one with low education and low bond to nature). The results could then be compared to each other and differences might be observable.

Moreover, the aspect of the differences between newcomers, returnees and locals might be interesting. For instance, some researches found that newcomers are more often active voluntarily because they try to get to know other people living in the region (e.g., DiEnno & Thompson, 2013; Wilson, 2000). It could only be identified that newcomers are more working in the area of animal protection and art, culture and museum than returnees or locals. It might be interesting to examine the differences of these groups in more diverse areas of volunteering. There might be some other differences between these groups in more social or political areas. This could be done with descriptive analyses.

Another interesting aspect could be to get a deeper analysis about the relationship between the amount of children and the perception of the environment. It might be the case that people having no child at all have a different picture of the landscapes than the people with children. This could be even deeper analyzed if the number of children would be divided because this offers the possibility to understand if the amount of children is relevant for the perception of the landscapes.

It could be interesting to examine the relationship between the WTP or WTV – or both – and the motives for this support. With the data, it was not possible to integrate the motives of the landscape support because these motives were not part of the used questionnaire. It might be interesting – especially for the regional management staff – to understand why people would pay or work for landscape protection. Certainly, they could use this to get more people to volunteer for work in their region and thereby exploit this potential.

10 Conclusive Summary of the Research

This part of the thesis describes the key results, the contributions to the previous literature and based on the results of the analyses some political recommendations. The chapter ends with some ideas for further research.

10.1 Key Results and the Contributions to Previous Research

The research gap that was identified showed a combination of cultural and economic values (Chapter 2.1.2) in a meso-level perspective with the integration of different disciplines (Chapter 2.1.3). These analyses combined different disciplines to identify the meaning of the landscape for its residents. The theoretical background of each analysis is that each was formed from theories out of the research area of psychology or environmental science. The meaning for each individual is ascertained with the environmental economics method of the WTP. The results are described on a meso-level perspective regarding the implications for the region instead of using each individual one. Using only an indicator instead of the WTP as a monetary value, the cultural values could be added to the economic values. In the end, there are recommendations for the region based on the results, created on the meso-level – albeit, some recommendations are lifted to a macro-level perspective.

The more detailed research gap would be closed for the following reasons. The first empirical work is going deeper into the topic of perception of regional cultural landscapes and its comparison between different regions. Such an analysis could not be identified in the literature before, even because the current value is determined instead of a future development, as prior research did (see Chapter 2.1.4). The second chapter analyses differences between everyday and symbolic landscapes and the perception of them by resident groups of locals, returnees and newcomers in the region of STB. The focus of this chapter is on the different perceptions of landscapes rather than on individual characteristics that are indicating moving or the motives themselves. Such an analysis could, as described in Chapter 2.1.4, not be identified before. The method of the WTP is extended by the WTV in Chapter 8 to analyze the support of the landscape. As described in Chapter 2.1.4, this process is usually done in developing countries and not for developed regions. This is the first analysis looking deeper in the support of landscapes by using the WTP and WTV approach in one analy-

J. Matloch, *The Assessment of German Cultural Landscapes*, RaumFragen: Stadt – Region – Landschaft, https://doi.org/10.1007/978-3-658-21416-6_10

sis. The last text chapter goes deeper into the topic if there are groups of individuals who are more involved in the special areas for the protection of the landscapes. It could not find any other examination that analyzed this topic (Chapter 2.1.4). By using a cluster analysis and an analysis of the WTP, WTV, life satisfaction and the use of the region, the literature is extended in more than one research area by a characterization of the people.

The descriptive results of the WTP show that less than 50 % of the residents value the landscapes monetary. It might be the case that the method of the landscape valuation was, as mentioned by the project partners of the regions, less understandable for respondents. But it could also be that the people in these regions are not aware of the features in their landscape. This result is fairly common in the literature and therefore not new. It could be argued that this result is (also because of other literature) probably similar to other regions, because the general public is not used to assigning a monetary value to a public good.

Additionally, with respect to the monetary contribution, volunteering seems to offer high potential for including residents in the support of landscapes. Almost 75 % of the respondents would be volunteering for at least four hours a month in one of the areas defined as related to landscape support. Combined with the results of WTP, it can be concluded that almost one-third of the respondents are willing to contribute financially and voluntary working hours. Only one-fourth would not be contributing at all. This potential seems to be really high, but it needs to be considered that these results are only hypothetical ones. It might be the case that the real potential with the right offers is lower. Therefore, the results might not be adequate to compare with other regions and thus not transferrable.

It could be found that the village landscape is perceived better if the analyzed region is more rural and, in this case, the perception of the city landscape garners less attention. The results suggest that even in bigger cities the more rural areas are better valued than built-up ones. Because of the differences between the regions, this result seems to be a generalized result for other regions as well – even if this result was not identified in the literature before. Other analyses also found a preference for water landscape as described in the corresponding chapter, so that result seems to be robust for other regions as well. It might be the case that water landscapes are something special to the people. It could be the case that regions more in the middle of the land have other features – like the mountains around them – that would be valued higher than other landscapes. The differences between the symbolic landscapes are only minor and these are located somewhere in between the result for the city and the village landscape. This result is further not generalizable because the symbolic landscapes differ between the regions. These landscapes are special for the region and even a region has similar landscapes, the perception of these might be differing because of different regional conditions or the work with these landscapes.

The statistical analysis shows that some variables are depending on the type of the landscape while others are depending on the region. Others are independent of the landscape type or the region. Because of the complexities of the dependence of the variables, it might be the case that the generalization is not given for every variable. The relationship to nature might be applicable to other regions, even because other research found similar results regarding the relationship to nature. These analyses also show more often that the bond to nature is an important factor for landscape support. This argument is confirmed by the used control variables that indicate the same as previous analyses. The others variables should be checked before using it, because the analyses did not show significance in some regions or landscape types. It could be the case that the directions of these variables change with a bigger sample.

One common finding in these statistical analyses and previous research is that the educational level has a positive impact on the perception of landscapes. This can be confirmed for each type of the landscape and seems to be an important driver for the perception of the landscapes in terms of money and volunteer work. This can also be verified with respect to the use of the region. Because of these, the positive impact of the education might be transferrable to other regions or landscapes.

Usually, the results display that people over 65 years are less willing to contribute volunteer work or money for the support of landscapes. This result is common and also showed up during the cluster analysis. The literature assumes a negative impact of the age on doing a contribution from and the theory explains this phenomenon. The results of these analyses confirm this finding as well. Therefore, it might be transferable to other regions.

For close-nature landscapes the number of children seems to have a negative influence on the perception of these landscape types – at least for the people moved to a region. This result should be considered carefully because the sample in this analysis is small and therefore, it might be the case that the results could be change with a bigger sample. Additionally, the result is only examined for one region that is characterized between an urban and rural area. For more urban or rural structured regions the results might be different because of the opportunities the people have in such regions.

The most discussed field of regional identity was identified for locals to be important to value their close-nature landscapes. For the other landscape types covered with build elements the coefficients did not show any significance values. Therefore, it might be the case that the relationship to the region is important for the other landscapes as well, but the current results cannot prove this. For the other two groups (newcomers and returnees), no significant impact could be identified. Regarding the cluster analysis, it can be concluded that this relationship might be also be important for returnees. Nevertheless, no impact on the groups of newcomers could be found. It might be the case that they need to build new ties in the region when they arrive, so the feeling is not as strong as for people born in the region. This result might also be transferable to other regions, but it should consider how much time the newcomers have lived in the area. There might be some differences for people living longer in a region and those living for there for a shorter time.

10.2 Political Recommendations

In addition to the added value of the research with regard to the topics above, the management of a region might benefit from the analyses as well. Some recommendations are made in this part of the analyses.

The results of the analyses of Chapter 9 show that the people are interested to get involved in different areas of volunteer work. The results are in some cases based on the individual characteristics, but also on some are based on the regional conditions. While the analyses examined the individual characteristics, the regional management probably knows about the regional conditions. If these two constructs were combined from the regional management, it could be used to address possible volunteers based on their personal interests. With this process, it might be the case that more people would get involved in voluntary activities for the region and the region could benefit from the high potential displayed by the volunteering in Chapter 8.

The questionnaire did ask about the obstacles to volunteer involvement, but since the barriers are pretty broad and not really analyzed in this thesis, it might be a good idea to have a deeper look into the barriers in order to understand why the potential shown in Chapter 8 and 9 is higher than the current involvement in the regions. If the barriers are known, the region can use the results to reduce them and thus increase the involvement of the region's residents. This would be probably a gain for the region as well as for the residents getting involved.

The little use of the region for educational or cultural purposes (Chapter 9) could be strengthening by the region in different ways. The regional management could offer more alternatives to use the region for educational or cultural purposes. Regarding the educational use, for instance, it might be a good idea to create some place on a bike path that informs riders about special plants or animals. To make these places more attractive for families, these places could be more interactive. In some museums the children are able to answer some questions and get the answers in a child-oriented way. But only creating such places seems be barely adequate – the regional management needs to promote them. It could be some events set around these places to advertise them. In the long run, there will be a need for someone who will keep these places in good condition. For this purpose, there could be used the potential of the volunteers described in the paragraph above. But this is only an example; other possibilities are also welcome to increasing the use of the region for educational purposes even because the education seems to be an important driver for the perception of landscape (Chapter 6). Regarding the cultural purposes, it might be the case that the activities regarding culture are not well known in the population or these events happen less often. In this case, the attractiveness of these events could be improved or the frequency could be increased. These two examples might incur some initial costs, but the region could benefit by having interested people in the region support the environment and the region as well, and by using these improvements to further boost the region's image.

Because of the importance of bond to nature for the perception of landscapes (Chapter 6, Chapter 7, Chapter 8), intensifying the knowledge about the nature in childhood (e.g., in school or kindergarten) should be thought about. This could be done spending some days outside discovering the environment or teaching some lessons outdoors to better understand our relationship with nature. This could be a benefit for the region and its landscape, but also for the environmental change that is present at this time. These discovery days or teaching lessons could be combined with volunteers, so that these engaged people could share their expertise and the educator or teacher could look after the pedagogic aspects.

The analyses of this thesis show that the population is interested in the landscapes and their preservation of protection (Chapter 7). Especially village landscapes seem to have a special meaning for more rural areas (Chapter 6). Therefore, it might be a good idea to incorporate the residents if the regional management plans changes in the landscapes. This could not only lead to higher satisfaction of the residents because of their involvement in decision processes, but could also generate new ideas how some problems could be solved. Certainly, the effort to integrate the population might be higher, but usually, this could also be profitable for the development of the region. Furthermore, it might also strengthen the bond to the region, since these people now feel like they have the possibility to change or help shape developments.

The last recommendation focuses on the meaning of the water landscapes. Chapter 6 shows that this landscape type is highest perceived by the residents. But, as in the analyses, many regions have water landscapes. So if the landscapes are used to improve the attractiveness of the region, it might be a good idea to combine these with other special aspects, because water in itself is not a unique factor. Only by combining others factors could regions be emphasized over other ones. But if there could be identified other important regional features, the water landscape could be used as an additional for the characterization and attractiveness of the region.

In a broader sense, the regional management might be use the results for designing an improved marketing for tourist, advertising of the region for newcomers or returnees, or to improve their image for firms or the region's value chain. The aims might be different for different fields of work in the region, but it could also be a basis for deeper analyses regarding the characterization of the region itself.

10.3 Further Research

For the analyses, only regions located in the north of Germany were used. As shown in the theoretical part of the different topics and some of the results, the cultural backgrounds might have an impact on landscape perception. Therefore, the regions in the South might have different perceptions regarding the landscape types. Especially the symbolic landscapes might differ if the analyses would be done with other regions – with respect to the south, they might have mountain landscapes instead of water. These were not considered in the current examination. The differences between East and West are covered to some extent because the region of LNWM and LPLD integrate regions of the eastern and western part of Germany.

This research was done even if the regional sample were small. Some of the coefficients might not display significant because of this small sample. Therefore, in some cases the statistical analyses could only give indications of the impact of the some variables. With a bigger sample the generalization of results might be higher and hence more robust. Some of the analyses should even be repeated with a bigger sample to confirm the results presented in this research.

The survey included only people older than 18 years of age. Younger people are not involved. It might be interesting to analyze the perceptions of this group of residents. It could be the case that the perceptions of the landscapes between the youngest people and the older ones (especially the retirees) differ. It could be interesting to compare the perception of landscape between these groups. This could indicate whether younger people are more aware of the worthiness of the landscapes, given that nowadays they are growing up with the threat of climate change. In the past, this was not as visible and therefore older people might not have understood how important the landscapes were.

In the literature, there is already a huge discussion about the importance of the regional identity (e.g., Anton & Lawrence, 2014; Gustafson, 2006; Kaltenborn & Williams, 2002), but it might be interesting to deeper analyze the influence the regional bond has in different regions and their surrounding environment. It could be said that the relationship to the region is important for different groups and for different regions, but its impact seems also to be different. For the region of STB and newcomers, these results indicate a negative impact, while its impact is positive for the other two regions and the group of locals. A more differentiated analysis could help to understand these different impacts of the relationship to the regions. Furthermore, it might be interesting to look deeper into the construct why some people would say they have a bond while others would not. It could be the case that the regional conditions or individual characteristics are important drivers for the development of a bond to the region.

Another aspect that is worth to look deeper into is this of the influences of the landscape quality on the quality of life. Most of the existing analyses looked into the environmental quality like the air pollution or the climate (e.g., Levinson, 2012; Moro, Brereton, Ferreira & Clinch, 2008; Rehdanz & Maddison, 2008). However, the relationship between the landscape perception and the quality of life or life satisfaction might be worth to examine. The only work that was identified regarding a connection of WTP and life satisfaction is from Ambrey and Fleming (2011) who already combined these two constructs. The authors could not find a linear relationship between the WTP and the amenity, but Mackerron and Mourato (2013) found that the life satisfaction is higher if people are in nature during the moment of questioning. This result could be an indication that there might further relationships between these constructs. Further research could examine the relationship of the environment in terms of the landscapes that are visible and currently accessible in the regions and the WTP for the regional landscapes. This topic might be interesting not only for scientist but also for regional managers. The managers could use the information to improve the life quality of its residents and therefore make the regions more attractive for potential newcomers.

Connected with the last research idea is the analysis of the relationship between the WTP and more detailed aspects of the quality of life. It might be interesting to have a look at other regional conditions – like public services or the social life – and how satisfaction with these factors might influence the WTP for landscapes. In particular, infrastructure could be an important driver for the WTP, as Upham et al. (2009) and Kim et al. (2015) concluded that the geographical proximity might influence the WTP positively, while the results of this work suggest a negative impact. There might also be other influences like the ones named.

The results and previous research suggest that landscape perception is influenced especially from the educational level and the bond to nature. It might be the cases that the educational level and the relationship to nature might be interact to build a perception, but it was not possible to examine this relationship deeper than done in the analyses. This might be an interesting field for further research, because if the educational level in terms of learning about the nature and the bond to nature are depending on each other, the regions might benefit of this awareness. The regions might then have the chance to work on this to strengthen their protection of the landscapes.

A more psychological approach would be the analysis of the motives for WTP or WTV. Even if the analyses show some differences between different people it might be the case that these people also differ in their motives. The motives behind any given behavior are based on well-known theories, but it might be interesting to compare the motives for these types of landscape support. Most of the research looked only at the motives for WTV, but the integration of WTP would be new.

The survey data could not include newcomers that have left the regions after a while. But it might be interesting to understand the reasons for them leaving. This would help to understand if the regional conditions are responsible for that or if they were not feeling welcome or other circumstances. In turn, this could help the regions motivate more newcomers to stay if the reasons for leaving are based on the regional conditions.

And still, as described in each chapter, there might some other further research questions that are of interest for other researchers and the people working in the regions. All topics might not be covered in this short part of the further research, especially because researcher specialized in a different research area or people in public administration might see other aspects of interest in their corresponding fields of work.

References

Abraham, A., Sommerhalder, K., & Abel, T. (2010). Landscape and well-being: A scoping study on the health-promoting impact of outdoor environments. *International Journal of Public Health, 55*(1), 59-69.

Abrams, J. B., Gosnell, H., Gill, N. J., & Klepeis, P. J. (2012). Re-creating the rural, reconstructing nature: An international literature review of the environmental implications of amenity migration. *Conservation and Society, 10*(3), 270-284. doi:10.4103/0972-4923.101837

Adams, H. (2016). Why populations persist: mobility, place attachment and climate change. *Population and Environment, 37*, 429-448. doi:10.1007/s11111-015-0246-3

Agahi, N., & Parker, M. (2005). Are today's older people more active than their predecessors? Participation in leisure-time activities in Sweden in 1992 and 2002. *Ageing & Society, 25*, 925-941. doi:10.1017/S0144686X05004058

Ahlheim, M., Frör, O., Heinke, A., Duc, H. M., & Dinh, P. V. (2010). Labour as a Utility Measure in Contingent Valuation Studies - How good is it really?. *FZID Discussion Papers, 13-2010*.

Ajzen, I. (1985). From Intentions to Actions: A Theory of Planned Behavior. In J. Kuhl & J. Beckmann (Eds.), *Action Control* (pp. 11-39). Berlin/ Heidelberg: Springer. doi:10.1007/978-3-642-69746-3_2

Álvarez-Farizo, B., Hanley, N., & Barberán, R. (2001). The Value of Leisure Time: A Contingent Rating Approach. *Journal of Environmental Planning and Management, 44*(5), 681-699. doi:10.1080/09640560120079975

Ambrey, C. L., & Fleming, C. M. (2011). Valuing scenic amenity using life satisfaction data. *Ecological Economics, 72*, 106-115. doi:10.1016/j.ecolecon.2011.09.011

Anderson, P. M., & Zimmerer, M. E. (2003). Dollar Value of Volunteer Time: A Review of Five Estimation Methods. *The International Journal of Volunteer Administration, 21*(2), 39-44.

Andreoni, J. (1989). Giving with Impure Altruism: Applications to Charity and Ricardian Equivalence. *Journal of Political Economy, 97*(6), 1447-1458. doi:10.1086/261662

Andreoni, J. (1990). Impure Altruism and Donations to Public Goods: A Theory of Warm-Glow Giving. *The Economic Journal, 100*(401), 464-477. doi:10.2307/2234133

Anton, C. E., & Lawrence, C. (2014). Home is where the heart is: The effect of place of residence on place attachment and community participation. *Journal of Environmental Psychology, 40*, 451-461. doi:10.1016/j.jenvp.2014.10.007

Antrop, M. (2005). Why landscapes of the past are important for the future. *Landscape and Urban Planning, 70*, 21-34. doi:10.1016/j.landurbplan.2003.10.002

Aoki, Y. (1999). Review Article: Trends in the study of the psychological evaluation of landscape. *Landscape Research, 24*(1), 85-94. doi:10.1080/01426399908706552

Appelton, J. (1975). *The Experience of Landscape*. New York, USA: Wiley.

Appelton, J. (1984). Prospects and Refuges Re-Visited. *Landscape Journal*, *3*(2), 91-103.

Arriaza, M., Cañas-Ortega, J., Cañas-Madueño, J., & Ruiz-Aviles, P. (2004). Assessing the visual quality of rural landscapes. *Landscape and Urban Planning, 69*(1), 115-125. doi:10.1016/j.landurbplan.2003.10.029

Arrow, K. J. (1972). Gifts and Exchanges. *Philosophy & Public Affairs, 1*(4), 343-362. doi:10.1515/9781400853564.139

Arrow, K., Solow, R., Portney, P., Leamer, E., Radner, R., & Schuman, H. (2001). *Report of the NOAA panel on contingent valuation.*

J. Matloch, *The Assessment of German Cultural Landscapes*, RaumFragen: Stadt – Region – Landschaft, https://doi.org/10.1007/978-3-658-21416-6

Asah, S. T., & Blahna, D. J. (2012). Motivational functionalism and urban conservation stewardship: Implications for volunteer involvement. *Conservation Letters, 5*, 470-477. doi:10.1111/j.1755-263x.2012.00263.x

Asah, S. T., Lenentine, M. M., & Blahna, D. J. (2014). Benefits of urban landscape eco-volunteerism: Mixed methods segmentation analysis and implications for volunteer retention. *Landscape and Urban Planning, 123*, 108-113. doi:10.1016/j.landurbplan.2013.12.011

Balling, J. D., & Falk, J. H. (1982). Development of Visual Preference for Natural Environments. *Environment and Behavior, 14*(1), 5-28. doi:10.1177/0013916582141001

Bamert, M., Ströbele, M., & Buchecker, M. (2016). Ramshackle farmhouses, useless old stables, or irreplaceable cultural heritage? Local inhabitants' perspectives on future uses of the Walser built heritage. *Land Use Policy, 55*, 121-129. doi:10.1016/j.landusepol.2016.03.016

Barcus, H. R., & Brunn, S. D. (2010). Place Elasticity: Exploring A New Conceptualization Of Mobility And Place Attachment In Rural America. *Geografiska Annaler: Series B, Human Geography, 92*(4), 281-295. doi:10.1111/j.1468-0467.2010.00353.x

Bateman, I. J., & Turner, R. K. (1992). Evaluation of the environment: The contingent valuation method. *CSERGE Working Paper GEC 92-18.*

Bauer, N., Wallner, A., & Hunziker, M. (2009). The change of European landscapes: Human-nature relationships, public attitudes towards rewilding, and the implications for landscape management in Switzerland. *Journal of Environmental Management, 90*(9), 2910-2920. doi:10.1016/j.jenvman.2008.01.021

Becker, G. (1974). A Theory of Social Interactions. *Journal of Political Economy, 82*(6), 1063-1093. doi:10.3386/w0042

Becker, G. S. (1965). A Theory of the Allocation of Time. *The Economic Journal, 75*(299), 493-517. doi:10.2307/2228949

Berrens, R. P., Jenkins-Smith, H., Bohara, A. K., & Silva, C. L. (2002). Further Investigation of Voluntary Contribution Contingent Valuation: Fair Share, Time of Contribution, and Respondent Uncertainty. *Journal of Environmental Economics and Management, 44*, 144-168. doi:10.1006/jeem.2001.1193

Bishop, R. C., & Heberlein, T. A. (1979). Measuring Values of Extramarket Goods: Are Indirect Measures Biased? *American Journal of Agricultural Economics, 61*(5), 926-930. doi:10.2307/3180348

Bockstael, N. E., Strand, I. E., & Hanemann, W. M. (1987). Time and the Recreational Demand Model. *American Journal of Agricultural Economics, 69*(2), 293-302. doi:10.2307/1242279

Bohnet, I. C., & Konold, W. (2015). New approaches to support implementation of nature conservation, landscape management and cultural landscape development: Experiences from Germany's southwest. *Sustainability Science, 10*(2), 245-255. doi:10.1007/s11625-015-0290-z

Boll, T., Haaren, C. Von., & Albert, C. (2014). How do urban dwellers react to potential landscape changes in recreation areas? A case study with particular focus on the introduction of dendromass in the Hamburg Metropolitan Region. *Forest - Biogeosciences and Forestry, 7*, 423-433. doi:10.3832/ifor1173-007

Boneva, B. S., & Frieze, I. H. (2001). Toward a Concept of a Migrant Personality. *Journal of Social Issues, 57*(3), 477-491. doi:10.1111/0022-4537.00224

Boyle, K. J. (2003). Contingent Valuation in Practice. In P. A. Champ, K. J. Boyle, & T. C. Brown (Authors), *A Primer on Nonmarket Valuation* (pp. 111-169). Dordrecht: Kluwer Academic.

Boyle, K. J., Johnson, F. R., McCollum, D. W., Desvousges, W. H., Dunford, R. W., & Hudson, S. P. (1996). Valuing Public Goods: Discrete versus Continuous Contingent-Valuation Responses. *Land Economics, 72*(3), 381-396. doi:10.2307/3147204

Bramston, P., Pretty, G., & Zammit, C. (2011). Assessing Environmental Stewardship Motivation. *Environment and Behavior, 43*(6), 776-788. doi:10.1177/0013916510382875

Brennan, K. M., & Cooper, C. A. (2008). Rural mountain natives, in-migrants, and the cultural divide. *The Social Science Journal, 45*, 279-295. doi:10.1016/j.soscij.2008.03.006

Brereton, F., Clinch, J., & Ferreira, S. (2008). Happiness, Geography And The Environment. *Ecological Economics, 65*, 386-396.

Brown, G., & Brabyn, L. (2012). An analysis of the relationships between multiple values and physical landscapes at a regional scale using public participation GIS and landscape character classification. *Landscape and Urban Planning, 107*, 317-331. doi:10.1016/j.landurbplan.2012.06.007

Bruyere, B., & Rappe, S. (2007). Identifying the motivations of environmental volunteers. *Journal of Environmental Planning and Management, 50*(4), 503-516. doi:10.1080/09640560701402034

Bundesministerium für Umwelt, Naturschutz, Bau und Reaktorsicherheit (BMUB). (2014). *Naturbewusstsein 2013: Bevölkerungsumfrage zu Natur und biologischer Vielfalt [Awareness of nature 2013. Survey conducted to nature and biodiversity]*. Berlin: BMUB.

Burt, O. R., & Brewer, D. (1971). Estimation of Net Social Benefits from Outdoor Recreation. *Econometrica, 39*(5), 813-827. doi:10.2307/1909581

Butler, A. (2016). Dynamics of integrating landscape values in landscape character assessment: The hidden dominance of the objective outsider. *Landscape Research, 41*(2), 239-252. doi:10.1080/01426397.2015.1135315

Campbell, D. (2007). Willingness to Pay for Rural Landscape Improvements: Combining Mixed Logit and Random-Effects Models. *Journal of Agricultural Economics, 58*(3), 467-483. doi:10.1111/j.1477-

Campbell, R. R., Stangler, G. J., & Dailey Jr, G. H. (1977). Migration to Nonmetropolitan Areas: The Ozarks. In *Annual meetings of the Rural Sociological Society.* Madison, Wisconsin.

Carson, R. T., Flores, N. E., & Meade, N. F. (2001). Contingent Valuation: Controversies and Evidence. *Environmental and Resource Economics, 19*, 173-210. doi:10.1023/A:1011128332243

Carstensen, L. L. (1995). Evidence for a Life-Span Theory of Socioemotional Selectivity. *Current Directions in Psychological Science, 4*(5), 151-156. doi:10.1111/1467-8721.ep11512261

Clary, E. G., & Snyder, M. (1999). The Motivations to Volunteer: Theoretical and Practical Considerations. *Current Directions in Psychological Science, 8*(5), 156-159. doi:10.1111/1467-8721.00037

Clary, E. G., Snyder, M., & Stukas, A. A. (1996). Volunteers' Motivations: Findings from a National Survey. *Nonprofit and Voluntary Sector Quarterly, 25*(4), 485-505. doi:10.1177/0899764096254006

Clary, E. G., Snyder, M., Ridge, R. D., Copeland, J., Stukas, A. A., Haugen, J., & Miene, P. (1998). Understanding and assessing the motivations of volunteers: A functional approach. *Journal of Personality and Social Psychology, 74*(6), 1516-1530. doi:10.1037//0022-3514.74.6.1516

Connelly, N. A., Brown, T. L., & Decker, D. J. (2003). Factors Affecting Response Rates to Natural Resource - Focused Mail Surveys: Empirical Evidence of Declining Rates Over Time. *Society & Natural Resources, 16*(6), 541-549. doi:10.1080/08941920309152

Cornes, R., & Sandler, T. (1986). *The theory of externalities, public goods and club goods*. Cambridge: Cambridge University Press.

Cosgrove, D. (1989). Geography is Everywhere: Culture and Symbolism in Human Landscapes. In D. Gregory & R. Walford (Eds.), *Horizons in Human Geography* (pp. 118-135). Barnes & Noble Books.

Council of Europe. (2000). *European Landscape Convention*. Strasbourg: Council of Europe Publications.

Cummings, R. G., & Harrison, G. W. (1994). Was the Ohio Court Well Informed in its Assessment of the Accuracy of the Contingent Valuation Method? *Natural Resources Journal, 34*, 1-36.

Dale, A., Ling, C., & Newman, L. (2008). Does Place Matter? Sustainable Community Development in Three Canadian Communities. *Ethics, Place & Environment, 11*(3), 267-281. doi:10.1080/13668790802559676

Dalenberg, D., Fitzgerald, J. M., Schuck, E., & Wicks, J. (2004). How Much Is Leisure Worth? Direct Measurement with Contingent Valuation. *Review of Economics of the Household, 2*, 351-365. doi:10.1007/s11150-004-5652-8

Daniel, T. C. (2001). Whither scenic beauty? Visual landscape quality assessment in the 21st century. *Landscape and Urban Planning, 54*, 267-281. doi:10.1016/s0169-2046(01)00141-4

Davis, R. K. (1963). Recreation planning as an economic problem. *Natural Resources Journal, 3*, 239-249.

Deller, S. C., Tsai, T., Marcouiller, D. W., & English, D. B. (2001). The Role of Amenities and Quality of Life in Rural Economic Growth. *American Journal of Agricultural Economics, 83*(2), 352-365. doi:10.1111/0002-9092.00161

DiEnno, C. M., & Thompson, J. L. (2013). For the Love of the Land: How Emotions Motivate Volunteerism in Ecological Restoration. *Emotion, Space and Society, 6*, 63-72. doi:10.1016/j.emospa.2012.02.002

Dishman, R. K., Sallis, J. F., & Orenstein, D. R. (1985). The Determinants of Physical Activity and Exercise. *Public Health Reports, 100*(2), 158-171.

Easterlin, R. A. (1995). Will raising the incomes of all increase the happiness of all? *Journal of Economic Behavior and Organization, 27*, 35-47. doi:10.1016/0167-2681(95)00003-B

Echessah, P. N., Swallow, B. M., Kamara, D. W., & Curry, J. J. (1997). Willingness to contribute labor and money to tsetse control: Application of contingent valuation in Busia District, Kenya. *World Development, 25*(2), 239-253. doi:10.1016/s0305-750x(96)00095-2

Einig, K., & Dosch, F. (2008). Raumordnungsplanung und Kulturlandschaft [Planning of regional development and cultural landscape]. *Informationen Zur Raumentwicklung, 5*, I-IV.

Enste, D., & Ewers, M. (2014). Lebenszufriedenheit in Deutschland: Entwicklungen und Einflussfaktoren [Life satisfaction in Germany: Developments and influencing factors]. In *IW-Trends* (Vol. 2, pp. 43-58). IW.

Enste, D., Neumann, M., & Schare, T. (2012). *First Civic Engagement Report 2012: For a culture of shared responsibility*. Berlin: Ministry of Family Affairs, Senior Citizens, Women and Youth (BMFSFJ).

Eom, Y., & Larson, D. M. (2006). Valuing housework time from willingness to spend time and money for environmental quality improvements. *Review of Economics of the Household, 4*, 205-227. doi:10.1007/s11150-006-0008-1

Farnum, J., Hall, T., & Kruger, L. (2005). *Sense of place in natural resource recreation and tourism: an evaluation and assessment of research findings*. Portland, OR: U.S. Department of Agriculture, Forest Service, Pacific Northwest Research Station.

Fielding, A. (1992). Migration and Culture. In A. Champion & A. Fielding (Eds.), *Migration Processes and Patterns: Research, progress and prospects* (pp. 201-212). London: Belhaven Press.

Fishbein, M., & Ajzen, I. (1975). *Belief, attitude, intention, and behavior: an introduction to theory and research*. Reading, Mass.: Addison-Wesley.

Fokusregion Griese Gegend-Elbe-Wendland. (n. d.). Retrieved August 04, 2016, from http://regiobranding.de/content/fokusregion-griese-gegend—-elbe—-wendland

Fokusregion Lübeck und Nordwestmecklenburg. (n. d.). Retrieved August 04, 2016, from http://regiobranding.de/content/fokusregion-lübeck-und-nordwestmecklenburg

Fokusregion Steinburg. (o. D.). (n. d.). Retrieved August 04, 2016, from http://regiobranding.de/content/fokusregion-steinburg

Fortmann, L., & Kusel, J. (1990). New Voices, Old Beliefs: Forest Environmentalism Among New and Long-Standing Rural Residents. *Rural Sociology, 55*(2), 214-232. doi:10.1111/j.1549-0831.1990.tb00681.x

Freeman, R. (1997). Working for Nothing: The Supply of Volunteer Labor. *Journal of Labor Economics, 5*(1), 140-166. doi:10.3386/w5435

Frey, B. S. (2008). *Happiness: A Revolution in Economics*. Cambridge, MA: MIT Press.

Frey, B., & Stutzer, A. (2005). Happiness Research: State and Prospects. *Review of Social Economy, 62*(2), 207-228.

Frey, W. H., & Liaw, K. (1998). Immigrant Concentration and Domestic Migrant Dispersal: Is Movement to Nonometropolitan Areas "White flight"? *The Professional Geographer, 50*(2), 215-232. doi:10.1111/0033-0124.00116

Fritz-Vietta, N. V., Vega-Leinert, A. C. De la., & Stoll-Kleemann, S. (2015). Local Perceptions and Preferences for Landscape and Land Use in the Fischland-Darß-Zingst Region, German Baltic Sea. *Greifswalder Geographische Arbeiten, 51*, 1-40.

Fürst, D., & Löb, S. (2004). *"Inwertsetzung von Kulturlandschaft" - Aktivierung der Entwicklungspotentiale von Kulturlandschaft im Umfeld ostdeutscher Städte - Studie im Auftrag des BBR/BMVBW [Volarization of cultural landscape. Activating potential for development of cultural landscape in the setting of eastern cities]*. Bonn: Bundesamt für Bauwesen und Raumordnung.

Gailing, L. (2014). *Kulturlandschaftspolitik: Die gesellschaftliche Konstituierung von Kulturlandschaft durch Institutionen und Governance [Policy of cultural landscape. The social constitution of cultural landscape through institutions and governance]*. Detmold: Verlag Dorothea Rohn.

Gailing, L., & Röhring, A. (2008). Kulturlandschaften als Handlungsräume der Regionalentwicklung. Implikationen des neuen Leitbildes zur Kulturlandschaftsgestaltung [Cultural landscape as space of action of regional development. Implications of the new model for the design of cultural landscape]. *RaumPlanung, 136*, 5-10.

Gailing, L., Kühn, M., & Vetter, A. (2008). Kulturlandschaftsgestaltung und Raumordnung [Design of cultural landscape and regional development]. *Informationen Zur Raumentwicklung, 5*, 261-270.

German Socio-Economic Panel Study (SOEP) Group. (2015a). SOEP 2013 – SOEPmonitor Household 1984-2013 (SOEP v30). *SOEP Survey Papers 283*. Series E. Berlin: DIW/SOEP

German Socio-Economic Panel Study (SOEP) Group. (2015b). SOEP 2013 – SOEPmonitor Individuals 1984-2013 (SOEP v30). *SOEP Survey Papers 284*. Series E. Berlin: DIW/SOEP

Gibson, J. M., Rigby, D., Polya, D. A., & Russell, N. (2015). Discrete Choice Experiments in Developing Countries: Willingness to Pay Versus Willingness to Work. *Environmental and Resource Economics*, 1-25. doi:10.1007/s10640-015-9919-8

Gmelch, G. (1980). Return Migration. *Annual Review of Anthropology, 9*, 135-159.

Gottlieb, P. D. (1994). Amenities as an Economic Development Tool: Is there Enough Evidence? *Economic Development Quarterly, 8*(3), 270-285. doi:10.1177/089124249400800304

Gouldner, A. W. (1960). The Norm of Reciprocity: A Preliminary Statement. *American Sociological Review, 25*(2), 161-178. doi:10.2307/2092623

Green, G. P. (2001). Amenities and Community Economic Development: Strategies for Sustainability. *Journal of Regional Analysis and Policy, 31*(2), 61-75.

Green, K. E. (1996). Sociodemographic factors and mail survey response. *Psychology and Marketing, 13*(2), 171-184. doi:10.1002/(sici)1520-6793(199602)13:23.0.co;2-c

Green, K. E., Boser, J. A. & Hutchinson, S. R. (1997). Effects of population type on mail survey response rates and on the efficacy of response enhancers. Paper presented at the Annual Meeting of the American Educational Research Association (Chicago, IL, March 24-28, 1997).

Gronau, R. (1977). Leisure, Home Production, and Work- the Theory of the Allocation of Time Revisited. *Journal of Political Economy, 85*(6), 1099-1123.

Groot, W. De, & Born, R. Van den (2003). Visions of nature and landscape type preferences: an exploration in the Netherlands. *Landscape and Urban Planning, 63*, 127-138. doi:10.1016/s0169-2046(02)00184-6

Groth, P. (1997). Frameworks for Cultural Landscape Study. In P. Groth & T. W. Bressi (Eds.), *Understanding Everyday Landscapes* (pp. 1-21). New Haven: Yale University.

Gustafson, P. (2006). Place Attachment and Mobility. In N. McIntyre, D. Williams, & K. McHugh (Eds.), *Multiple Dwelling and Tourism: Negotiating Place, Home and Identity* (pp. 17-31). Oxford/ Cambridge: CAB International.

Haab, T. C., & McConnell, K. E. (2002). *Valuing Environmental and Natural Resources: The Econometrics of Non-market Valuation*. Northhampton, MA: Edward Elgar.

Haan, A. De. (1999). Livelihood and Poverty: The Role of Migraion - A Review of the Migration Literature. *Journal of Development Studies, 36*(2), 1-47. doi:10.1080/00220389908422619

Hackl, F., & Pruckner, G. J. (1999). On the gap between payment card and closed-ended CVM-answers. *Applied Economics, 31*, 733-742. doi:10.1080/000368499323940

Halfacree, K. H., & Boyle, P. J. (1993). The challenge facing migration research: the case for a biographical approach. *Progress in Human Geography, 17*(3), 333-348. doi:10.1177/030913259301700303

Halfacree, K. H., & Rivera, M. J. (2011). Moving to the Countryside ... and Staying: Lives beyond Representations. *Sociologia Ruralis, 52*(1), 92-114. doi:10.1111/j.1467-9523.2011.00556.x

Hall, P. A., & Soskice, D. (2001). Varieties of Capitalism: The Institutional Foundations of Comparative Advantage. Oxford: Oxford University Press.

Hanemann, W. M. (2001). The economic theory of WTP and WTA. In I. J. Bateman & K. G. Willis (Authors), *Valuing Environmental Preferences: Theory and Practice of the Contingent Valuation Method in the US, EU, and developing Countries* (2nd ed., pp. 42-96). Oxford: Oxford University Press.

Hansen, A. J., Rasker, R., Maxwell, B., Rotella, J. J., Johnson, J. D., Parmenter, A. W., Langner, U., Cohen, W. B., Lawrence, R. L., Kraska, M. P. (2002). Ecological Causes and Consequences of Demographic Change in the New West. *BioScience, 52*(2), 151-162. doi:10.1641/0006-3568(2002)052[0151:ecacod]2.0.co;2

Harbaugh, W. T., Mayr, U., & Burghart, D. R. (2007). Neural Responses to Taxation and Voluntary Giving Reveal Motives for Charitable Donations. *Science, 316*, 1622-1625. doi:10.1126/science.1140738

Harding, S. (1985). Values and the Nature of Psychological Well-Being. In: M. Abrams, D. Gerard, & N. Timms (Eds.), *Values and social change in Britain* (pp. 227-252). London: Macmillan in association with the European Value Systems Study Group.

Hardner, J. J. (1996). Measuring The Value Of Potable Water In Partially Monetized Rural Economies. *Journal of the American Water Resources, 32*(6), 1361-1366. doi:10.1111/j.1752-1688.1996.tb03504.x

Hartje, V., Degenhardt, J., & Dehnhardt, A. (2003). *Monetäre Bewertung einer nachhaltigen Entwicklung der Stromlandschaft Elbe: Endbericht Dezember 2003 [Monetary valuation of a sustainable development of the river landscape Elbe. Final report December 2003]*. Berlin: Technische Universität Berlin, Inst. für Landschafts- und Umweltplanung.

Haugen, K. (2015). Contested Lands? Dissonance and Common Ground in Stakeholder Views on Forest Values. *Tijdschrift Voor Economische En Sociale Geografie*. doi:10.1111/tesg.12165

Haumann, W. (2014). *Motive des bügerschaftlichen Engagements: Kernergebnisse einer bevölkerungsrepräsentativen Befragung durch das Institut für Demoskopie Allensbach im August 2013 [Motives of civic engagement: central results of a representative survey]*. Berlin: Ministry of Family Affairs, Senior Citizens, Women and Youth (BMFSFJ).

Hausenblas, H., Carron, A., & Mack, D. (1997). Application of the Theories of Reasoned Action and Planned Behavior to Exercise Behavior: A Meta-Analysis. *Journal of Sport & Exercise Psychology, 19*, 36-51. doi:10.1123/jsep.19.1.36

Heberle, R. (1938). The Causes of Rural-Urban Migration a Survey of German Theories. *American Journal of Sociology, 43*(6), 932-950. doi:10.1086/217875

Heberlein, T. A., & Baumgartner, R. (1978). Factors Affecting Response Rates to Mailed Questionnaires: A Quantitative Analysis of the Published Literature. *American Sociological Review, 43*(4), 447-462. doi:10.2307/2094771

Heyne, M., Meannig, W., & Süßmuth, B. (2009). Die intangiblen Effekte der Fußball-WM 2006 in Deutschland - Eine Bewertung mit der Contingent-Valuation-Methode [The intangible effects of the soccer world championship 2006 in Germany – A valuation using contingent valuation]. In: *Organisation und Folgewirkung von Großveranstaltungen - interdisziplinäre Studien zur FIFA Fussball-WM 2006* (pp. 83-100). Wiesbaden: Springer Fachmedien.

Hicks, J. (1939). *Value and capital; an inquiry into some fundamental principles of economic theory* (2nd ed., Reprinted 2001). Oxford: Oxford University Press.

Hossack, F., & An, H. (2014). Does payment type affect willingness-to-pay? Valuing new seed varieties in India. *Environment and Development Economics, 20*(03), 407-423. doi:10.1017/s1355770x14000503

Houle, B. J., Sagarin, B. J., & Kaplan, M. F. (2005). A Functional Approach to Volunteerism: Do Volunteer Motives Predict Task Selection? *Basic and Applied Social Psychology, 27*(4), 337-344. doi:10.1037/e413802005-076

Howley, P. (2011). Landscape aesthetics: Assessing the general publics' preferences towards rural landscapes. *Ecological Economics, 72*, 161-169. doi:10.1016/j.ecolecon.2011.09.026

Howley, P., Hynes, S., & Donoghue, C. (2012). Countryside Preferences: Exploring Individuals' Willingness to Pay for the Conservation of the Traditional Farm Landscape. *Landscape Research, 37*(6), 703-719. doi:10.1080/01426397.2011.637619

Hung, L. T., Loomis, J. B., & Thinh, V. T. (2007). Comparing money and labour payment in contingent valuation: The case of forest fire prevention in Vietnamese context. *Journal of International Development, 19*, 173-185. doi:10.1002/jid.1294

Hurley, P. T., & Halfacre, A. C. (2011). Dodging alligators, rattlesnakes, and backyard docks: A political ecology of sweetgrass basket-making and conservation in the South Carolina Lowcountry, USA. *GeoJournal, 76*(4), 383-399. doi:10.1007/s10708-009-9276-7

Hutter, M. (1997). Economic Perspectives on Cultural Heritage: An Introduction. In M. Hutter & I. Rizzo (Eds.), *Economic Perspectives on Cultural Heritage* (pp. 3-10). MacMillan Press Ltd.

Ipsen, D. (2006). *Ort und Landschaft [Location and landscape]*. Wiesbaden: Verlag für Sozialwissenschaften.

Jessop, B. (2011). Rethinking the diversity of capitalism: varieties of capitalism, variegated capitalism, and the world market. In G. Wood & C. Lane (Eds.), *Capitalist Diversity and Diversity within Capitalism* (pp. 209-237). London: Routledge.

Job, H. (2003). *Der ökonomische Wert der Kulturlandschaft. Die Anwendung der Zahlungsbereitschaftsanalyse auf szenariohafte Landschaftsbild-Simulationen [The economic value of cultural landscape. Implementation of the analysis of willingness to pay using simulated scenarios of landscape pictures]*. München: Ludwig-Maximilian-Universität München.

Job, H., & Knies, S. (2001). Der Wert der Landschaft: Ansätze zur Quantifizierung der Schutzwürdigkeit von Kulturlandschaften [The value of the landscape: Basic approaches for quantification of worthiness of protection of cultural landscapes]. *Raumforschung Und Raumordnung*, 1, 19-28.

Jobes, P. C. (2000). *Moving nearer to heaven: the illusions and disillusions of migrants to scenic rural places*. Westport, CT: Praeger.

Johnson, B. K., & Whitehead, J. C. (2000). Value of public goods from sports stadiums: The CVM approach. *Contemporary Economic Policy, 18*(1), 48-58. doi:10.1111/j.1465-7287.2000.tb00005.x

Johnson, H. P. (2000). Movin′ Out: Domestic Migration to and from California in the 1990s. *California Counts - Population Trends and Profiles, 2*(1), 1-15.

Johnson, K. M., & Beale, C. L. (1998). The Rural Rebound. *The Wilson Quarterly, 22*(2), 16-27.

Jorgensen, B. S., & Stedman, R. C. (2001). Sense Of Place As An Attitude: Lakeshore Owners Attitudes Toward Their Properties. *Journal of Environmental Psychology, 21*, 233-248. doi:10.1006/jevp.2001.0226

Jorgensen, B., Syme, G., Bishop, B., & Nancarrow, B. (1999). Protest Responses in Contingent Valuation. *Environmental and Resource Economics, 14*(1), 131-150. doi:10.1023/A:1008372522243

Kahneman, D., & Knetsch, J. L. (1992). Valuing public goods: The purchase of moral satisfaction. *Journal of Environmental Economics and Management, 22*(1), 57-70. doi:10.1016/0095-0696(92)90019-S

Kahneman, D., & Krueger, A. (2006). Developments In The Measurement Of Subjective Well-Being. *Journal of Economic Perspectives, 20*(1), 3-24.

Kahneman, D., Diener, E., & Scharz, N. (1999). *Well-Being: Foundations of Hedonic Psychology*. New York: Russell Sage Foundation.

Kaltenborn, B. P. (1998). Effects of sense of place on responses to environmental impacts: A study among residents in Svalbard in the Norwegian high Arctic. *Applied Geography, 18*(2), 169-189. doi:10.1016/s0143-6228(98)00002-2

Kaltenborn, B. P., & Williams, D. R. (2002). The meaning of place: Attachments to Femundsmarka National Park, Norway, among tourists and locals. *Norsk Geografisk Tidsskrift - Norwegian Journal of Geography, 56*, 189-198. doi:10.1080/00291950260293011

Kaltenborn, B., & Bjerke, T. (2002). Associations between Environmental Value Orientations and Landscape Preferences. *Landscape and Urban Planning, 59*, 1-11. doi:10.1016/S0169-2046(01)00243-2

Kämmerer, S., Schmitz, P., & Wiegand, S. (1996). Monetäre Bewertung der Kulturlandschaft in Baden-Württemberg — Bürger bewerten ihre Umwelt [Monetary valuation of the cultural landscape in Baden-Wuerttemberg – Inhabitants are valuing their environment]. In: *Nachhaltige Land- und Forstwirtschaft: Expertisen* (pp. 503-523). Berlin: Springer.

Kaplan, R. (1973). Some Psychological Benefits of Gardening. *Environment and Behavior, 5*(2), 145-162. doi:10.1177/001391657300500202

Kaplan, S. (1987). Aesthetics, Affect, and Cognition: Environmental Preference from an Evolutionary Perspective. *Environment and Behavior, 19*(1), 3-32. doi:10.1177/0013916587191001

Kaplan, S., Kaplan, R., & Wendt, J. S. (1972). Rated preference and complexity for natural and urban visual material. *Perception & Psychophysics, 12*(4), 354-356. doi:10.3758/bf03207221

Karkow, K., & Gronemann, S. (2005). Akzeptanz und Zahlungsbereitschaft bei Besuchern der Ackerlandschaft [Acceptance and willingness to pay of visitors of the agricultural landscape]. In: *Ackerlandschaften Nachhaltigkeit und Naturschutz auf ertragsschwachen Standorten* (pp. 115-128). Berlin: Springer.

Kempa, D. (Dezember 2014). Zum Projekt (About the project). *Regiobranding Newsletter, 1*.

Kempa, D., & Herrmann, S. (2014). Regiobranding: Kulturlandschaft als Identitätsträger und Marke einer Region [Regiobranding: cultural landscapes as carrier of identity and brand of a region]. *Unimaganzin - Forschungsmagazin Der Leibniz Universität Hannover, 3 | 4*, 64-67.

Kent, R. L. (1993). Attributes, features and reasons for enjoyment of scenic routes: A comparison of experts, residents, and citizens. *Landscape Research, 18*(2), 92-102. doi:10.1080/01426399308706398

Kim, J. H., Kim, S., & Doh, S. (2015). The distance decay of willingness to pay and the spatial distribution of benefits and costs for the ecological restoration of an urban branch stream in Ulsan, South Korea. *The Annals of Regional Science, 54*(3), 835-853. doi:10.1007/s00168-015-0688-7

King, R. (2002). Towards a new map of European migration. *International Journal of Population Geography, 8*, 89-106. doi:10.1002/ijpg.246

Klamer, A. (2003). Social, cultural and economic value of cultural goods. *Cultural Economics, Japanese Association for Cultural Economics, 3*(3), 17-38.

Knapp, T. A., & Graves, P. E. (1989). On The Role Of Amenities In Models Of Migration And Regional Development. *Journal of Regional Science, 29*(1), 71-87. doi:10.1111/j.1467-9787.1989.tb01223.x

Kolk, M. (2017). A Life-Course Analysis of Geographical Distance to Siblings, Parents, and Grandparents in Sweden. *Population, Space and Place, 23*(3). doi:10.1002/psp.2020

Kondo, M. C., Rivera, R., & Rullman, S. (2012). Protecting the idyll but not the environment: Second homes, amenity migration and rural exclusion in Washington State. *Landscape and Urban Planning, 106*, 174-182. doi:10.1016/j.landurbplan.2012.03.003

Kowalewski, J., & Schulze, S. (2010). *Die Struktur der Landwirtschaft in der Metropolregion Hamburg [The structure of the agriculture sector of the metropolitan area of Hamburg]*. Hamburg: Hamburgisches WeltWirtschaftsInstitut(HWWI).

Kuechler, S; (1993) *Landscape as Memory*. In: Bender, B, (ed.) Landscape: Politics and Perspectives. (pp. 85-106). Berg: Oxford.

Kühne, O. (2011). Heimat und sozial nachhaltige Landschaftsentwicklung [Home and socail sustainable development of landscapes]. *Raumforschung und Raumordnung, 69*, 291-301. doi:10.1007/s13147-011-0108-0

Kupiszewskiw, M., & Rey, V. (1996). Future of East-West migration in Europe. In F. W. Carter & P. Jordan (Eds.), *Central Europe after the fall of the Iron Curtain: Geopolitical perspectives, spatial patterns and trends* (pp. 247-268). Frankfurt am Main, Germany: Peter Lang, Europäischer Verlag der Wissenschaften.

Kyle, G., Graefe, A., Manning, R., & Bacon, J. (2004). Effects of place attachment on users' perceptions of social and environmental conditions in a natural setting. *Journal of Environmental Psychology, 24*, 213-225. doi:10.1016/j.jenvp.2003.12.006

Lalli, M. (1992). Urban-Related Identity: Theory, Measurement, and Empirical Findings. *Journal of Environmental Psychology, 12*, 285-303. doi:10.1016/s0272-4944(05)80078-7

Lankia, T., Neuvonen, M., Pouta, E., & Sievänen, T. (2014). Willingness to contribute to the management of recreational quality on private lands in Finland. *Journal of Forest Economics, 20*(2), 141-160. doi:10.1016/j.jfe.2014.04.001

Lansing, J., & Mueller, E. (1967). *The geographic mobility of labor*. Michigan: Ann Arbor, Survey Research Center, University of Michigan.

Lee, E. (1966). A Theory of Migration. *Demography, 3*(1), 47-57.

Levinson, A. (2012). Valuing Public Goods Using Happiness Data: The Case of Air Quality. *Journal of Public Economics, 96*, 869-880. doi:10.1016/j.jpubeco.2012.06.007

Lin, C., & Lockwood, M. (2014). Forms and sources of place attachment: Evidence from two protected areas. *Geoforum, 53*, 74-81. doi:10.1016/j.geoforum.2014.02.008

Lorah, P. (2000). Population growth, economic security, and cultural change. In S. F. McCool, D. N. Cole, W. T. Borrie, & J. O'Loughlin (Eds.), *Wilderness science in a time of change conference - Volume 2: Wilderness within the context of larger systems* (pp. 230-237). Missoula, MT: U.S. Department of Agriculture, Forest Service, Rocky Mountain Research Station.

Lothian, A. (2014). Theories of Landscape Aesthetics. Retrieved August 15, 2017, from http://www.scenicsolutions.com.au/Attached%20PDFs/Theory%20of%20landscape%20quality.pdf

Mabogunje, A. L. (1970). Systems Approach to a Theory of Rural-Urban Migration. *Geographical Analysis, 2*(1), 1-18. doi:10.1111/j.1538-4632.1970.tb00140.x

Maciá, A. (1979). Visual Perception of Landscape: Sex and Personality Differences. In *Our National Landscape* (pp. 279-285). Berkely Pacific Southwest Faorest and Range Experiement Station.

Mackerron, G., & Mourato, S. (2013). Happiness is Greater in Natural Environments. *Global Environmental Change, 23*(5), 992-1000. doi:10.1016/j.gloenvcha.2013.03.010

Madden, T. J., Scholder Ellen, P., & Ajzen, I. (1992). A Comparison of the Theory of Planned Behavior and the Theory of Reasoned Action. *Personality and Social Psychology Bulletin, 18*(1), 3-9.

Mäler, K. (1974). *Environmental Economics: A Theoretical Inquiry*. New York, Oxan: Resources for the Future by the Johns Hopkins University Press.

Malinowski, B. (1926). *Crime and custom in savage society*. New York: Harcourt, Brace & Company, Inc.

Marangon, F., & Visintin, F. (2007). Rural landscape valuation in a cross-border region. *Cahiers D'économie Et Sociologie Rurales, 84-85*, 114-132.

Massey, D. S., Arango, J., Hugo, G., Kouaouci, A., Pellegrino, A., & Taylor, J. E. (1993). Theories of International Migration: A Review and Appraisal. *Population and Development Review, 19*(3), 431-466. doi:10.2307/2938462

Matloch, J., C. Blaumann, B. Böhm, M. Ferretti, S. Herrmann, D. Kempa, F. Knaps, L. Lange, I. Lüder, M. Schaffert, D. Schiller, P. Steffenhagen-Koch & F. Wenger (2016). Bevölkerungsbefragung zur Kulturlandschaft im Projekt Regiobranding - Beschreibende Ergebnisse aus den Fokusregionen Lübeck-Nordwestmecklenburg, Steinburger Elbmarschen und Griese Gegend-Elbe-Wendland [Residents´ survey about the cultural landscapes in the project Regionbranding – Descriptive results for the regions Lübeck-Nordwestmecklenburg, Steinburger Elbmarschen and Griese Gegend-Elbe-Wendland] - Farbversion, Arbeitspapier Nr. 01a.

Mayer, M., & Job, H. (2014). The Economics of Protected Areas - A European Perspective. *Zeitschrift Für Wirtschaftsgeographie, 58*(2-3), 73-97. doi:10.1515/zfw.2014.0006

McElhinny, B. (2006). Written in Sand: Language and landscape in an environmental dispute in southern Ontario. *Critical Discourse Studies, 3*(2), 123-152. doi:10.1080/17405900600908087

McFadden, D. (1998). Measuring Willingness-to-Pay for Transportation Improvements. *Theoretical Foundations of Travel Choice Modeling*, 339-364.

McGranahan, D. A. (1999). *Natural amenities drive rural population change*. Washington, D.C.: U.S. Department of Agriculture, Food and Rural Economics Division, Economic Research Service.

Measham, T. G., & Barnett, G. B. (2008). Environmental Volunteering: Motivations, Modes and Outcomes. *Australian Geographer, 39*(4), 537-552. doi:10.1080/00049180802419237

Meier, C., Bucher, A., & Hagenbuch, R. (2010). Landschaft, Landschaftsbewusstsein und landschaftliche Identität als Potenziale für die regionale Entwicklung - Eine empirische Fallstudie in Glarus Süd, Schweiz [Landscape, Landscape Awareness, and Landscape Identity as potential for Regional Development – An Empirical Case Study in Southern Glarus, Switzerland]. *GAIA - Ecological Perspectives for Science and Society, 19*(3), 213-222.

Meinig, D. W. (1979). The Beholding Eye: Ten Versions of the Same Scene. In: Meinig, D. W., & Jackson, J. B. (Eds.). *The Interpretation of everyday landscapes: Geographical essays* (pp. 33-48). New York: Oxford University Press.

Menchik, P. L., & Weisbrod, B. A. (1987). Volunteer Labor Supply. *Journal of Public Economics, 32*, 159-183. doi:10.1016/0047-2727(87)90010-7

Mendham, E., & Curtis, A. (2010). Taking Over the Reins: Trends and Impacts of Changes in Rural Property Ownership. *Society & Natural Resources, 23*(7), 653-668. doi:10.1080/08941920801998893

Miles, I., Sullivan, W. C., & Kuo, F. E. (1998). Ecological Restoration Volunteers: The Benefits of Participation. *Urban Ecosystems, 2*, 27-41. doi:10.1023/A:1009501515335

Mincer, J., & Polachek, S. (1974). Family Investments in Human Capital: Earnings of Women. *Journal of Political Economy, 82*(2, Part 2), 76-103. doi:10.1086/260293

Mitchell, R. C., & Carson, R. T. (1989). *Using surveys to value public goods: The contingent valuation method.* Washington, D.C.: Resources for the Future.

Molina, J. R., Silva, F. R., & Herrera, M. Á. (2016). Integrating economic landscape valuation into Mediterranean territorial planning. *Environmental Science & Policy, 56*, 120-128. doi:10.1016/j.envsci.2015.11.010

Moro, M., Brereton, F., Ferreira, S., & Clinch, J. P. (2008). Ranking quality of life using subjective well-being data. *Ecological Economics, 65*, 448-460. doi:10.1016/j.ecolecon.2008.01.003

Morrison, M., & Dowell, D. (2013). Sense of Place and Willingness to Pay: Complementary Concepts When Evaluating Contributions of Cultural Resources to Regional Communities. *Regional Studies, 49*(8), 1374-1386. doi:10.1080/00343404.2013.827335

Naudiet, R., Arlt, K., Jansen, U., & Maiwald, D. (1994). *Atlas des Kreises Steinburg: Innenansichten einer Region (Atlas of Steinburg: internal point of views of a region).* Münsterdorf: Hansen und Hansen.

Neill, H. R., Cummings, R. G., Ganderton, P. T., Harrison, G. W., & McGuckin, T. (1994). Hypothetical Surveys and Real Economic Commitments. *Land Economics, 70*(2), 145-154. doi:10.2307/3146318

Nettle, D. (2005). *Happiness: the science behind your smile*. Oxford: Oxford University Press.

Ní Laoire, C. (2000). Conceptualising Irish Rural Youth Migration: A Biographical Approach. *International Journal of Population Geography, 6*(3), 229-243. doi:10.1002/1099-1220(200005/06)6:3<229::aid-ijpg185>3.0.co;2-r

Ní Laoire, C. (2007). The "Green Green Grass of Home"? Return Migrants to Rural Ireland. *Journal of Rural Studies, 23*, 332-344. doi:10.1016/j.jrurstud.2007.01.005

Nohl, W. (2006). Heimat als symbolischer Aneignungsprozess - Konzeptionelle Überlegungen und empirsche Untersuchungen ["Home" as Symbolic Process of Acquisition - Conceptual considerations and empirical investigations]. *Naturschutz und Landschaftsplanung, 38*, 140-145.

Nohl, W. (2009). Landschaftsästhetisches Erleben - Grundformen und ihre nachhaltige Wirkung [Landscape aesthetic experiences – Basic forms and its sustainable impact]. Retrieved January 9, 2017, from http://www.wanderforschung.de/files/nohl-landschaftssaethetik1375881539.pdf

Organisation for Economic Co-operation and Development (OECD) Better Life Index. (n.d.). Life Satisfaction. Retrieved August 14, 2017, from http://www.oecdbetterlifeindex.org/topics/life-satisfaction/

Organisation for Economic Co-operation and Development (OECD). (1994). *The Contribution of Amenities to Rural Development*. Paris.

Overdevest, C., Orr, C. H., & Stepenuck, K. (2004). Volunteer Stream Monitoring and Local Participation in Natural Resource Issues. *Research in Human Ecology, 11*(2), 177-185.

Paasi, A. (2002). Bounded spaces in the mobile world: Deconstructing 'regional identity' *Tijdschrift Voor Economische En Sociale Geografie, 93*(2), 137-148. doi:10.1111/1467-9663.00190

Palang, H., Printsmann, A., Gyuró, É K., Urbanc, M., Skowronek, E., & Woloszyn, W. (2006). The Forgotten Rural Landscapes of Central and Eastern Europe. *Landscape Ecology, 21*(2), 347-357. doi:10.1007/s10980-004-5034-x

Palmore, E., & Luikart, C. (1972). Health and Social Factors Related to Life Satisfaction. *Journal of Health and Social Behavior, 13*, 68-80.

Peacock, A. (1992). Economics, Cultural Values and Cultural Policies. In R. Towse & A. Khakee (Eds.), *Cultural Economics* (pp. 9-20). Berlin-Heidelberg: Springer Verlag.

Peck, J., & Thoedore, N. (2007). Variegated Capitalism. *Progress in Human Geography, 31*(6), 731-772. doi:10.1177/0309132507083505

Petersen, W. (1958). A General Typology of Migration. *American Sociological Review, 23*(3), 256-266.

Price, D., & Sikes, M. (1975). *Rural-Urban Migration Research in the United States*. Washington, D.C.: U.S. Government Printing Office.

Projektgebiet. (n. d.). Retrieved August 04, 2016, from http://regiobranding.de/content/das-projektgebiet

Proshansky, H. M., Fabian, A. K., & Kaminoff, R. (1982). Place-identity: Physical world socialization of the self. *Journal of Environmental Psychology, 3*(1), 57-83. doi:10.1016/S0272-4944(83)80021-8

Pruckner, G. J. (1995). Der kontingente Bewertungsansatz zur Messung von Umweltgütern: Stand der Debatte und umweltpolitische Einsatzmöglichkeiten [The contingent valuation method to measure environmental goods:state-of-the-art of the discussion and possible ecopolitical applications]. *Zeitschrift für Umweltpolitik & Umweltrecht: ZfU ; Beiträge zur rechts-, wirtschafts- und sozialwissenschaftlichen Umweltforschung, 18*(4), 1-31.

Purcell, A. T. (1992). Abstract and specific physical attributes and the experience of landscape. *Journal of Environmental Management, 34*, 159-177. doi:10.1016/s0301-4797(05)80149-5

Rai, R. K., & Scarborough, H. (2013). Economic value of mitigation of plant invaders in a subsistence economy: Incorporating labour as a mode of payment. *Environment and Development Economics, 18*(02), 225-244. doi:10.1017/s1355770x1200037x

Rai, R. K., & Scarborough, H. (2014). Nonmarket valuation in developing countries: Incorporating labour contributions in environmental benefits estimates. *Australian Journal of Agricultural and Resource Economics, 59*, 479-498. doi:10.1111/1467-8489.12071

Randall, A. (1998). Beyond the crucial experiment: Mapping the performance characteristics of contingent valuation. *Resource and Energy Economics, 20*, 197-206. doi:10.1016/s0928-7655(97)00036-5

Rasker, R., & Hansen, A. (2000). Natural amenities and population growth in the greater yellowstone region. *Human Ecology Review, 7*(2), 30-40.

Ray, C. (1998). Culture, Intellectual Property and Territorial Rural Development. *Sociologia Ruralis, 38*(1), 3-20. doi:10.1111/1467-9523.00060

Rehdanz, K., & Maddison, D. (2008). Local environmental quality and life-satisfaction in Germany. *Ecological Economics, 64*, 787-797. doi:10.1016/j.ecolecon.2007.04.016

Reichert, C. Von., Cromartie, J. B., & Arthun, R. O. (2012). Reasons for Returning and Not Returning to Rural U.S. Communities. *The Professional Geographer, 66*(1), 58-72. doi:10.1080/00330124.2012.725373

Reichert, C. Von., Cromartie, J. B., & Arthun, R. O. (2014). Impacts of Return Migration on Rural U.S. Communities. *Rural Sociology, 79*(2), 200-226. doi:10.1111/ruso.12024

Ritchey, P. N. (1976). Explnations of Migration. *Annual Review of Sociology, 2*, 363-404.

Ritter, J. (1962). *Landschaft : zur Funktion der Aesthetik in der modernen Gesellschaft [Landscape: the function of aesthetic in the modern society]*. Speech presented at the assumption of the rector position of the Westfaelische Wilhelms-University in Muenster, Germany.

Roesch-McNally, G. E., & Rabotyagov, S. S. (2015). Paying for Forest Ecosystem Services: Voluntary Versus Mandatory Payments. *Environmental Management, 57*(3), 585-600. doi:10.1007/s00267-015-0641-7

Roschewitz, A. (1999). *Der monetäre Wert der Kulturlandschaft: Eine Contingent-Valuation-Studie [The monetary value of tha culture landscape: A ontingent valuation study]*. Kiel: Wissenschaftsverlag. Vauk.

Rudzitis, G. (1999). Amenities Increasingly Draw People to Rural West. *Rural Development Perspectives, 14*(2), 9-13.

Ryan, A., & Spash, C. (2011). Is WTP an attitudinal measure? Empirical analysis of the psychological explanation for contingent values. *Journal of Economic Psychology, 32*, 674-687. doi:10.1016/j.joep.2011.07.004

Ryan, R., & Deci, E. (2001). On Happiness and Human Potentials: A Review of Research on Hedonic and Eudaimonic Well-Being. *Annual Review of Psychology, 52*, 141-166.

Sajardo, A., & Serra, I. (2011). The Economic Value of Volunteer Work: Methodological Analysis and Application to Spain. *Nonprofit and Voluntary Sector Quarterly, 40*(5), 873-895. doi:10.1177/0899764010371233

Salamon, L. M., Sokolowski, S. W., & Haddock, M. A. (2011). Measuring the Economic Value Of Volunteer Work Globally: Concepts, Estimates, And A Roadmap to the Future. *Annals of Public and Cooperative Economics, 82*(3), 217-252. doi:10.1111/j.1467-8292.2011.00437.x

Samuelson, P. A. (1954). Pure Theory of Public Expenditure and Taxation. *The Review of Economics and Statistics, 36*(4), 387-389. doi:10.2307/1925895

Santos, J. M. (2001). *A Synthesis of country Reports on Demand Measurement of Non-Commodity Outputs in OECD Agriculture*. Organisation for Economic Co-operation and Development (OECD).

Sauer, C. O. (1962). Seashore - Primitive Home of Man? *Proceedings of the American Philosophical Society, 106*(1), 41-47.

Scheer, C. (2014). Bauernhäuser in den holsteinischen Elbmarschen als Ausdruck der Siedlungs- und Agrargeschichte [Farmhouses in the Holstein Elbe Marsh as expression of the settlement and agricultural history]. In C. Heuter, M. Schimek, & C. Vorwig (Eds), *Bauern-, Herren-, Fertighäuser - Hausforschung als Sozialgeschichte (Farm, manor, prefab houses – Research of houses a spart of the social history)* (pp. 123-140). Münster: Waxmann.

Schefold, B. (1994). *Wirtschaftsstile (Economic styles)*. Frankfurt am Main: Fischer Taschenbuch Verlag.

Schein, R. H. (1997). The Place of Landscape: A Conceptual Framework for Interpreting an American Scene. *Annals of the Association of American Geographers, 87*(4), 660-680. doi:10.1111/1467-8306.00072

Schervish, P. G., & Havens, J. J. (1997). Social participation and charitable giving: A multivariate analysis. *Voluntas: International Journal of Voluntary and Nonprofit Organizations, 8*(3), 235-260. doi:10.1007/bf02354199

Schiappacasse, I., Vásquez, F., Nahuelhual, L., & Echeverría, C. (2013). Labor as a Welfare Measure in Contingent Valuation: The Value of a Forest Restoration Project. *Ciencia E Investigación Agraria, 40*(1), 69-84. doi:10.4067/s0718-16202013000100006

Schindler-Rainman, E., & Lippitt, R. (1971). *The volunteer community: creative use of human resources*. Michigan: Center for a Voluntary Society.

Schkade, D., & Payne, J. (1994). How People Respond to Contingent Valuation Questions: A Verbal Protocol Analysis of Willingness to Pay for an Environmental Regulation. *Journal of Environmental Economics and Management, 26*(1), 88-109. doi:10.1006/jeem.1994.1006

Schmitz, K., Schmitz, P. M., & Wronka, T. W. (2003). Bewertung von Landschaftsfunktionen mit Choice Experiments [Valuation of landscape functions using choice experiments]. *Argarwirtschaft, 52*(8), 379-389.

Schultz, P. W., & Zelezny, L. (1999). Values As Predictors Of Environmental Attitudes: Evidence For Consistency Across 14 Countries. *Journal of Environmental Psychology, 19*, 255-265. doi:10.1006/jevp.1999.0129

Schwandt, H. (2013). *Unmet aspirations as an explanation for the age U-shape in human wellbeing* (No. 7604). Bonn, Germany: IZA.

Schwarck, J. (2010). Empörung über Regenwasser-Gebühr | shz.de. Retrieved September 09, 2016, from http://www.shz.de/lokales/wilstersche-zeitung/empoerung-ueber-regenwasser-gebuehr-id1918971.html

Shaw, R. P. (1975). *Migration theory and fact: a review and bibliography of current literature*. Philadelphia, PA: Regional Science Research Institute.

Sheppard, E. (2011). Geographical Political Economy. *Journal of Economic Geography, 11*, 319-331. doi:10.1093/jeg/lbq049

Shumway, J. M., & Otterstrom, S. M. (2001). Spatial Patterns of Migration and Income Change in the Mountain West: The Dominance of Service-Based, Amenity-Rich Counties. *The Professional Geographer, 53*(4), 492-502. doi:10.1111/0033-0124.00299

Simmons, R. G. (1991). Presidential Address on Altruism and Sociology. *The Sociological Quarterly, 32*(1), 1-22. doi:10.1111/j.1533-8525.1991.tb00342.x

Simonson, J., Vogel, C., & Tesch-Römer, C. (2016). *Freiwilliges Engagement in Deutschland: Zentrale Ergebnisse des Deutschen Freiwilligensurveys 2014 [Voluntary Civic Engagement in Germany: central results of the German Volunteering survey 2014]*. Berlin: Ministry of Family Affairs, Senior Citizens, Women and Youth (BMFSFJ).

Smith, M. B., Bruner, J. S., & White, R. W. (1956). *Opinions and Personality*. New York: Wiley.

Smith, N. (1984). *Uneven Development: Nature, Capital, and the Production of Space*. Athens, GA: University of Georgia Press.

Sniehotta, F., Presseau, J., & Araújo-Soares, V. (2014). Editorial: Time to retire the theory of planned behaviour. *Health Psycholoy Review, 8*(1), 1-7. doi:10.1080/17437199.2013.869710

Soini, K., Vaarala, H., & Pouta, E. (2012). Residents' sense of place and landscape perceptions at the rural–urban interface. *Landscape and Urban Planning, 104*, 124-134. doi:10.1016/j.landurbplan.2011.10.002

Sørensen, J. (2014). Rural–Urban Differences in Life Satisfaction: Evidence from the European Union. *Regional Studies, 48*(9), 1451-1466.

Southey, G. (2011). The Theories of Reasoned Action and Planned Behavior Applied to Business Decisions: A Selective Annotated Bibliography. *Journal of New Business Ideas and Trends, 9*(1), 43-50.

Spash, C. L. (2008). Contingent valuation design and data treatment: If you can't shoot the messenger, change the message. *Environment and Planning C: Government and Policy, 26*, 34-53. doi:10.1068/cav4

Spring, A., Ackert, E., Crowder, K., & South, S. J. (2017). Influence of Proximity to Kin on Residential Mobility and Destination Choice: Examining Local Movers in Metropolitan Areas. *Demography*, 1-28. doi:10.1007/s13524-017-0587-x

Stamps, A. E. (2004). Mystery, complexity, legibility and coherence: A meta-analysis. *Journal of Environmental Psychology, 24*, 1-16. doi:10.1016/s0272-4944(03)00023-9

Stephenson, J. (2008). The Cultural Values Model: An integrated approach to values in landscapes. *Landscape and Urban Planning, 84*(2), 127-139. doi:10.1016/j.landurbplan.2007.07.003

Stone, K., Bhat, M., Bhatta, R., & Mathews, A. (2008). Factors influencing community participation in mangroves restoration: A contingent valuation analysis. *Ocean & Coastal Management, 51*, 476-484. doi:10.1016/j.ocecoaman.2008.02.001

SurveyMonkey Germany. (2014). An welchem Wochentag Du Deine Umfrage verschicken solltest? | SurveyMonkey Blog [At what day should you mail your survey? | SurveyMonkey Blog]. Retrieved August 04, 2016, from https://de.surveymonkey.com/blog/de/2014/08/12/welchem-wochentag-du-deine-umfrage-verschicken-solltest/

Swanwick, C. (2009). Society's attitudes to and preferences for land and landscape. *Land Use Policy, 26*, 62-75. doi:10.1016/j.landusepol.2009.08.025

Swanwick, C., Hanley, N., & Termansen, M. (2007). Scoping Study on Agricultural Landscape Valuation. Final Report to DEFRA. London.

Szerszynski, B., & Urry, J. (2006). Visuality, mobility and the cosmopolitan: inhabiting the world from afar. *The British Journal of Sociology, 57*(1), 113-131. doi:10.1111/j.1468-4446.2006.00096.x

Throsby, D. (2001). Introduction. In *Economics and Culture* (pp. 1-18). Cambridge: Cambridge University Press.

Throsby, D. (2003). Determining the Value of Cultural Goods: How Much (or How Little) Does Contingent Valuation Tell Us? *Journal of Cultural Economics, 27*, 275-285. doi:10.1023/A:1026353905772

Throsby, D. (2007). Regional Aspects of Heritage Economics: Analytical and Policy Issues. *Australasian Journal of Regional Studies, 13*(1), 21-30.

Tidwell, L. S., & Brunson, M. W. (2008). Volunteering to Manage Rangeland Weeds: Results of a Citizen Survey in the Southwestern United States. *Rangelands, 30*(4), 19-24. doi:10.2111/1551-501x(2008)30[19:vtmrw]2.0.co;2

Titmuss, R. M. (1971). *The gift relationship: from human blood to social policy*. London and New York: Pantheon Books a Division of Random House.

Tselios, V., Noback, I., McCann, P., & Dijk, J. V. (2015). Local social engagement, satisfaction, and embeddedness in the Netherlands: which effects matter and for whom? *Environment and Planning A, 47*, 1132-1154. doi:10.1177/0308518x15592301

Tuan, Y.-F. (1990). Topophillia - A Study of Environmental Perceptions, Attitudes and Values. New York, USA: Columbia University Press.

Tveit, M. S. (2009). Indicators of visual scale as predictors of landscape preference; a comparison between groups. *Journal of Environmental Management, 90*(9), 2882-2888. doi:10.1016/j.jenvman.2007.12.021

Tweed, C., & Sutherland, M. (2007). Built cultural heritage and sustainable urban development. *Landscape and Urban Planning, 83*, 62-69. doi:10.1016/j.landurbplan.2007.05.008

Ulrich, R. S. (1981). Natural Versus Urban Scenes: Some Psychophysiological Effects. *Environment and Behavior, 13*(5), 523-556. doi:10.1177/0013916581135001

Ulrich, R. S. (1983). Aesthetic and Affective Response to Natural Environment (I. Altman & J. Wohlwill, Eds.). *Human Behavior and Environment, 6*, Behavior and Natural Environment, 85-125. doi:10.1007/978-1-4613-3539-9_4

Unwin, K. I. (1975). The Relationship of Observer and Landscape in Landscape Evaluation. *Transactions of the Institute of British Geographers,* No. 66, 130-134. doi:10.2307/621627

Upham, P., Whitmarsh, L., Poortinga, W., Purdam, K., Darnton, A., McLachan, C., & Devine-Wright, P. (2009). Public Attitudes to Environmental Change: a selective review of theory and practice. United Kingdom: Living With Environmental Change Research Programme.

Vanderheyden, V., Horst, D. V., Rompaey, A. V., & Schmitz, S. (2013). Perceiving the Everyday: A Study of Everyday Landscapes in Belgium. *Tijdschrift Voor Economische En Sociale Geografie, 105*(5), 591-603. doi:10.1111/tesg.12066

Vaske, J. J., & Kobrin, K. C. (2001). Place Attachment and Environmentally Responsible Behavior. *Journal of Environmental Education, 32*(4), 16-21. doi:10.1080/00958960109598658

Vásquez, W. F. (2014). Willingness to pay and willingness to work for improvements of municipal and community-managed water services. *Water Resources Research, 50*(10), 8002-8014. doi:10.1002/2014wr015913

Veenhoven, R. (1996). Developments in Statisfaction-Research. *Social Indicators Research (Soc Indic Res), 37*, 1-46.

Veenhoven, R. (2004). Subjective measures of well-being. Wider Discussion Papers, 7.

Veenhoven, R. (2011). Glück als subjektives Wohlbefinden: Lehren aus der empirischen Forschung [Luck as subjective well-being: doctrines oft he empirical research]. In D. Thomä, C. Henning, & O. Mitscherlich-Schönherr (Eds.), *Glück: Ein interdisziplinäres Handbuch* (pp. 396-404). Stuttgart: Metzler.

Venkatachalam, L. (2004). The contingent valuation method: A review. *Environmental Impact Assessment Review*, 24, 89-124. doi:10.1016/S0195-9255(03)00138-0

Verbič, M., & Slabe-Erker, R. (2009). An Econometric Analysis Of Willingness-to-pay For Sustainable Development: A Case Study Of The Volčji Potok Landscape Area. *Ecological Economics, 68*, 1316-1328. doi:10.1016/j.ecolecon.2008.09.002

Vohs, K. D., Mead, N. L., & Goode, M. R. (2006). The Psychological Consequences of Money. *Science, 314*, 1154-1156. doi:10.1126/science.1132491

Völckner, F. (2006). Methoden zur Messung individueller Zahlungsbereitschaften: Ein Überblick zum State of the Art [Methods of measuring the individual Willingness to pay: an overview of the state-of-the art]. *Journal Für Betriebswirtschaft*, 56, 33-60.

Vondolia, G. K., Eggert, H., Navrud, S., & Stage, J. (2014). What do Respondents Bring to Contingent Valuation? A Comparison of Monetary and Labour Payment Vehicles. *Journal of Environmental Economics and Policy, 3*(3), 253-267. doi:10.1080/21606544.2014.892034

Vos, W., & Meekes, H. (1999). Trends in European cultural landscape development: perspectives for a sustainable future. *Landscape and Urban Planning, 46*, 3-14. doi:10.1016/s0169-2046(99)00043-2

Warnes, A. (1992). Migration and the life course. In A. Champion & A. Fielding (Eds.), *Migration processes and patterns: Research progress and prospects* (pp. 175-187). London: Belhaven Press.

Waterman, A. (1993). Two conceptions of happiness: Contrasts of personal expressiveness (eudaimonia) and hedonic enjoyment. *Journal of Personality and Social Psychology, 64*(4), 678-691.

Webb, D. J., Green, C. L., & Brashear, T. G. (2000). Development and Validation of Scales to Measure Attitudes Influencing Monetary Donations to Charitable Organizations. *Journal of the Academy of Marketing Science, 28*(2), 299-309. doi:10.1177/0092070300282010

Weichhart, P., Weiske, C., & Werlen, B. (2006). *Place Identity und Images: Das Beispiel Eisenhüttenstadt [Place identiy and images: the example of Eisenhuettenstadt]*. Wien: Institut für Geographie und Regionalforschung der Universität Wien.

Welsch, H. (2006). Environment and happiness: Valuation of air pollution using life satisfaction data. *Ecological Economics, 58*, 801-813.

Welsch, H. (2007). Environmental welfare analysis: A life satisfaction approach. *Ecological Economics, 62*, 544-551.

Welsch, H. (2009). Implications of happiness research for environmental economics. *Ecological Economics, 68*, 2735-2742.

Wilcox, S., Castro, C., King, A. C., Hausmann, R., & Brownson, R. C. (2000). Determinants of leisure time physical activity in rural compared with urban older and ethnically diverse women in the United States. *Journal of Epidemiology and Community Health, 54*, 667-672. doi:10.1136/jech.54.9.667

Williams, J. D., & Sofranko, A. J. (1979). Motivations for the Inmigration Component of Population Turnaround in Nonmetropolitan Areas. *Demography, 16*(2), 239-255. doi:10.2307/2061141

Wilson, J. (2000). Volunteering. *Annual Review of Sociology, 26*, 215-240. doi:10.1146/annurev.soc.26.1.215

Wolpert, J. (1965). Behavioral Aspects Of The Decision To Migrate. *Regional Science, 15*(1), 159-169. doi:10.1111/j.1435-5597.1965.tb01320.x

World Cultural and Natural Heritage. (2015). *Operational Guidelines for the Implementation of the World Heritage Convention*. Paris: World Heritage Centre.

Yu, K. (1995). Cultural variations in landscape preference: comparisons among Chinese sub-groups and Western design experts. *Landscape and Urban Planning, 32*, 107-126. doi:10.1016/0169-2046(94)00188-9

Zäch, C., Schulz, T., Waltert, F., & Pütz, M. (2015). *Landscape and Amenity Migration*. Birmensdorf: Eidg. Forschungsanstalt für Wald, Schnee und Landschaft (WSL).

Zajonc, R. B. (1980). Feeling and thinking: Preferences need no inferences. *American Psychologist, 35*(2), 151-175. doi:10.1037//0003-066x.35.2.151

Ziele. (n. d.). Retrieved August 04, 2016, from http://regiobranding.de/content/ziele

Zube, E. H., Sell, J. L., & Taylor, J. G. (1982). Landscape Perception: Research, Application and Theory. *Landscape Planning*, *9*(1), 1-33. doi:10.1016/0304-3924(82)90009-0

Appendix: Further Tables and Figures

Tab. A 1: Coefficients of the estimation of the logit model for the water landscape (coefficients, standard errors in parentheses)

	All	LNWM	LPLD	STB
Gender	-0.07	0.30	-0.05	-0.40
	(0.14)	(0.34)	(0.21)	(0.27)
Age	**-0.02*** **	**-0.03*** **	-0.01	**-0.03*** **
	(0.00)	(0.01)	(0.01)	(0.01)
Education	0.05	0.05	0.04	0.08
	(0.07)	(0.16)	(0.10)	(0.12)
Children	0.03	0.07	0.04	-0.00
	(0.06)	(0.14)	(0.08)	(0.10)
Income	**-0.00***	-0.00	-0.00	-0.00
	(0.00)	(0.00)	(0.00)	(0.00)
Daily relaxing	0.11	0.10	0.20	0.09
	(0.15)	(0.36)	(0.21)	(0.27)
Daily sports	0.15	-0.05	0.26	-0.06
	(0.17)	(0.42)	(0.25)	(0.31)
Occupational use	**-0.38** **	0.54	-0.34	**-0.83** **
	(0.19)	(0.52)	(0.25)	(0.38)
Rural environment	0.13	0.07	0.22	0.18
	(0.14)	(0.36)	(0.21)	(0.25)
Returnee	0.22	0.38	0.30	0.12
	(0.21)	(0.56)	(0.29)	(0.41)
Newcomer	**0.43** **	0.55	**0.59** **	0.01
	(0.17)	(0.40)	(0.24)	(0.32)
Bond to nature	**0.47*** **	0.09	0.34	**0.95*** **
	(0.15)	(0.34)	(0.22)	(0.28)
Bond to region	0.11	0.56	0.24	-0.32
	(0.15)	(0.35)	(0.22)	(0.28)
Constant	0.39	0.75	-0.39	**1.45** **
	(0.34)	(0.80)	(0.51)	(0.61)
Pseudo R-squared	0.03	0.06	0.03	0.06
N	921.00	179.00	438.00	304.00

Note. Stars indicate significance levels: * p<0.1, ** p<0.05, *** p<0.01. N displays the number of cases.
Source: Resident survey, own calculations.

J. Matloch, *The Assessment of German Cultural Landscapes*, RaumFragen: Stadt – Region – Landschaft, https://doi.org/10.1007/978-3-658-21416-6

Tab. A 2: Coefficients of the estimation of the logit model for the city landscape (coefficients, standard errors in parentheses)

	All	LNWM	LPLD	STB
Gender	-0.09	0.34	-0.04	**-0.46***
	(0.15)	(0.35)	(0.21)	(0.27)
Age	**-0.01***	-0.01	-0.00	-0.01
	(0.00)	(0.01)	(0.01)	(0.01)
Education	0.03	0.13	-0.04	0.05
	(0.07)	(0.16)	(0.10)	(0.12)
Children	0.06	**0.29***	0.05	-0.03
	(0.06)	(0.15)	(0.09)	(0.10)
Income	**-0.00***	-0.00	-0.00	-0.00
	(0.00)	(0.00)	(0.00)	(0.00)
Daily relaxing	0.04	0.17	0.33	-0.35
	(0.15)	(0.36)	(0.21)	(0.28)
Daily sports	0.24	0.37	0.17	0.08
	(0.17)	(0.41)	(0.25)	(0.32)
Occupational use	**-0.36***	0.59	-0.41	**-1.01****
	(0.20)	(0.50)	(0.27)	(0.45)
Rural environment	**-0.50*****	-0.59	**-0.56*****	-0.35
	(0.14)	(0.38)	(0.21)	(0.26)
Returnee	**0.50****	**1.05***	**0.63****	-0.03
	(0.22)	(0.58)	(0.30)	(0.44)
Newcomer	**0.53*****	0.58	**0.80*****	-0.01
	(0.17)	(0.44)	(0.25)	(0.33)
Bond to nature	**0.44*****	0.22	0.25	**0.92*****
	(0.15)	(0.34)	(0.22)	(0.28)
Bond to region	0.03	0.10	**0.39***	**-0.55****
	(0.15)	(0.36)	(0.23)	(0.27)
Constant	-0.25	-0.85	-0.75	0.88
	(0.34)	(0.81)	(0.51)	(0.58)
Pseudo R-squared	0.03	0.08	0.05	0.07
N	921.00	179.00	438.00	304.00

Note. Stars indicate significance levels: * p<0.1, ** p<0.05, *** p<0.01. N displays the number of cases.

Source: Resident survey, own calculations.

Tab. A 3: Coefficients of the estimation of the logit model for the village landscape (coefficients, standard errors in parentheses)

	All	LNWM	LPLD	STB
Gender	-0.13	0.21	-0.15	**-0.46***
	(0.14)	(0.35)	(0.21)	(0.27)
Age	**-0.02*****	**-0.02***	**-0.02*****	**-0.03*****
	(0.00)	(0.01)	(0.01)	(0.01)
Education	-0.05	-0.07	-0.13	0.07
	(0.07)	(0.16)	(0.10)	(0.12)
Children	0.07	**0.30****	0.08	-0.04
	(0.06)	(0.15)	(0.09)	(0.10)
Income	**-0.00****	**-0.00***	-0.00	-0.00
	(0.00)	(0.00)	(0.00)	(0.00)
Daily relaxing	0.01	0.44	0.07	-0.19
	(0.15)	(0.36)	(0.21)	(0.27)
Daily sports	0.19	0.19	0.28	-0.05
	(0.17)	(0.41)	(0.26)	(0.32)
Occupational use	-0.24	0.66	-0.30	**-0.67***
	(0.19)	(0.50)	(0.25)	(0.38)
Rural environment	**0.45*****	0.04	0.33	**0.77*****
	(0.14)	(0.37)	(0.21)	(0.26)
Returnee	0.19	0.34	0.20	0.08
	(0.21)	(0.55)	(0.30)	(0.41)
Newcomer	**0.31***	0.17	0.31	0.34
	(0.17)	(0.41)	(0.24)	(0.32)
Bond to nature	**0.59*****	0.11	**0.46****	**1.15*****
	(0.15)	(0.35)	(0.22)	(0.27)
Bond to region	0.11	0.16	0.28	-0.23
	(0.15)	(0.36)	(0.23)	(0.28)
Constant	0.50	0.26	0.69	0.71
	(0.34)	(0.81)	(0.52)	(0.59)
Pseudo R-squared	0.04	0.06	0.04	0.09
N	921.00	179.00	438.00	304.00

Note. Stars indicate significance levels: * p<0.1, ** p<0.05, *** p<0.01. N displays the number of cases.

Source: Resident survey, own calculations.

Tab. A 4: Coefficients of the estimation of the logit model for at least one of the regional landscapes (coefficients, standard errors in parentheses)

WTP	All	Locals	Returnees	Newcomer
Gender	0.08	0.16	-0.18	0.14
	(0.25)	(0.47)	(0.85)	(0.36)
Under 25 years	0.22	0.64	0.00	0.57
	(0.62)	(1.01)	(.)	(1.03)
Between 25 and 65 years	0.33	0.21	-0.81	0.63
	(0.34)	(0.62)	(1.06)	(0.47)
Number of children	**-0.20****	-0.10	-0.22	**-0.34****
	(0.10)	(0.18)	(0.41)	(0.14)
Higher education	**0.90*****	0.45	**2.02****	**1.05*****
	(0.28)	(0.59)	(0.87)	(0.38)
Bond to region	-0.25	0.56	-0.31	-0.58
	(0.25)	(0.54)	(0.83)	(0.36)
Use of nature	-0.02	-0.07	0.97	-0.33
	(0.25)	(0.45)	(0.81)	(0.35)
Volunteering	-0.06	-0.10	0.44	0.14
	(0.26)	(0.47)	(0.84)	(0.38)
Constant	-0.11	-0.70	-0.26	0.02
	(0.42)	(0.86)	(1.31)	(0.58)
Pseudo R-squared	0.06	0.03	0.18	0.11
N	293.00	92.00	36.00	163.00

Note. Stars indicate significance levels: * p<0.1, ** p<0.05, *** p<0.01. N displays the number of cases.
Source: Resident survey, own calculations.

Tab. A 5: Coefficients of the estimation of the logit model for the farmhouses (coefficients, standard errors in parentheses)

Farmhouses	All	Locals	Returnees	Newcomer
Gender	-0.05	-0.34	-1.24	0.28
	(0.27)	(0.52)	(1.04)	(0.36)
Under 25 years	0.24	0.97	0,00	0.89
	(0.62)	(1.07)	*(omitted)*	(1.04)
Between 25 and 65 years	0.04	0.13	-0.72	0.11
	(0.37)	(0.69)	(1.26)	(0.49)
Number of children	-0.14	-0.06	-0.27	-0.22
	(0.10)	(0.20)	(0.43)	(0.14)
Higher education	**0.98*****	0.40	**2.55****	**1.15*****
	(0.28)	(0.62)	(1.07)	(0.37)
Bond to region	-0.34	0.69	-0.74	-0.57
	(0.26)	(0.61)	(0.93)	(0.37)
Use of nature	-0.09	-0.46	0.99	-0.17
	(0.26)	(0.48)	(0.91)	(0.36)
Volunteering	-0.02	-0.05	1.15	-0.01
	(0.27)	(0.51)	(0.96)	(0.39)
Constant	-0.49	-1.13	-0.86	-0.48
	(0.44)	(0.95)	(1.38)	(0.59)
Pseudo R-squared	0.06	0.05	0.25	0.10
N	293.00	92.00	36.00	163.00

Note. Stars indicate significance levels: * p<0.1, ** p<0.05, *** p<0.01. N displays the number of cases.
Source: Resident survey, own calculations.

Tab. A 6: Coefficients of the estimation of the logit model for the marshland (coefficients, standard errors in parentheses)

Marshland	All	Locals	Returnees	Newcomer
Gender	0.37	-0.17	0.39	**0.69***
	(0.26)	(0.51)	(0.83)	(0.37)
Under 25 years	0.49	1.61	0,00	0.91
	(0.62)	(1.14)	*(omitted)*	(1.02)
Between 25 and 65 years	0.19	0.51	-0.35	0.09
	(0.36)	(0.70)	(1.10)	(0.48)
Number of children	**-0.17***	-0.07	-0.16	**-0.25***
	(0.10)	(0.19)	(0.39)	(0.14)
Higher education	**0.51***	0.30	**1.65****	0.44
	(0.28)	(0.64)	(0.83)	(0.37)
Bond to region	0.24	**1.78****	-0.18	-0.25
	(0.26)	(0.72)	(0.82)	(0.37)
Use of nature	-0.25	-0.43	0.53	-0.43
	(0.25)	(0.48)	(0.79)	(0.36)
Volunteering	-0.25	-0.44	-0.03	-0.18
	(0.27)	(0.51)	(0.81)	(0.39)
Constant	-0.69	**-1.99***	-0.77	-0.45
	(0.44)	(1.03)	(1.34)	(0.59)
Pseudo R-squared	0.04	0.09	0.12	0.07
N	293.00	92.00	36.00	163.00

Note. Stars indicate significance levels: * p<0.1, ** p<0.05, *** p<0.01. N displays the number of cases.
Source: Resident survey, own calculations.

Tab. A 7: Coefficients of the estimation of the logit model for the water landscape (coefficients, standard errors in parentheses)

Water	All	Locals	Returnees	Newcomer
Gender	-0.01	-0.39	-0.22	0.28
	(0.25)	(0.49)	(0.82)	(0.35)
Under 25 years	0.47	1.06	0,00	0.65
	(0.62)	(1.06)	*(omitted)*	(1.02)
Between 25 and 65 years	0.29	0.56	-0.88	0.40
	(0.35)	(0.66)	(1.05)	(0.47)
Number of children	-0.16	-0.08	-0.04	**-0.28****
	(0.10)	(0.18)	(0.38)	(0.14)
Higher education	**0.91*****	**1.03***	**1.61***	**0.97*****
	(0.28)	(0.62)	(0.82)	(0.37)
Bond to region	-0.07	**0.99***	-0.56	-0.41
	(0.25)	(0.59)	(0.81)	(0.36)
Use of nature	0.03	0.12	0.70	-0.25
	(0.25)	(0.46)	(0.77)	(0.34)
Volunteering	-0.21	-0.67	0.01	0.12
	(0.26)	(0.50)	(0.80)	(0.37)
Constant	-0.30	-1.17	0.04	-0.25
	(0.43)	(0.93)	(1.29)	(0.57)
Pseudo R-squared	0.06	0.07	0.12	0.09
N	293.00	92.00	36.00	163.00

Note. Stars indicate significance levels: * p<0.1, ** p<0.05, *** p<0.01. N displays the number of cases.
Source: Resident survey, own calculations.

Tab. A 8: Coefficients of the estimation of the logit model for the city landscape (coefficients, standard errors in parentheses)

City	All	Locals	Returnees	Newcomer
Gender	-0.22	0.19	-1.10	-0.26
	(0.28)	(0.53)	(1.14)	(0.37)
Under 25 years	0.44	0.51	0,00	1.21
	(0.63)	(1.07)	*(omitted)*	(1.04)
Between 25 and 65 years	-0.20	-0.17	-0.07	-0.25
	(0.38)	(0.71)	(1.57)	(0.49)
Number of children	-0.16	-0.07	-0.41	-0.16
	(0.11)	(0.21)	(0.49)	(0.15)
Higher education	**0.82*****	0.77	1.48	**0.97****
	(0.29)	(0.61)	(1.10)	(0.38)
Bond to region	-0.39	0.14	-0.13	-0.60
	(0.28)	(0.60)	(0.99)	(0.40)
Use of nature	-0.19	-0.33	-1.57	0.01
	(0.28)	(0.50)	(1.06)	(0.37)
Volunteering	-0.04	0.60	0.87	-0.35
	(0.28)	(0.52)	(1.15)	(0.41)
Constant	-0.37	-1.22	-0.52	-0.24
	(0.46)	(0.95)	(1.61)	(0.60)
Pseudo R-squared	0.06	0.08	0.15	0.09
N	293.00	92.00	36.00	163.00

Note. Stars indicate significance levels: * p<0.1, ** p<0.05, *** p<0.01. N displays the number of cases.
Source: Resident survey, own calculations.

Tab. A 9: Coefficients of the estimation of the logit model for the village landscape (coefficients, standard errors in parentheses)

Village	All	Locals	Returnees	Newcomer
Gender	0.02	0.24	0.64	-0.19
	(0.26)	(0.50)	(0.86)	(0.35)
Under 25 years	0.42	0.69	0,00	0.85
	(0.62)	(1.06)	*(omitted)*	(1.03)
Between 25 and 65 years	0.20	-0.08	0.63	0.30
	(0.36)	(0.66)	(1.29)	(0.47)
Number of children	**-0.19***	-0.07	-0.11	**-0.27***
	(0.10)	(0.19)	(0.39)	(0.14)
Higher education	**0.87*****	0.89	0.51	**1.06*****
	(0.28)	(0.63)	(0.80)	(0.37)
Bond to region	0.05	**1.30****	0.31	-0.29
	(0.26)	(0.66)	(0.87)	(0.36)
Use of nature	-0.23	-0.26	-0.00	-0.18
	(0.25)	(0.47)	(0.80)	(0.35)
Volunteering	-0.05	0.35	-0.66	0.11
	(0.26)	(0.49)	(0.82)	(0.38)
Constant	-0.44	**-1.78***	-1.48	-0.14
	(0.44)	(0.96)	(1.57)	(0.57)
Pseudo R-squared	0.06	0.08	0.08	0.08
N	293.00	92.00	36.00	163.00

Note. Stars indicate significance levels: * p<0.1, ** p<0.05, *** p<0.01. N displays the number of cases.

Source: Resident survey, own calculations.

Tab. A 10: Results of the willingness to volunteer for each cluster divided into areas of volunteering and regional areas (in percent)

Cluster		Place development			Conservation of the environment			Animal protection			Art/ culture/ museum			Monument conservation			Local history		
		LNWM	LPLD	STB	LNWM	LPLD	STB	LNWM	LPLD	STB	LNWM	LPLD	STB	LNWM	LPLD	STB	LNWM	LPLD	STB
1	middle-aged men without any bond	23.81	22.00	15.22	***9.52***	28.00	***19.57***	***9.52***	28.00	***19.57***	4.76	10.00	8.70	**14.29**	8.00	**6.52**	***4.76***	8.00	6.52
2	highly educated young women	***14.29***	17.14	***5.26***	28.57	***22.86***	23.68	35.71	**34.29**	28.95	***0.00***	20.00	**18.42**	3.57	***5.71***	5.26	10.71	8.57	2.63
3	people bonded to the region	22.22	22.97	19.15	16.67	29.73	27.66	22.22	27.03	**29.79**	11.11	4.05	2.13	5.56	6.76	4.26	5.56	**13.51**	***0.00***
4	bonded high educated people	**31.58**	**34.38**	**22.73**	26.32	**40.63**	27.27	21.05	***9.38***	22.73	21.05	**21.88**	9.09	10.53	**9.38**	***0.00***	5.26	9.38	**13.64**
5	returnees bonded to the region	18.18	27.08	14.81	**50.00**	22.92	29.63	27.27	22.92	22.22	18.18	8.33	3.70	13.64	8.33	3.70	18.18	10.42	7.41
6	retired incomers	30.43	10.91	15.38	21.74	25.45	28.21	39.13	23.64	23.08	13.04	12.73	12.82	8.70	7.27	2.56	8.70	***3.64***	5.13
7	bonded to the region	20.00	***8.00***	12.50	20.00	24.00	**31.25**	**40.00**	16.00	25.00	**40.00**	***4.00***	***0.00***	***0.00***	8.00	6.25	**20.00**	12.00	6.25
All		22.60	19.72	15.59	26.03	28.06	25.86	28.08	25.00	23.57	12.33	10.28	9.51	8.22	7.22	9.51	9.59	9.17	6.46

Note: Bold values display the highest percentage of the region and the italic and bold ones the lowest ones.

Tab. A 11: Coefficients of the estimation of the cluster for the willingness to pay (coefficients, standard errors in parentheses)

	Cluster 1	Cluster 2	Cluster 3	Cluster 4	Cluster 5	Cluster 6	Cluster 7
	Middle-aged men without any bond	Highly educated young women	Low-educated people bonded to the region	Bonded highly educated people	Returnees bonded to the region	Retired newcomers	Retired locals bonded to the region
Gender	-0.18	-0.15	-0.14	**-1.18***	0.02	0.18	0.44
	(0.53)	(0.98)	(0.38)	(0.68)	(0.48)	(0.39)	(0.55)
Age	**-0.04****	**-0.07*****	-0.01	-0.04	-0.01	**-0.04*****	-0.01
	(0.02)	(0.03)	(0.02)	(0.03)	(0.02)	(0.01)	(0.02)
Higher Education	0.35	0.00	**1.76****	0.00	0.80	0.13	0.19
	(0.45)	(omitted)	(0.74)	(omitted)	(0.53)	(0.39)	(0.67)
Employed	0.00	1.18	-0.21	0.00	-0.58	0.00	0.00
	(omitted)	(1.27)	(1.38)	(omitted)	(0.76)	(omitted)	(omitted)
Income	0.03	-0.16	-0.04	-0.25	-0.11	0.14	-0.15
	(0.10)	(0.12)	(0.09)	(0.18)	(0.12)	(0.10)	(0.18)
Children	0.18	0.32	-0.23	-0.21	-0.06	0.13	-0.08
	(0.16)	(0.25)	(0.17)	(0.19)	(0.21)	(0.14)	(0.23)
Rural Environment	-0.07	-0.00	-0.25	**1.67*****	-0.11	0.17	-0.13
	(0.39)	(0.53)	(0.38)	(0.59)	(0.44)	(0.38)	(0.54)
Returnee	0.00	0.29	0.00	0.00	0.00	0.00	0.83
	(omitted)	(0.81)	(omitted)	(omitted)	(omitted)	(omitted)	(0.92)
Newcomer	0.26	0.54	-0.35	0.63	0.00	1.82	0.00
	(0.52)	(0.66)	(0.44)	(0.64)	(omitted)	(1.17)	(omitted)
Bond to Nature	-0.00	**1.16****	**1.08*****	**1.58***	-0.23	**0.96****	0.02
	(0.48)	(0.58)	(0.37)	(0.91)	(0.49)	(0.39)	(0.65)
Bond to Region	-0.29	0.13	-0.29	0.00	-0.19	0.07	-0.02
	(0.64)	(0.83)	(0.66)	(omitted)	(0.49)	(0.38)	(1.68)
Constant	1.39	2.05	1.02	2.26	1.59	-0.94	1.18
	(1.00)	(1.73)	(1.51)	(1.76)	(1.58)	(1.48)	(1.80)
Pseudo R-squared	0.05	0.11	0.07	0.21	0.03	0.12	0.05
N	136.00	97.00	165.00	79.00	101.00	149.00	69.00

Note. Stars indicate significance levels: * p<0.1, ** p<0.05, *** p<0.01. N displays the number of cases.

Source: Resident survey, own calculations.

Tab. A 12: Coefficients of the estimation of the cluster for the willingness to volunteer (coefficients, standard errors in parentheses)

	Cluster 1	Cluster 2	Cluster 3	Cluster 4	Cluster 5	Cluster 6	Cluster 7
	Middle-aged men without any bond	Highly educated young women	Low-educated people bonded to the region	Bonded highly educated people	Returnees bonded to the region	Retired newcomers	Retired locals bonded to the region
Gender	0.86	-0.64	**-0.73***	-1.14	-0.75	**-1.06*****	**-1.58****
	(0.55)	(0.97)	(0.38)	(0.74)	(0.52)	(0.38)	(0.72)
Age	0.01	-0.01	0.00	-0.03	-0.01	**-0.03****	**-0.04****
	(0.02)	(0.02)	(0.02)	(0.03)	(0.02)	(0.01)	(0.02)
Higher Education	0.36	0.00	-0.11	0.00	-0.26	0.09	-0.35
	(0.47)	(omitted)	(0.64)	(omitted)	(0.56)	(0.37)	(0.74)
Employed	0.00	1.21	0.08	0.00	0.27	0.00	0.00
	(omitted)	(1.26)	(1.31)	(omitted)	(0.79)	(omitted)	(omitted)
Income	0.14	-0.01	0.04	-0.32	0.00	0.03	0.16
	(0.10)	(0.12)	(0.09)	(0.21)	(0.12)	(0.09)	(0.20)
Children	-0.19	-0.19	-0.04	-0.10	-0.17	-0.05	-0.16
	(0.17)	(0.25)	(0.16)	(0.19)	(0.22)	(0.13)	(0.26)
Rural Environment	-0.51	0.23	0.14	0.86	0.09	-0.16	-0.17
	(0.41)	(0.53)	(0.37)	(0.62)	(0.46)	(0.36)	(0.61)
Returnee	0.00	-1.11	0.00	0.00	-0.40	0.00	-0.08
	(omitted)	(0.79)	(omitted)	(omitted)	(1.60)	(omitted)	(0.93)
Newcomer	0.41	-0.39	-0.32	**1.54****	0.00	-0.45	0.00
	(0.53)	(0.68)	(0.44)	(0.71)	(omitted)	(0.85)	(omitted)
Bond to Nature	**1.90*****	**1.19****	**0.88****	0.77	0.63	**0.72****	-0.66
	(0.56)	(0.56)	(0.37)	(0.97)	(0.52)	(0.37)	(0.68)
Bond to Region	0.17	-1.00	0.20	0.00	**0.91***	-0.20	0.00
	(0.67)	(0.81)	(0.62)	(omitted)	(0.50)	(0.36)	(omitted)
Constant	-1.61	0.71	-0.60	3.09	0.76	**2.30***	**3.20***
	(1.03)	(1.72)	(1.46)	(1.98)	(2.18)	(1.28)	(1.88)
Pseudo R-squared	0.10	0.10	0.06	0.14	0.05	0.07	0.14
N	136.00	97.00	165.00	79.00	103.00	149.00	67.00

Note. Stars indicate significance levels: * p<0.1, ** p<0.05, *** p<0.01. N displays the number of cases.

Source: Resident survey, own calculations.

Fig. A 1: Questionnaire

Regiobranding

Logo der entsprechenden Praxispartner

FONA

Innovationsgruppen für ein Nachhaltiges Landmanagement

GEFÖRDERT VOM Bundesministerium für Bildung und Forschung

Sind meine Daten denn bei Ihnen sicher?

Die Erhebung und Auswertung der Daten erfolgt **anonym**. Es wird kein persönlicher Bezug zwischen den Befragten und den erhobenen Daten hergestellt. Die Daten dienen lediglich dem Forschungszweck und werden **nicht an Dritte weitergegeben**. Die Ergebnisse werden nur in zusammengefasster Form in Tabellen und Abbildungen dargestellt und veröffentlicht. Die ausgefüllten Fragebögen werden nach Ablauf der Aufbewahrungsfrist vernichtet.

Wir wünschen Ihnen viel Spaß beim Ausfüllen und sind gespannt auf Ihren Blickwinkel!

1. FRAGENBLOCK: MEINE NUTZUNG MEINER REGION

1.1 Zu welchem Zweck und wie häufig nutzen Sie persönlich Ihre Region?

	(fast) täglich	mind. 1x pro Woche	mind. 1x pro Monat	seltener	gar nicht
Ich nutze sie in meiner Freizeit...					
...zur Entspannung (z. B. Spazierengehen, Hund ausführen).	☐	☐	☐	☐	☐
...für sportliche Aktivitäten (z. B. Reiten, Angeln, Radfahren, Nordic Walking).	☐	☐	☐	☐	☐
...als Treffpunkt und Kommunikationsort.	☐	☐	☐	☐	☐
...als Lern- und Bildungsort (z. B. Natur- und Tierbeobachtung, Besichtigung historischer Stätten, Museen).	☐	☐	☐	☐	☐
...für Besuche von Einkaufsstätten innerhalb der Stadt- und Ortskerne.	☐	☐	☐	☐	☐
...für Besuche von Einkaufsstätten außerhalb der Stadt- und Ortskerne.	☐	☐	☐	☐	☐
...für kulturelle / gemeinschaftliche Aktivitäten (z. B. Dorffeste).	☐	☐	☐	☐	☐
...für andere Aktivitäten und zwar ____________________.	☐	☐	☐	☐	☐
Ich nutze sie beruflich als...					
...Land-, Forst-, Energiewirt/in.	☐	☐	☐	☐	☐
...Angestellte/r oder Beamte/r im öffentlichen Dienst mit Bezug zum Erhalt, Schutz und / oder Gestaltung der Landschaft, der Städte und Dörfer sowie der Denkmale.	☐	☐	☐	☐	☐
...andere/r Angestellte/r oder Unternehmer/in mit Bezug zum Erhalt, Schutz und / oder Gestaltung der Landschaft, der Städte und Dörfer sowie der Denkmale und zwar ____________________.	☐	☐	☐	☐	☐

- 1 -

1.2 In welchen Städten / Gemeinden finden die folgenden Aktivitäten in Ihrem Leben überwiegend statt?
(Falls Sie einer Aktivität nicht nachgehen, lassen Sie die Zeile bitte frei.)

Wohnen	______________________
Arbeiten	______________________
Besuch der Schule / Bildungsstätte	______________________
Einkaufen / Besorgungen / sonstige Erledigungen	______________________
Freizeit (Kultur, Sport, Erholung oder Ähnliches)	______________________
Besuch(e) von Familienangehörigen und Freunden	______________________
andere Aktivitäten und zwar ______________	______________________

1.3 Welche Verkehrsmittel nutzen Sie überwiegend in Ihrer Region?
(Mehrfachnennungen möglich)

☐ Auto
☐ Zug / Bahn
☐ Bus
☐ Mofa / Roller / Motorrad
☐ E-Bike
☐ Fahrrad
☐ zu Fuß
☐ Boot / Fähre
☐ andere und zwar ______________
____

1.4 Bitte bewerten Sie die Lebensqualität in Ihrer Region.
(Falls Sie zu einem Aspekt keine Antwort geben können, lassen Sie die Zeile bitte frei.)

1.4.1 Wie wichtig sind folgende Aspekte für Ihre Lebensqualität?

1.4.2 Wie zufrieden sind Sie mit den folgenden Angeboten in Ihrer Region?

sehr unwichtig		unent-schieden		sehr wichtig		sehr unzufrieden		unent-schieden		sehr zufrieden
--	-	0	+	++		--	-	0	+	++
☐	☐	☐	☐	☐	Gesundheitliche Versorgung	☐	☐	☐	☐	☐
☐	☐	☐	☐	☐	„Gesunde" Umgebung (z. B. Luftqualität)	☐	☐	☐	☐	☐
☐	☐	☐	☐	☐	Naherholung	☐	☐	☐	☐	☐
☐	☐	☐	☐	☐	Landschaftliche Schönheit	☐	☐	☐	☐	☐
☐	☐	☐	☐	☐	Nahversorgung mit Supermärkten	☐	☐	☐	☐	☐
☐	☐	☐	☐	☐	Regionaltypische Gebäude und Denkmale	☐	☐	☐	☐	☐
☐	☐	☐	☐	☐	Miet- bzw. Immobilienpreise	☐	☐	☐	☐	☐
☐	☐	☐	☐	☐	Straßenverkehrsinfrastruktur	☐	☐	☐	☐	☐
☐	☐	☐	☐	☐	Öffentliche Verkehrsangebote (z. B. Bus, Bahn)	☐	☐	☐	☐	☐
☐	☐	☐	☐	☐	Kulturelle Angebote (z. B. Museen, Theater)	☐	☐	☐	☐	☐
☐	☐	☐	☐	☐	Tradition, Brauchtum und Geschichte	☐	☐	☐	☐	☐
☐	☐	☐	☐	☐	Regionaltypische Produkte	☐	☐	☐	☐	☐
☐	☐	☐	☐	☐	Nachbarschaftliche bzw. dörfliche Gemeinschaft, Vereinsleben	☐	☐	☐	☐	☐
☐	☐	☐	☐	☐	Bildungsangebote	☐	☐	☐	☐	☐
☐	☐	☐	☐	☐	andere Aspekte und zwar ______________	☐	☐	☐	☐	☐

2. FRAGENBLOCK: MEINE WAHRNEHMUNG MEINER REGION

2.1 Was schätzen Sie an Ihrer Region?
(Dies können sowohl Landschaften, Gebäude, Siedlungen, archäologische und historische Denkmale oder Ähnliches sein. Bitte geben Sie maximal 5 Orte / Plätze an.

- ______________________________
- ______________________________
- ______________________________
- ______________________________
- ______________________________

2.2 Bitte bewerten Sie folgende Aussagen zu Ihrer Bindung an Ihre Region.

	trifft nicht zu		unent-schieden		trifft zu
	--	-	0	+	++
Ich fühle mich in meiner Region wirklich zu Hause.	☐	☐	☐	☐	☐
Mit meiner Region verbinde ich viele persönliche Erinnerungen.	☐	☐	☐	☐	☐
Meine persönliche Zukunft ist eng an die Region geknüpft.	☐	☐	☐	☐	☐

2.3 Bitte bewerten Sie, ob es Ihrer Meinung nach eine Veränderung der Landnutzung / Landbedeckung in Ihrer Region gibt und wie Sie diese wahrnehmen.

(Falls Sie zu einer Landnutzung / -bedeckung keine Antwort geben können, lassen Sie die Zeile bitte frei.)

2.3.1 Wie nehmen Sie die Veränderung der <u>Flächengrößen</u> seit dem Jahr 2000 wahr?

2.3.2 Wie empfinden Sie diese Veränderung?

starke Abnahme		keine		starke Zunahme		sehr negativ		unent-schieden		sehr positiv
--	-	0	+	++		--	-	0	+	++
☐	☐	☐	☐	☐	Ackerbauflächen	☐	☐	☐	☐	☐
☐	☐	☐	☐	☐	Grünland (z. B. Weiden, Wiesen)	☐	☐	☐	☐	☐
☐	☐	☐	☐	☐	anderes Offenland (z. B. Heide, Dünen, Deichvorland)	☐	☐	☐	☐	☐
☐	☐	☐	☐	☐	Wald- und Gehölzflächen	☐	☐	☐	☐	☐
☐	☐	☐	☐	☐	Feuchtgebiete (z. B. Moore, Sümpfe)	☐	☐	☐	☐	☐
☐	☐	☐	☐	☐	Wasser (z. B. Seen, Flüsse, Meer)	☐	☐	☐	☐	☐
☐	☐	☐	☐	☐	Brachland	☐	☐	☐	☐	☐
☐	☐	☐	☐	☐	Verkehrs-, Siedlungs- und Gewerbeflächen	☐	☐	☐	☐	☐
☐	☐	☐	☐	☐	Flächen für erneuerbare Energien (z. B. Biomasse für Biogas und -kraftstoffe, Windkraftanlagen)	☐	☐	☐	☐	☐
☐	☐	☐	☐	☐	andere Nutzung / Bedeckung und zwar ____________	☐	☐	☐	☐	☐

2.4 Welche allgemeinen Beeinträchtigungen bezogen auf die zukünftige Entwicklung befürchten Sie für Ihre Region?
(Mehrfachnennungen möglich)

☐ Leerstand und / oder Verfall
☐ Abwanderung der Bevölkerung
☐ behördliche Auflagen
☐ Zerstörung durch den Menschen
☐ Verlust von handwerklichen Techniken / Fähigkeiten
☐ andere und zwar ____________________
____
☐ Zerstörung durch Naturereignisse und -gefahren
☐ allgemeine Kostensteigerungen
☐ erneuerbare Energien (z. B. Windräder, Biogasanlagen)
☐ Ausbau von Verkehrs-, Siedlungs- und Gewerbeflächen
☐ Ich sehe keine zukünftigen Beeinträchtigungen.

3. FRAGENBLOCK: MEIN EHRENAMTLICHES ENGAGEMENT IN MEINER REGION

Unter dem Begriff ehrenamtliches Engagement werden im Folgenden sämtliche freiwilligen Aktivitäten verstanden, die unabhängig vom Beruf und in der Regel unentgeltlich (mit Ausnahme von Aufwandsentschädigungen) ausgeführt werden. Dazu zählen sowohl Vereinsengagement als auch freie oder phasenweise, projektbezogene Tätigkeiten.

3.1 Engagieren Sie sich in irgendeiner Weise ehrenamtlich?

☐ ja
☐ nein → *Bitte gehen Sie zu Frage **3.6.***

3.2 Wie viele Stunden pro Monat sind Sie im Durchschnitt ehrenamtlich tätig?

☐ 1 – 2 Stunden
☐ 3 – 5 Stunden
☐ 6 – 10 Stunden
☐ 11 – 20 Stunden
☐ mehr als 20 Stunden

3.3 Wofür engagieren Sie sich ehrenamtlich?
(Mehrfachnennungen möglich)

☐ Ortsentwicklung
☐ Politik
☐ Natur- und Umweltschutz
☐ Tierschutz
☐ Feuerwehr / Rettungsdienst
☐ Sport / Bewegung
☐ Bildung / Kindergarten / Schule
☐ Religion / Kirche
☐ Kunst / Kultur / Museen
☐ Pflege von Denkmalen und Gebäuden
☐ Heimatkunde und -forschung
☐ Soziales / Gesellschaftliches (z. B. Flüchtlingshilfe, Betreuung beeinträchtigter Menschen)
☐ anderes und zwar ____________________
____

3.4 Was bringen Sie ehrenamtlich ein?
(Mehrfachnennungen möglich)

☐ Arbeitskraft / Zeit
☐ Geld
☐ Material
☐ Ideen / Wissen / Planung
☐ Räumlichkeiten
☐ anderes und zwar ____________________
____

3.5 Bitte führen Sie Ihre ehrenamtliche Tätigkeit kurz aus. Falls Sie in Vereinen tätig sind, nennen Sie diese bitte namentlich.
(Bitte antworten Sie in Stichpunkten)

____
____
____
____

→ *Bitte gehen Sie zu Frage **3.7.***

3.6 Was hält Sie bisher davon ab, sich in Ihrer Region ehrenamtlich zu engagieren?
(Mehrfachnennungen möglich)

☐ keine Zeit
☐ zu unflexible Zeiteinteilung
☐ kein passendes Angebot
☐ keine Lust / kein Interesse
☐ anderes und zwar ____________________
____

3.7 In welchen der folgenden Bereiche könnten Sie sich ein ehrenamtliches Engagement in der Zukunft (erstmalig, verstärkt oder zusätzlich) am ehesten vorstellen?
(Mehrfachnennungen möglich)

☐ Ortsentwicklung
☐ Politik
☐ Natur- und Umweltschutz
☐ Tierschutz
☐ Feuerwehr / Rettungsdienst
☐ Sport / Bewegung
☐ Bildung / Kindergarten / Schule
☐ Religion / Kirche
☐ Kunst / Kultur / Museen
☐ Pflege von Denkmalen und Gebäuden
☐ Heimatkunde und -forschung
☐ Soziales / Gesellschaftliches (z. B. Flüchtlingshilfe, Betreuung beeinträchtigter Menschen)
☐ anderes und zwar ____________________
____

3.8 Wie viele Stunden pro Monat wären Sie generell bereit, sich ehrenamtlich für Ihre Region zu engagieren?

☐ 1 – 2 Stunden
☐ 3 – 5 Stunden
☐ 6 – 10 Stunden
☐ 11 – 20 Stunden
☐ mehr als 20 Stunden
☐ kein Interesse

4. FRAGENBLOCK: MEINE WERTSCHÄTZUNG MEINER REGION

Erhalt, Schutz und Gestaltung der Landschaft, der Städte und Dörfer sowie der Denkmale werden durch Ihre Steuern und andere öffentliche Mittel finanziert. Hierzu gehören u. a. Anlegen und Instandhalten von (Be-)Pflanzungen und Gewässern, Bau und Erhalt von Wegen sowie Schutz vor und Beseitigung von Verschmutzungen. Möglicherweise könnte sich die Bevölkerung mit einem zusätzlichen Beitrag an weiteren Investitionen für diesen Bereich beteiligen.

4.1 Sollte Ihrer Meinung nach die gesamte Bevölkerung grundsätzlich einen direkten finanziellen Beitrag für Erhalt, Schutz und Gestaltung der Landschaft, der Städte und Dörfer sowie der Denkmale zusätzlich zu den bereits gezahlten Steuern leisten?
(Es besteht die Annahme, dass diese zusätzlichen Mittel sinnvoll und effizient eingesetzt werden.)

☐ ja ☐ nein

4.2 Welchen Geldbetrag wären Sie persönlich bereit, <u>monatlich</u> zusätzlich zu den bereits von Ihnen gezahlten Steuern für Erhalt, Schutz und Gestaltung der jeweiligen, auf den folgenden Fotos abgebildeten Landschaften zu zahlen?
Bitte bewerten Sie die Landschaften einzeln und unabhängig voneinander.
(Denken Sie dabei bitte an Ihren persönlichen Nutzen z. B. im Vergleich zu Besuchen von Restaurants, Theatern, Sportveranstaltungen.)

- 7 -

Lübeck

Strandlandschaft

☐ 0 Euro
☐ 1 – 5 Euro
☐ 6 – 10 Euro
☐ 11 – 20 Euro
☐ 21 – 50 Euro
☐ 51 – 100 Euro
☐ mehr als 100 Euro

Picture: Hansestadt Lübeck

Wälder und Gehölze

☐ 0 Euro
☐ 1 – 5 Euro
☐ 6 – 10 Euro
☐ 11 – 20 Euro
☐ 21 – 50 Euro
☐ 51 – 100 Euro
☐ mehr als 100 Euro

Picture: Hansestadt Lübeck

Wasserfläche

☐ 0 Euro
☐ 1 – 5 Euro
☐ 6 – 10 Euro
☐ 11 – 20 Euro
☐ 21 – 50 Euro
☐ 51 – 100 Euro
☐ mehr als 100 Euro

Picture: Hansestadt Lübeck

Städtische Umgebung

☐ 0 Euro
☐ 1 – 5 Euro
☐ 6 – 10 Euro
☐ 11 – 20 Euro
☐ 21 – 50 Euro
☐ 51 – 100 Euro
☐ mehr als 100 Euro

Picture: Hansestadt Lübeck

Dörfliche Umgebung

☐ 0 Euro
☐ 1 – 5 Euro
☐ 6 – 10 Euro
☐ 11 – 20 Euro
☐ 21 – 50 Euro
☐ 51 – 100 Euro
☐ mehr als 100 Euro

Picture: Hansestadt Lübeck

Ludwigslust-Parchim

Dünenlandschaft

- ☐ 0 Euro
- ☐ 1 – 5 Euro
- ☐ 6 – 10 Euro
- ☐ 11 – 20 Euro
- ☐ 21 – 50 Euro
- ☐ 51 – 100 Euro
- ☐ mehr als 100 Euro

Picture: Dirk Foitlänger, Biosphärenreservatsamt Schaalsee-Elbe

Hecken und Wälder

- ☐ 0 Euro
- ☐ 1 – 5 Euro
- ☐ 6 – 10 Euro
- ☐ 11 – 20 Euro
- ☐ 21 – 50 Euro
- ☐ 51 – 100 Euro
- ☐ mehr als 100 Euro

Picture: Ingrid Herrmann

Wasserfläche

- ☐ 0 Euro
- ☐ 1 – 5 Euro
- ☐ 6 – 10 Euro
- ☐ 11 – 20 Euro
- ☐ 21 – 50 Euro
- ☐ 51 – 100 Euro
- ☐ mehr als 100 Euro

Picture: Ingrid Herrmann

Städtische Umgebung

- ☐ 0 Euro
- ☐ 1 – 5 Euro
- ☐ 6 – 10 Euro
- ☐ 11 – 20 Euro
- ☐ 21 – 50 Euro
- ☐ 51 – 100 Euro
- ☐ mehr als 100 Euro

Picture: Sylvia Wegener

Dörfliche Umgebung

- ☐ 0 Euro
- ☐ 1 – 5 Euro
- ☐ 6 – 10 Euro
- ☐ 11 – 20 Euro
- ☐ 21 – 50 Euro
- ☐ 51 – 100 Euro
- ☐ mehr als 100 Euro

Picture: Ingrid Herrmann

Lüchow-Dannenberg

Heidelandschaft

☐ 0 Euro
☐ 1 – 5 Euro
☐ 6 – 10 Euro
☐ 11 – 20 Euro
☐ 21 – 50 Euro
☐ 51 – 100 Euro
☐ mehr als 100 Euro

Picture: Ute Ostermann

Wälder und Gehölze

☐ 0 Euro
☐ 1 – 5 Euro
☐ 6 – 10 Euro
☐ 11 – 20 Euro
☐ 21 – 50 Euro
☐ 51 – 100 Euro
☐ mehr als 100 Euro

Picture: Umweltministerium / Untere Naturschutzbehörde, Landkreis Lüchow-Dannenberg

Wasserfläche

☐ 0 Euro
☐ 1 – 5 Euro
☐ 6 – 10 Euro
☐ 11 – 20 Euro
☐ 21 – 50 Euro
☐ 51 – 100 Euro
☐ mehr als 100 Euro

Picture: Jenny Raeder

Städtische Umgebung

☐ 0 Euro
☐ 1 – 5 Euro
☐ 6 – 10 Euro
☐ 11 – 20 Euro
☐ 21 – 50 Euro
☐ 51 – 100 Euro
☐ mehr als 100 Euro

Picture: Jenny Raeder

Dörfliche Umgebung

☐ 0 Euro
☐ 1 – 5 Euro
☐ 6 – 10 Euro
☐ 11 – 20 Euro
☐ 21 – 50 Euro
☐ 51 – 100 Euro
☐ mehr als 100 Euro

Picture: Jenny Raeder

Steinburg

Hauslandschaft

☐ 0 Euro
☐ 1 – 5 Euro
☐ 6 – 10 Euro
☐ 11 – 20 Euro
☐ 21 – 50 Euro
☐ 51 – 100 Euro
☐ mehr als 100 Euro

Picture: Melanie Herrmann

Marschlandschaft

☐ 0 Euro
☐ 1 – 5 Euro
☐ 6 – 10 Euro
☐ 11 – 20 Euro
☐ 21 – 50 Euro
☐ 51 – 100 Euro
☐ mehr als 100 Euro

Picture: Archäologisches Landesamt Schleswig-Holstein

Wasserfläche

☐ 0 Euro
☐ 1 – 5 Euro
☐ 6 – 10 Euro
☐ 11 – 20 Euro
☐ 21 – 50 Euro
☐ 51 – 100 Euro
☐ mehr als 100 Euro

Picture: Christina Gieb

Städtische Umgebung

☐ 0 Euro
☐ 1 – 5 Euro
☐ 6 – 10 Euro
☐ 11 – 20 Euro
☐ 21 – 50 Euro
☐ 51 – 100 Euro
☐ mehr als 100 Euro

Picture: Glückstadt Destination Management (GDM)

Dörfliche Umgebung

☐ 0 Euro
☐ 1 – 5 Euro
☐ 6 – 10 Euro
☐ 11 – 20 Euro
☐ 21 – 50 Euro
☐ 51 – 100 Euro
☐ mehr als 100 Euro

Picture: Beate von Malottky

4.3 Angenommen eine zusätzliche finanzielle Unterstützung für Erhalt, Schutz und Gestaltung der Landschaft, der Städte und Dörfer sowie der Denkmale durch die Bevölkerung würde notwendig werden, welche Art der Zahlung würden Sie bevorzugen?
(Mehrfachnennungen möglich)

- ☐ Eintrittspreise für den Zutritt zu bestimmten Landschaften / Denkmalen
- ☐ Erhebung eines Landschaftsbeitrages mit prozentualem Anteil des Einkommens
- ☐ Gebühren für (Besucher-)Parkplätze
- ☐ Erhöhung der Verkaufspreise für regionale Produkte
- ☐ freiwillige Spenden / Stiftungen / Vereine
- ☐ andere und zwar ____________________

4.4 Bitte bewerten Sie folgende Aussagen zu Ihrer Beziehung zur Natur im Allgemeinen.

	trifft nicht zu		unent-schieden		trifft zu
	--	-	0	+	++
Ich versuche, so oft wie möglich in der Natur zu sein.	☐	☐	☐	☐	☐
Natur spielt in meinem Leben eine untergeordnete Rolle.	☐	☐	☐	☐	☐
Ich fühle mich mit Natur und Landschaft in meiner Region eng verbunden.	☐	☐	☐	☐	☐
Ich glaube, der Mensch profitiert von einer intakten Natur (z. B. durch Nahrung, sauberes Wasser, Erholung).	☐	☐	☐	☐	☐
Ich sehe eine Verpflichtung, die Natur zu bewahren und zu schützen.	☐	☐	☐	☐	☐
Ich fühle mich als Teil der Natur.	☐	☐	☐	☐	☐

5. FRAGENBLOCK: MERKMALE ZU MEINER PERSON

5.1 Welchem Geschlecht gehören Sie an? ☐ weiblich ☐ männlich

5.2 Wie alt sind Sie?

☐ 0 – 17 Jahre	☐ 45 – 54 Jahre
☐ 18 – 24 Jahre	☐ 55 – 64 Jahre
☐ 25 – 34 Jahre	☐ 65 – 74 Jahre
☐ 35 – 44 Jahre	☐ 75 Jahre und älter

5.3 Welche Staatsangehörigkeit(en) haben Sie?

☐ deutsch ☐ andere und zwar ____________________

5.4 Welchen höchsten Bildungsabschluss haben Sie erreicht?

☐ Volksschule / Hauptschulabschluss
☐ Realschulabschluss / Mittlere Reife
☐ (Fach-)Abitur / (Fach-)Hochschulreife
☐ (Fach-)Hochschulabschluss
☐ keinen Schulabschluss

5.5 Haben Sie eine Berufsausbildung / Lehre abgeschlossen? ☐ ja ☐ nein

5.6 Welchen Umfang hat Ihre derzeitige berufliche Tätigkeit?

☐ Vollzeit (Beruf, Studium, Ausbildung)
☐ Teilzeit (inkl. Mini-Job)
☐ in Mutterschutz / Elternzeit
☐ arbeitssuchend
☐ in Rente
☐ aus anderen Gründen nicht berufstätig

5.7 Wie hoch ist das Nettoeinkommen aller Mitglieder in Ihrem Haushalt?
(Bitte kreuzen Sie den (geschätzten) monatlichen Nettobetrag an; also den Betrag nach Abzug von Steuern und Sozialabgaben. Zum Nettobetrag zählen auch Kindergeld, Rente, Sozialleistungen usw.)

☐ unter 1.000 Euro
☐ 1.001 – 1.500 Euro
☐ 1.501 – 2.000 Euro
☐ 2.001 – 2.500 Euro
☐ 2.501 – 3.000 Euro
☐ 3.001 – 4.000 Euro
☐ 4.001 – 5.000 Euro
☐ mehr als 5.000 Euro

5.8 Wie viele Kinder haben Sie? ____________

5.9 Wie wohnen Sie?

☐ alleine
☐ mit Partner(in)
☐ mit Partner(in) und Kind(ern)
☐ ohne Partner(in) mit Kind(ern)
☐ mit meinen Eltern
☐ insgesamt mit drei Generationen
☐ in einer Wohngemeinschaft
☐ anders und zwar ________________________
____

5.10 In was für einem Haus wohnen Sie derzeit?

☐ Einfamilienhaus
☐ Doppel- bzw. Zweifamilienhaus
☐ Reihenhaus
☐ Mehrfamilienhaus
☐ anderem und zwar ______________________
____

5.11 Wie wohnen Sie derzeit? ☐ zur Miete ☐ im Eigentum

- 13 -

5.12 In welcher Lage liegt Ihr Wohnhaus?

☐ in einer Stadt ☐ in einem Dorf

☐ am Stadtrand ☐ außerhalb einer Stadt / eines Dorfes

5.13 Seit wann wohnen Sie in Ihrer Region?

☐ seit meiner Geburt ohne Unterbrechungen → *Bitte gehen Sie zu Frage 5.15.*

☐ seit meiner Geburt mit Unterbrechungen

☐ Ich bin zugezogen.

5.14 In welchem Jahr sind Sie zurückgekehrt / zugezogen? ___________

5.15 Warum sind Sie in Ihrer Region geblieben bzw. zurückgekehrt / zugezogen?
(Mehrfachantworten möglich)

☐ berufliche Gründe ☐ Wohnumgebung

☐ familiäre Gründe ☐ anderer Grund und zwar ________________

☐ finanzielle Gründe ______________________________

5.16 Wie zufrieden sind Sie derzeit, alles in allem, mit Ihrem Leben?
(Bitte bewerten Sie auf einer Skala von 0 (ganz und gar unzufrieden) bis 10 (ganz und gar zufrieden).)

☐	☐	☐	☐	☐	☐	☐	☐	☐	☐	☐
0	1	2	3	4	5	6	7	8	9	10

5.17 Was wünschen Sie sich für Ihre Region und ihre Landschaften?
(Bitte antworten Sie in Stichpunkten)

____

____

____

____

Jetzt haben Sie es fast geschafft!

Bitte geben Sie den ausgefüllten Bogen in einer der folgenden **Sammelstellen** in Ihrer Nähe ab:

Oder nutzen Sie den beigelegten Rückumschlag und senden Sie uns den Bogen für Sie portofrei zurück.

An wen kann ich mich wenden, wenn ich noch Rückfragen habe?

Gern stehen wir Ihnen im Befragungszeitraum für Rückfragen zur Verfügung. Sie erreichen uns unter der Emailadresse regiobranding@umwelt.uni-hannover.de oder auf dem Postweg unter der Anschrift Gottfried Wilhelm Leibniz Universität Hannover, Institut für Umweltplanung, Projekt Regiobranding, Herrenhäuser Straße 2, 30419 Hannover.

Wie kann ich weiterhin aktiv an dem Projekt teilhaben?

Im Nachgang zu der Befragung möchten wir mit einigen Bürgerinnen und Bürgern Gespräche führen. Wir würden uns freuen, wenn Sie uns Ihre Bereitschaft dazu durch die Weitergabe Ihrer Kontaktdaten (Adresse, Telefonnummer oder auch gern E-Mailadresse) auf einem separaten Blatt (nicht beiliegend) mitteilen und dieses dem Umschlag mit dem Fragebogen beifügen würden. Wir würden bei Bedarf mit Ihnen Kontakt aufnehmen. **Die Kontaktdaten werden nicht mit Ihren Angaben verknüpft.**

Herzlichen Dank für Ihre Teilnahme!

- 15 -

Source: Resident survey.

Fig. A 2: Decision tree of the nested logit model

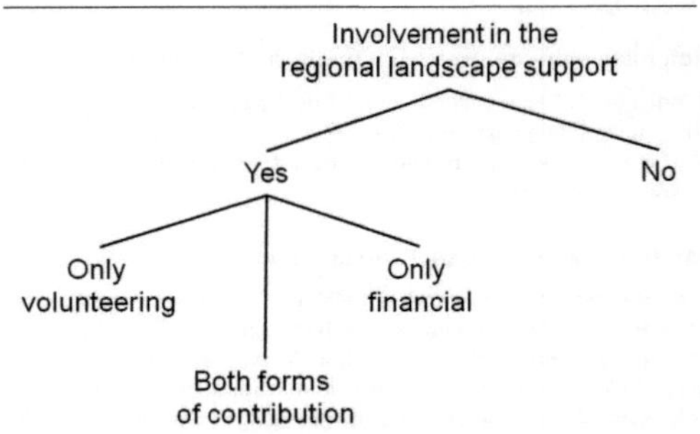

Source: Own illustration.

MIX
Papier aus verantwortungsvollen Quellen
Paper from responsible sources
FSC® C105338

If you have any concerns about our products,
you can contact us on
ProductSafety@springernature.com

In case Publisher is established outside the EU,
the EU authorized representative is:
Springer Nature Customer Service Center GmbH
Europaplatz 3, 69115 Heidelberg, Germany

Printed by Libri Plureos GmbH
in Hamburg, Germany

Labour Mobility in the European Union as an Example of the Transnationalization of Employment

Torben Krings

Labour Mobility in the European Union as an Example of the Transnationalization of Employment

Springer

Torben Krings
Johannes Kepler Universität Linz
Linz, Austria

ISBN 978-3-658-43976-7 ISBN 978-3-658-43977-4 (eBook)
https://doi.org/10.1007/978-3-658-43977-4

Translation from the German language edition: “Die Transnationalisierung der Arbeitswelt am Beispiel von Erwerbsmobilität in der Europäischen Union” by Torben Krings, © Springer Fachmedien Wiesbaden GmbH, ein Teil von Springer Nature 2022. Published by Springer Fachmedien Wiesbaden. All Rights Reserved.

This book is a translation of the original German edition “Die Transnationalisierung der Arbeitswelt am Beispiel von Erwerbsmobilität in der Europäischen Union” by Krings, Torben, published by Springer Fachmedien Wiesbaden GmbH in 2022. The translation was done with the help of an artificial intelligence machine translation tool. A subsequent human revision was done primarily in terms of content, so that the book will read stylistically differently from a conventional translation. Springer Nature works continuously to further the development of tools for the production of books and on the related technologies to support the authors.

This Springer imprint is published by the registered company Springer Fachmedien Wiesbaden GmbH, part of Springer Nature.
The registered company address is: Abraham-Lincoln-Str. 46, 65189 Wiesbaden, Germany

Paper in this product is recyclable.

Contents

1 Introduction

Sometimes the significance of an institutional regulation only becomes clear when it no longer applies. When the COVID-19 pandemic brought public life to a standstill in large parts of Europe in the spring of 2020, cross-border traffic also largely came to a halt. To hinder the spread of the virus, EU member states severely restricted the free movement of individuals, effectively suspending one of the four fundamental freedoms of the EU. This caught some economic sectors unprepared, in which migrants are an indispensable part of the workforce. In Austria, a "collapse" of 24-hour mobile care was feared, in Germany an absence of "entire columns of construction workers", and in the UK, where farmers still rely on seasonal workers from Southeast Europe despite Brexit, a "crop failure" in agriculture was feared (FAZ, 2020; ORF, 2020). Necessity is the mother of invention. To "save the broccoli" (The Economist, 2020), farm workers from Southeast Europe were flown in to England, something that was also observed in Germany. In Austria, special trains were chartered and a "train corridor" was set up to bring personal caregivers from Romania into the country. In other words, the crisis revealed the "systemic importance" of cross-border labour mobility, to use a much-quoted term at the time.

This book examines the transnationalization of work using the example of intra-European labour mobility. The focus is on the analysis of a new dynamic of cross-border mobility in the 21st century, as it has developed mainly in the course of EU enlargement to the East. Of particular interest are the socio-economic conditions, the extent and diversity of this mobility, as well as the labour market effects and the occupational position of EU migrants. The question of the regulation of this mobility in the enlarged EU, where social and institutional diversity has increased, is also raised. From a conceptual

T. Krings, *Labour Mobility in the European Union as an Example of the Transnationalization of Employment*,
https://doi.org/10.1007/978-3-658-43977-4_1

perspective, the challenges that the transnationalization of labour markets poses for social science labour research are examined more closely.

The central thesis developed in this book is that a transnational mobility space has emerged in the EU, which structurally promotes mobility and facilitates entry into the labour market. Since a comparable migration regime does not exist for persons from third countries, and given the still considerable wage gap within the enlarged EU, it can be assumed that intra-European mobility will remain the dominant form of employment-related migration in the EU for the foreseeable future.

Intra-European mobility, in contrast to migration movements from third countries, has long been scarcely addressed in academic research (SVR, 2013, p. 67). This may be due to the fact that this mobility has only increased to a significant extent since the 2000s, especially since enlargement in 2004 and 2007. Another reason could be that the distinction postulated by the European Commission between intra-European *mobility* and extra-European *migration* is also reproduced in the academic debate. In addition to different free movement rights, the duration of the migration project and questions about the center of life of mobile individuals (destination or country of origin) are occasionally used to distinguish between migration and mobility (Wagner & Hassel, 2016, p. 132). In fact, however, the boundaries between migration and mobility are often blurry as regards European cross-border movements, not least because the individual orientations and aspirations of the "free movers" (Favell, 2008) can change over time. Therefore, it seems more fitting to arrange intra-European migration on a continuum of short and longer-term mobility. Against this background, "mobility" and "migration" are used synonymously in this book. The following section introduces the four central themes of the book, which run through all subsequent chapters.

A New Transnational Mobility Space

International labour migration is not a new phenomenon, but has a centuries-long tradition, whether as seasonal mobility or as permanent migration (Bade, 2000). However, what is new is that intra-European migration is no longer regulated by nation-states.[1] While in the 20th century the nation-state was the central actor controlling borders and regulating migration movements, within the framework of the European integration process "the European nation-state's supreme early-

[1] Mind you, in non-crisis times, as the experiences with the Corona crisis teach us.

twentieth-century control over migration and population dynamics was being voluntarily dislodged" (Favell & Hansen, 2002, p. 585). Although access to the labour market can be *temporarily* restricted for new member states, as happened most recently in the EU enlargements of 2004, 2007, and 2013, this does not alter the fact that a transnational mobility space has emerged in the EU/EFTA, which is based on the principle of free movement and now includes 31 countries.[2] Freedom of movement is considered one of the most attractive features of the EU, as repeatedly shown in surveys. Most recently, in a Eurobarometer survey, over 80% of respondents supported "the freedom of movement of EU citizens who can live, work, study and do business anywhere in the EU" (European Commission, 2019).

The increased importance of intra-European labour mobility can be illustrated by some numbers. Between 2004 and 2018, the number of mobile EU/EFTA citizens who are employed in another EU/EFTA state increased by over six million to almost ten million people. These figures, based on the European Labour Force Survey, do not include short-term cross-border mobility such as posted work, seasonal work, or circular migration, and thus underestimate the actual extent of work-related migration in Europe (Fries-Tersch et al., 2020). If the latter are taken into account, it becomes clear that in 2018 approximately 12 million EU/EFTA citizens, and thus five percent of all workers in the European Economic Area, worked across borders in one form or another.

The existence of a transnational mobility space should not lead to the assumption that work-related migration in Europe takes place without barriers and obstacles. In addition to linguistic and cultural differences, the diversity of national education, social security, and tax systems can be a hindrance to intra-European mobility (Krause et al., 2014). Less visible borders also appear in transnational workplaces when foreign workers are not employed under the same conditions as domestic

---

[2] In addition to the 27 EU member states, the European mobility space also includes Iceland, Liechtenstein, Norway, and Switzerland, which form the European Free Trade Association (EFTA). Of the EFTA member states, Iceland, Liechtenstein, and Norway (as well as all EU member states) belong to the European Economic Area (EEA). Switzerland has regulated freedom of movement in a bilateral agreement with the EU. For reasons of data pragmatism and for better readability, EFTA citizens are not always separately identified in the book.

The UK, which left the EU at the beginning of 2020, no longer applies the four fundamental freedoms of the European single market after a transition phase. Nevertheless, many EU citizens still work in the UK, which is why the island state also plays an important role in this book.

workers, which can exacerbate segmentation by origin and employment status (Wagner, 2018).

It should also be borne in mind that the creation of an intra-European mobility space is accompanied by new tendencies towards closure from the outside. In a globalized world with selective border openings, mobility opportunities are unevenly distributed (Mau et al., 2012). Intra-European mobility in the "age of migration" (Castles & Miller, 2009) is only one facet of international migration. However, it differs from other migration movements primarily in terms of free movement rights, as the Italian sociologist Ettore Recchi (2015, p. 19) pointedly argues: "To a large extent, European citizens are international migrants with the same rights as internal migrants."

Intra-European migration is not evenly distributed. Between 2008 and 2018, almost 12% of the Romanian population of working age moved to another EU member state, but only just under one percent of the Spanish population, despite the economic crisis on the Iberian Peninsula at the beginning of the second decade (Eurostat, 2020). Despite reports about the "new guest workers" from Southern Europe (Becker et al., 2013), the increase in intra-European labour mobility is to a considerable extent due to migration movements from the new to the old member states.

A New Dynamic of Labour Mobility in Europe

After many years of lamenting the "immobile European", a new dynamic of labour mobility in the EU has developed in the wake of Eastern enlargement in 2004 and 2007 (and partly also since Croatia's accession in 2013). Since 2004, the number of migrants from the new member states (NMS) who are employed in an old member state (OMS) has increased by almost four million. It therefore does not seem exaggerated to say that Eastern enlargement "spurred the largest European movements of labour in modern times" (Døvik, 2016, p. 1).

Recent rounds of enlargement were by no means the first in the EU. However, the interstate income differences have never been as high as at the beginning of the 21st century. In 2007, the per capita income of the eight accession countries from Central and Eastern Europe in 2004 was less than 60% of the EU-15 average, and in the case of Bulgaria and Romania, only a third (in purchasing power standards) (European Commission, 2008, p. 124). Never before in the history of modern nation-states has a similar free movement regime been implemented against the backdrop of comparable income differences. This wealth gap is an important factor that majorly contributed to the new dynamic of intra-European migration.

A correlation between wage differentials and willingness to migrate applies less to migration within the "old" EU-15, where interstate income differences have tended to decrease at least until the economic and financial crisis of 2008/09 and where an increase in mobility was nevertheless recorded (Recchi, 2008). This is explained, among other things, by the fact that the introduction of EU citizenship in 1993 has reduced mobility costs and non-economic migration motives have gained greater importance (King, 2002). Different migration motives between mobile Europeans from the new and old member states seems to be partly reflected in survey data (European Commission, 2012, p. 264). However, a distinction between "lifestyle migration" (OMS) versus "labour migration" (NMS) would be highly simplistic. For many migrants from the OMS, labour market-related motives are of great importance, especially with regard to their own career, while the migration decision of individuals from the NMS often involves more than just economic considerations (Krings et al., 2013). It seems more useful to assume a continuum between economic and non-economic motives, which often overlap (King, 2002). Gainful employment remains of central importance in structuring intra-European mobility, even if the latter no longer takes place as part of organized "guest worker programs", but as part of a transnational free movement regime.

The Diversity of Intra-European Labour Mobility

Employment-related mobility in Europe differs not only in terms of occupational level and the duration of stay, but also in terms of the employment relationship. The most significant form of intra-European labour migration takes place within the regulatory framework of worker mobility and includes EU/EFTA citizens who pursue gainful employment in another member state and have at least temporarily their center of life there (Fries-Tersch et al., 2020). Commuter migration (in official EU parlance "cross-border workers") and seasonal work also take place within the framework of worker mobility, and the host country principle applies (with some exceptions in social security and taxation). Especially as regards short-term forms of employment-related mobility, frequent border crossings are quite widespread, creating "transnational social spaces" (Pries, 2010) characterized by relatively permanent cross-border interconnections that can extend over the territorial space of several nation-states.

Another example of a temporary form of mobility is cross-border posted work. In this employer-driven mobility, employees are sent to another member state for a limited period to perform an economic activity there. Unlike commuter migration or seasonal work, this mobility is regulated within the framework of the EU Posting of Workers Directive, and the host country principle applies only to some extent. The importance of posted work for intra-European employment mobility should not be underestimated. For example, in 2014 more posted workers crossed national borders in the EU than regular workers (Barslund & Busse, 2016, p. 5).

The mobility of self-employed persons, which takes place within the framework of freedom of establishment, is also an example of the diversity of employment-related migration. In 2018, it accounted for 10% of all mobile EU citizens of working age (Fries-Tersch et al., 2020, pp. 69–70). Collective agreements do not apply to self-employed persons, and most labour law provisions also do not apply. There are some indications that in the context of cross-border mobility, this often involves a form of "dependent self-employment" (Mühlberger, 2007) as work is carried out under supervision. An increase in posted work and self-employment can be seen as evidence that employment-related migration in the 21st century differs not only by qualification, but also by employment status (Bosch, 2013). This raises questions about the regulation of these movements in a transnational mobility space.

The (Re-)Regulation of Transnational Labour Mobility

Intra-European migration illustrates the challenges associated with a transnational mobility space characterized by significant income differences. This became particularly evident during the COVID-19 crisis, when the precarious working and living conditions of contract workers from Southeast Europe were revealed (Hecking & Klawitter, 2020). One challenge is that regulatory gaps in European multi-level governance offer an opportunity for "regime shopping" (Dølvik et al., 2014) between different labour and social systems. In addition, there are recurrent difficulties in enforcing existing legal norms in cross-border contexts (Staples et al., 2013). This underlines that European unification is dominated by a logic of "negative integration" (Scharpf, 1998). While barriers to cross-border goods and people mobility have largely been removed, the creation of cross-border regulatory institutions is "still in its infancy" (Gagawczuk, 2018). Against this background, the question arises to what extent a trend towards re-regulation of national labour and wage systems can be observed, taking into account the constraints of European law (Krings, 2016). To explore this further, the role of collective labour market actors such as trade unions and employers' associations, but also state agencies need to be considered.

Methodological Approach and Structure of the Book
From a methodological perspective, in addition to a wealth of scientific literature and policy studies, microdata from the European Labour Force Survey (LFS) was used to investigate the scope and employment patterns of intra-European mobility. The LFS is the largest representative household sample survey in Europe, conducted quarterly by national statistical offices, collecting information on the labour market situation and the origin of the respondents. In 2018, the quarterly sample size of the LFS comprised approximately 1.5 million people across the EU. An analysis of LFS data is intended to provide not only an overview of the extent of work-related mobility in the EU, but also of qualification levels and occupational status of EU migrants, with country-specific differences also to be highlighted.[3] In principle, it can be assumed that in sample surveys like the LFS, the number of migrant respondents, especially those newly arrived, is underestimated (European Commission, 2012, p. 283). In addition, the LFS only captures migrants who have lived in their country of residence for at least one year (or intend to do so). Therefore, it can be assumed that while the LFS can provide a reliable overview of the development and direction of migration movements, it underestimates the actual extent of work-related mobility, as short-term mobility such as posted work or seasonal work are not taken into account (ibid.). To also consider these, additional data from national labour market authorities was accessed.

Worker mobility is a political field in which a multitude of different actors operate at national and supranational level (Bernhard & Bernhard, 2014). To shed more light on the power relations among these actors especially with regard to the regulation of the labour market, policy documents, public statements and press releases from labour market actors at national and European level were additionally accessed and evaluated from a content-analytical perspective.

The empirical focus of the book is on the main destination countries of intra-European migration in Western Europe. Of course, reference is repeatedly made to the countries of origin, as cross-border mobility can only be understood in a relational perspective between country of origin and destination country. It should also be noted that many countries of origin in Central and Eastern Europe have now themselves been transformed into immigration countries. Obviously, cross-border mobility in the "age of migration" is not a one-way street. However, the focus on the destination countries in Western Europe arises from the fact that the

[3] For more information on the EU-LFS: https://ec.europa.eu/eurostat/de/web/lfs/overview.

employment situation of mobile EU citizens and the (re-)regulation of work and mobility is a central focus of investigation.

Chapter 2 outlines the conceptual framework of the book. The transnationalization of employment is analysed using classical labour market theories. It is argued that the interaction of supply and demand continues to significantly influence the dynamics of labour markets. However, it remains theoretically underdevleoped how this relationship unfolds in a transnational mobility space. For a sufficient understanding of cross-border labour markets, it is therefore crucial how economic push and pull factors interact with the European free movement regime and the institutional regulation of employment. In Chap. 3, the extent, diversity and labour market impact of intra-European labour mobility is examined empirically. It is shown that a new dynamic of cross-border migration has developed in the 21st century. This mobility has met a demand for skilled and less skilled workers, but also posed a challenge to workers' rights in a transnational migration space.

The following two chapters conduct two country case studies on the impact of intra-European labour mobility. Chapter 4 shows that a booming and open labour market in Germany provided the context for an exceptional rise in inward migration. Many EU migrants found employment relatively quickly, which helped to alleviate a shortage of skilled and unskilled labour in some sectors. However, this migration was also accompanied by dumping practices. This experience was a key factor in fuelling the debate about the introduction of a statutory minimum wage. In Chap. 5, it is shown that in Austria similar challenges existed with regard to cross-border labour mobility, especially in connection with "atypical" employment. Unlike in Germany, however, a re-regulation of the labour market was negotiated within the social partnership framework. This shows that there is considerable scope for national regulatory measures in the European system of multi-level governance.

In Chap. 6 a closer look is taken at the current situation of freedom of movement in view of the COVID-19 pandemic and Brexit. Using three case studies on German agriculture, Austrian 24-hour mobile care, and the British transport sector, the "systemic importance" of transnational labour mobility is demonstrated. The demand for migrant workers may be socially constructed, but it is nonetheless socially real. In the concluding Chap. 7 the main results of the book are summarized and a look into the future is taken. It is argued that intra-European mobility will continue to be the most significant form of employment-related migration in the EU. This is likely to be due to the still existing substantial wage differences between states, but also to the European free movement regime, which reduces the "transaction costs" for cross-border migration and enables relatively frictionless mobility.

References

Bade, K. J. (2000). *Europa in Bewegung. Migration vom späten 18. Jahrhundert bis zur Gegenwart*. Beck.

Barslund, M., & Busse, M. (2016). Labour mobility in the EU: Addressing challenges and ensuring 'fair mobility'. CEPS Special Report Nr. 139. Retrieved October 2, 2018, from https://www.ceps.eu/system/files/SR139%20MB%20and%20MB%20LabourMobility.pdf

Becker, S., et al. (February 2013). Der deutsche Traum. *Der Spiegel, 25*.

Bernhard, S., & Bernhard, S. (2014). Arbeitnehmerfreizügigkeit und Diskriminierung auf dem deutschen Arbeitsmarkt – Europäische Politik, transnationaler Mobilitätsraum und nationales Feld. *Berliner Journal für Soziologie, 24*(2), 169–199.

Bosch, G. (2013). Grenzüberschreitende Arbeitsmärkte und nationale Beschäftigungssysteme in der EU. In H. G. Soeffner (Ed.), *Transnationale Vergesellschaftungen* (pp. 279–292). Springer VS.

Castles, S., & Miller, M. J. (2009). *The age of migration: International population movements modern world* (4th ed.). Palgrave Macmillan

Dølvik, J. E. (2016). Introduction: Transnational labour mobility – Engine for social convergence or divergence in Europe? In J. E. Dølvik & L. Eldring (Hrsg.), *Labour mobility in the enlarged Single European Market* (Comparative Social Research) (Vol. 32, pp. 1–17). Emerald Group.

Dølvik, J. E., Eldring, L., & Visser, J. (2014). Setting wage floors in open labour markets: The social partners in Europe's multilevel governance. In S. Evju (Ed.), *Regulating transnational labour in Europe: The quandaries of multilevel governance* (pp. 131–186). University of Oslo, Department of Private Law.

Europäische Kommission. (2019). Standard-Eurobarometer 92. Herbst 2019. Retrieved April 21, 2021, from https://europa.eu/eurobarometer/surveys

Europäische Kommission, Generaldirektion Beschäftigung, soziale Angelegenheiten und Chancengleichheit. (2008). *Employment in Europe 2008*. Amt für amtliche Veröffentlichungen der Europäischen Gemeinschaften.

Europäische Kommission, Generaldirektion Beschäftigung, Soziales und Integration. (2012). *Employment and social developments in Europe 2011*. Amt für amtliche Veröffentlichungen der Europäischen Gemeinschaften.

Eurostat. (2020). EU citizens living in another Member State – Statistical overview. Retrieved April 15, 2020, from https://ec.europa.eu/eurostat/statistics-explained/index.php?title=EU_citizens_living_in_another_Member_State_-_statistical_overview

Favell, A. (2008). *Eurostars and Eurocities: Free movement and mobility in an integrating Europe*. Blackwell.

Favell, A., & Hansen, R. (2002). Markets against politics: Migration, EU enlargement and the idea of Europe. *Journal of Ethnic and Migration Studies, 28*(4), 581–601.

FAZ. (2020). Baustopp wegen des Coronavirus: Viele Kräne stehen still. Retrieved April 27, 2020, from https://www.faz.net/aktuell/gesellschaft/gesundheit/coronavirus/baustopp-wegen-corona-folgen-fuer-hessisches-baugewerbe-16685554.html

Fries-Tersch, E., Jones, M., Böök, B., de Keyser, L., & Tugran, T. (2020). *2019 Annual report on Intra-EU labour mobility*. Publications Office of the European Union.

Gagawczuk, W. (2018). Die Europäische Arbeitsbehörde – Lohndumping entgegenwirken. Infobrief EU & International 1/2018. Kammer für Arbeiter und Angestellte für Wien.

Retrieved February 22, 2020, from https://media.arbeiterkammer.at/wien/EU_Infobrief_1_2018.pdf

Hecking, C., & Klawitter, N. (2020). Corona-Infektionen in Schlachthöfen: Eine verseuchte Branche. Retrieved May 15, 2020, from https://www.spiegel.de/wirtschaft/unternehmen/corona-krise-deutschlands-schlachthoefe-sind-zu-hotspots-geworden-a-00212d07-5546-4edd-9040-22ec848409c9

King, R. (2002). Towards a new map of European migration. *International Journal of Population Geography, 8*(2), 89–106.

Krause, A., Rinne, U., & Zimmermann, K. F. (2014). *How far away is a single European labour market? IZA Discussion Paper Nr. 8383*. Forschungsinstitut zur Zukunft der Arbeit.

Krings, T. (2016). East-West mobility and the (re-)regulation of employment in transnational labour markets. In J. E. Dølvik & L. Eldring (Eds.), *Labour mobility in the enlarged Single European Market* (Comparative Social Research) (Vol. 32, pp. 183–213). Emerald Group.

Krings, T., Moriarty, E., Wickham, J., Bobek, A., & Salamońska, J. (2013). *New mobilities in Europe: Polish migration to Ireland post-2004*. Manchester University Press.

Mau, S., Brabandt, H., Laube, L., & Roos, C. (2012). *Liberal states and the freedom of movement: Selective borders, unequal mobility*. Palgrave Macmillan.

Mühlberger, U. (2007). *Dependent self-employment: Workers on the border between employment and self-employment*. Palgrave Macmillan.

ORF. (2020). Mediziner warnen: 24-Stunden-Betreuung vor Zusammenbruch. Retrieved April 27, 2020, from https://www.orf.at/stories/3162384/

Pries, L. (2010). *Transnationalisierung: Theorie und Empirie grenzüberschreitender Vergesellschaftung*. Springer.

Recchi, E. (2008). Cross–state mobility in the EU: Trends, puzzles and consequences. *European Societies, 10*(2), 197–224.

Recchi, E. (2015). *Mobile Europe: The theory and practice of free movement in the EU*. Palgrave Macmillan.

Sachverständigenrat deutscher Stiftungen für Integration und Migration (SVR). (2013). *Erfolgsfall Europa? Folgen und Herausforderungen der EU-Freizügigkeit für Deutschland Jahresgutachten 2013 mit Migrationsbarometer*. SVR.

Scharpf, F. W. (1998). Balancing positive and negative integration: The regulatory options for Europe. In D. Dettke (Ed.), *The challenge of globalization for Germany's Social Democracy: A policy agenda for the twenty-first century* (pp. 29–57). Berghahn.

Staples, R., Trinczek, R., & Whittall, M. (2013). Posted Workers: Zwischen Regulierung und Invisibilisierung. *Arbeit, 22*(4), 271–286.

The Economist. (2020, April 16). Agricultural labour: Not enough Britons want to work on farms.

Wagner, I. (2018). *Workers without borders: Posted work and precarity in the EU*. Cornell University Press.

Wagner, B., & Hassel, A. (2016). Move to work, move to stay? Mapping atypical labour migration into Germany. In J. E. Dølvik & L. Eldring (Eds.), *Labour mobility in the enlarged Single European Market* (Comparative Social Research) (Vol. 32, pp. 125–158). Emerald Group.

2 Labour Migration as an Exampe of the Transnationalization of Work

This chapter develops the conceptual framework of the book. The transnationalization of work is examined using the example of labour migration. In the 21st century, there is an increase in cross-border mobility within and outside of employment organizations. To a significant extent, this mobility is demand-driven, as there is still a need for workers with different qualifications in "post-industrial" societies. However, it is also supply-driven, as the transaction costs for cross-border mobility have been reduced, especially in the enlarged EU. Thus, drawing on classic labour market theories, it is argued that the interaction of supply and demand continues to be central to the dynamics of labour markets. However, this interaction increasingly takes place in a transnational mobility space, where labour market-related factors interact with the institutional architecture of a free movement regime.

2.1 The Transnationalization of Work

Labour market sociology, like other sub-areas of sociology, has mostly analyzed social structures and processes *within* nation-states in the past. This "methodological nationalism" was justified to a certain extent. One of the characteristics of European societies were (and are) distinctive *national* labour markets with historically grown traditions of labour relations, vocational education systems, and patterns of employment regulation, which are also shaped by cultural norms and values (Wickham, 2016). These national labour markets have not necessarily lost their importance in the face of economic globalization and internationalization. Thus, the negotiation of working conditions and collective agreements still

T. Krings, *Labour Mobility in the European Union as an Example of the Transnationalization of Employment*, https://doi.org/10.1007/978-3-658-43977-4_2

largely takes place at the national (or regional) level (Baccaro & Howell, 2017). The persistence of national employment models is also evident in the field of vocational training, the gender order, personnel management, and work organization (Bosch et al., 2009; Flecker, 2017). However, and this is central to the argument of this book, nationally framed employment increasingly takes place under conditions that transcend the borders of nation-states.

To analytically capture cross-border linkages in the world of work, the concept of transnationalization, as developed by, among others, Ludger Pries, is useful (Pries, 2010a, 2013). The theoretical advantages of this concept lie in its emphasis on the "continuing formative power of spatial references and national societies" (Pries, 2010a, p. 10), unlike the globalization perspective. This is of particular relevance for the analysis of the transformation of work. Employment is still mostly performed in specific workplaces and is spatially and locally bound.[1] At the same time, work is increasingly embedded in cross-border linkages that go beyond the nation-state. To capture these various socio-spatial references, the transnationalization perspective offers a helpful analytical framework, which opposes methodological nationalism in social science labour research, without advocating a de-territorialization and liquefaction of the social (Pries, 2013, p. 882).

Economic internationalization is primarily shaped by multinational corporations (MNCs). Supported by new information and communication technologies (ICTs), MNCs have reorganized themselves as cross-border "network enterprises" (Castells, 2017) since the 1980s. These companies are still anchored in their home countries and are shaped by them, but they increasingly have a global orientation in their structures, strategies, and identities. As part of transnational value chains, they transcend the borders of nation-states and rely on the outsourcing of parts of production and service functions (Hirsch-Kreinsen, 2010).

Internationalization strategies are by no means limited to larger companies. Many smaller companies are also part of cross-border corporate networks, whether as highly specialized supplier companies in the automotive industry or as personnel service providers in the placement of workers across national borders (Quack et al., 2018). Moreover, the phenomenon of cross-border subcontracting is particularly noticeable in the European construction industry. The former

[1] This point was somewhat qualified during the COVID-19 pandemic, as mobile work and home office have gained in importance.

mainly includes smaller companies that either offer highly specialized construction services ("cooperative subcontracting") or cost-effective, labour-intensive activities ("competitive subcontracting") (Harvey, 2003).

Economic internationalization often goes hand in hand with an increase in employment-related mobility within and outside of organizations (Maletzky & Pries, 2014). This mobility ranges from intra-corporate assignments of highly qualified "expatriates" to "posted workers" in the European construction or meat industry, to individual labour migration, commuting migration, seasonal work, and "work and travel", where the boundaries between economic and non-economic motives become blurry (Conradson & Latham, 2005; King, 2002). An increase in transnational labour migration is facilitated by new transport opportunities, but also by new communication technologies, through which information about possible jobs circulate in virtual networks (Urry, 2007). Especially in a transnational mobility space like the EU, where an institutional architecture promotes cross-border mobility, more and more people have work experience from abroad, ranging from a few months to several years (European Commission, 2010).

The transnationalization of work does not necessarily require the relocation of the production site or the mobility of employees. New ICTs enable the cross-border outsourcing of services in areas such as accounting or customer service (Quack et al., 2018, p. 4). Also in knowledge-intensive services such as software development, employees are increasingly involved in transnational work contexts (Schulz-Schaeffer & Bottel, 2018). Moreover, the transnationalization of employment is also reflected in a "transformation of everyday work and collaboration" (Platzer et al., 2020, p. 14). This is evident, for example, in virtual communication or in multicultural work teams from different regions of the world. As a rule, "all forms of employment - even the seemingly most 'down-to-earth' and traditional ones - are increasingly shaped by conditions that transcend the boundaries of national societies" (Pries, 2010b, p. 740).

In addition to the cross-border mobility and networking of companies, people, and work activities, transnationalization also includes the emergence of international patterns of employment regulation (Quack et al., 2018). As already mentioned, the negotiation of labour and social standards still primarily takes place at the national level. At the same time, mechanisms of cross-border employment regulation have emerged since the 1990s, ranging from the minimum labour standards of the International Labour Organization (ILO), through European Works Councils, to international framework agreements between trade union federations and multinational companies (Pries, 2017). Also in the field of cross-border labour mobility, international agreements such as the UN Migrant Workers

Convention or the ILO Equal Treatment Convention have been adopted. However, these agreements have only had limited legal effectiveness, as the wealthy destination countries of the North have mostly not signed them (Ruhs, 2013).

The most advanced system of international employment regulation is in the EU. Above all, EU secondary law comprises a considerable corpus of directives on labour and employment policy, which must be implemented by the individual member states. The aim is less the standardization of labour and social standards, but rather the establishment of European-wide minimum standards. This is considered necessary to take into account the various national employment systems (Rhodes, 2010). Also, in the 1990s, the Social Dialogue was institutionalized at the European level, which allows employers' associations and trade unions to negotiate framework agreements that are then transposed into EU directives. Despite these developments, it should be noted that collective interest policy at European level is only weakly developed. This is due, on the one hand, to the fact that the "hard" topics of wage policy continue to remain a "national prerogative" (Pernicka, 2015, p. 3). On the other hand, there is often no interest on the part of employers' associations to negotiate binding European-wide regulations. Nevertheless, a transnational regime of employment regulation has developed in the EU, in which state and non-state actors work together and which is characterized by the entanglement of the local, national, and supranational level (Pries, 2017, pp. 236–250).

In the course of the internationalization of work, transnational labour markets have emerged, especially in Europe, in which relations of labour supply and demand are mediated in a cross-border context (Mense-Petermann, 2020). This poses a challenge for social science research, which has so far analyzed the structure and functioning of labour markets mostly within national societies (Köhler & Krause, 2010; Sengenberger, 1987). This research tradition has worked out how the interaction of supply and demand shapes the dynamics of labour markets. However, it remains theoretically underdeveloped how this relationship unfolds in a transnational mobility space, where the inflow of workers is no longer regulated by the nation-state. Thus, an analytical framework is required that goes beyond the "container" of the nation-state (Pries, 2010b).

Understanding cross-border labour markets is central to how economic push factors (unemployment, low wages) and pull factors (demand for labour, higher wages) interact with processes of opening and closure in the European mobility space and how this influences the strategies of employers and workers. The various institutional settings in the multi-level system of the EU are also relevant, as they can both enable and constrain market actors (Mense-

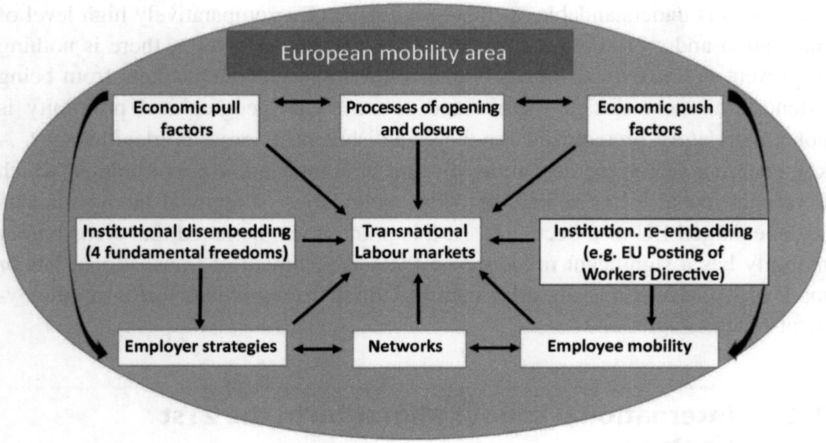

Fig. 2.1 The constitution of transnational labour markets in the European mobility space. (Source: Own representation)

Petermann, 2020). In this respect, new opportunity structures for employers have developed with the completion of the single market and Eastern enlargement. The former appear as "market makers" (ibid.), often with the support of cross-border intermediaries such as temporary employment agencies and subcontractors.

Thus, in the analysis of the transnationalization of labour markets, the "primacy of the demand side" (Krause & Köhler, 2010) can be maintained. However, the supply side also plays an important role in the European mobility space, as workers and service providers from an EU member state can freely offer themselves elsewhere, regardless of whether there is a shortage of labour. Thus, supply and demand in transnational labour markets can develop in a dynamic fashion, especially when employers develop a preference for "flexible" workers from abroad (Wiesböck, 2016). A crucial question is to what extent a "re-embedding" of cross-border mobility takes place in a transnational context, where different regulatory levels often stand in a tense relationship to each other (Staples et al., 2013). Of crucial importance is to what extent local labour market regulations are applied to cross-border mobility. The above Fig. 2.1 attempts to visualize the different factors that shape transnational labour markets in the European mobility space.

Most studies on transnational labour markets refer to regions in adjacent neighboring states (Gijsel et al., 1999; Verwiebe et al., 2014; Wiesböck, 2016).

This focus is understandable, as these regions have a comparatively high level of integration and cross-border commuting is important. However, there is nothing to prevent the analytical framework of transnational labour markets from being extended to more distant regions and countries, as geographical proximity is not a mandatory prerequisite for the establishment of cross-border linkages. A key enabling factor are new transport and communication opportunities, which have contributed to the redrawing of the boundaries of regional labour markets in the enlarged EU (see Sect. 2.4). At this point, it can be noted that cross-border mobility has a significant influence on the establishment of transnational labour markets. To what extent does this mobility differ from previous forms of employment-related migration?

2.2 International Labour Migration in the 21st Century

Even though non-economic forms of international migration have gained in importance, a clear majority of cross-border movements is still work-related (Pries, 2020, p. 20). The main regions of this mobility are Asia, North and Central America, and Europe, with countries of origin and destination becoming more diverse in the face of a "globalization of migration" (Castels & Miller, 2009). The qualifications of international migrants has also become more diverse. Especially since the 1990s, the importance of skilled migration has increased against the backdrop of a shift towards an "information society" (Castells, 2017). At the same time, there is still a demand to fill low-skilled jobs in the highly developed OECD countries, so that to a certain extent a polarization of employment among international migrants can be observed (Emmenegger & Careja, 2012; Pries, 2010b).

A demand-oriented approach to labour migration was developed by Michael Piore within the framework of dual labour market theory (DLMT) (1979, 1986). According to Piore, the labour markets in highly developed economies are split into a "primary", capital-intensive sector with high productivity and a less productive, labour-intensive "secondary" sector. Since domestic workers strive for the primary sector, where relatively stable, well-paid jobs can be found, there is a structural demand for labour in the secondary sector, where jobs are poorly paid, have a low social status, and offer few opportunities for advancement (Piore, 1979, pp. 26–45).

Piore's analysis mainly referred to the (ending) era of Fordist mass production. Authors such as Saskia Sassen or Roger Waldinger and Michael Lichter have further developed a demand-oriented perspective for "post-industrial" societies.

Sassen (1988, 1991) argues that structural change in the core economic countries and the integration of formerly peripheral regions into the world economy produces new migration movements. In particular, "global cities" are a pull factor for labour migrants, as here a growth of knowledge-intensive services (e.g., in finance or IT) goes hand in hand with a rising demand for low-skilled services (e.g., cleaning or household-related services). The latter are mostly performed by women and are considered a paradigmatic example of the "feminization of migration" (Spindler, 2011).

While Sassen pursues a macro-theoretical approach in the tradition of world system theory, Waldinger and Lichter (2003) focus on the micro level of companies. They investigate why, despite a generally rising level of education, there is still a demand for labour to fill low-skilled jobs. Part of their answer is that formal qualifications are of little importance for many precarious service jobs. Rather, it is about the "right attitude" to the job, which employers suspect more among new migrants than among long-term residents. The former would compare the working conditions at the destination with those in the region of origin ("dual frame of reference"), with the new destination usually scoring better and thus even a low-wage job can appear attractive (cf. also Anderson et al., 2006).

Even though the dividing lines in the labour market are more complex than assumed by DLMT, segmentation theories still represent a promising approach to explain why there is still a demand for migrant labour in "post-industrial" societies. This also includes a demand for irregular workers, which are both a cause and a consequence of a trend towards flexibilization and informalization of labour markets (Wallace, 2001, p. 55–56; Wickham, 2016, p. 192). Above all, the main assumption of segmentation theory that certain groups of workers can be assigned to distinctive job clusters, which differ in terms of pay, employment stability and mobility opportunities, proves to be correct, especially with regard to intra-European labour mobility (Felbo-Kolding et al., 2019; McCollum & Findlay, 2015).

However, DLMT with its focus on migrant inflows into secondary labour markets cannot explain why an increasing differentiation of migration can be observed. While a majority of migrants are still employed in low-skilled positions globally, the segment of skilled migration has grown since the 1990s, for which many OECD countries have lowered the barriers (Martin et al., 2006). This development is to be seen against the background of the change towards an information society, in which the demand for specialists in areas such as IT, research and development, financial services and consulting has increased. The mobility of these "highly qualified professionals" (Castells, 2017) takes place within the internal labour markets of multinational companies, but also externally as part of international company networks or within the framework of state recruitment policies (Koser & Salt, 1997).

Another factor is demographic change in Western societies, which has increased the demand for skilled workers, especially in the health sector. Most OECD countries have lowered the barriers for these migrants, who are often employed by the state (Adel et al., 2004). In this context, ethical questions about the social consequences of recruiting medical professionals from less developed regions ("brain drain") keep arising (Martin et al., 2006, pp. 70–71). Especially the new EU member states experienced significant emigration of medical staff post-2004, which they could only partially compensate for through migration from third countries (Fóti, 2013).

How the demand for labour is specifically designed depends not least on the institutional environment, the collective training system and employer strategies in different "varieties of capitalism" (Afonso & Devitt, 2016). While, for example, in Scandinavia there is mainly a demand for skilled workers due to a "high road strategy" and relatively high wages, migrants in Southern Europe are mainly found in low-skilled jobs (Sciortino, 2013). In English-speaking countries and in continental Europe, there is mainly a demand for skilled migration but also for workers in low-wage jobs, the number of which has grown especially in Germany since the mid-1990s (Bosch & Kalina, 2008).

Overall, there is little doubt that there continues to be a demand for migrants with different qualifications in Western societies. However, it must be taken into account that this demand is socially constructed. Often, there is less of a shortage of workers per se, but rather a shortage of workers who are willing to work under certain conditions. At least theoretically, there is the possibility to make these 'bad' jobs more attractive to domestic workers by increasing wages or improving working conditions (Castles, 2006, p. 761).

The situation is different in more skilled segments of the labour market, where there seems to be an absolute shortage of qualified workers in industries such as healthcare or IT. But here too, this shortage is not a given, but is directly related to the (lack of) willingness of companies to train and further qualify their employees (Ulrich, 2019). Thus, a shortage of skilled workers cannot be considered separately from the system of vocational training and the willingness of companies to invest in training. Especially in liberal market economies, companies tend to have a stronger "buy not make" orientation in terms of skills and qualifications (Wickham & Bruff, 2008), but even in continental Europe, a shortage of skilled workers is sometimes "home-made" (Ulrich, 2019).

The argument that a demand for labour is always socially constructed is more of a theoretical nature. The social reality of Western European labour markets seems to be closer to Piore's model of structural demand for foreign labour. However, it would be too simplistic to explain labour migration only with the demand side. While the latter is of central importance especially at the beginning of the migration process, migration movements develop a certain dynamic of their own especially through the role of social networks. In addition, demand-side factors usually cannot be considered separately from supply-side factors such as the search for employment opportunities and higher income (Martin et al., 2006, pp. 7–8). Of particular relevance for the argument of this book is how these economic push and pull factors interact with the EU free movement regime. How has this regime developed in historical perspective?

2.3 A Transnational Mobility Space in Europe

In the EU, a transnational mobility space has emerged, based on the principle of freedom of movement. The main feature of this regime, which has been characterized as "the most open cross-state movement policy worldwide" (Recchi, 2006, p. 71), is the freedom of movement for workers. Access to the labour market can temporarily be restricted for new accession countries, as happened most recently during EU enlargement in 2004, 2007, and 2013. After a transition period, mobile EU/EFTA citizens are largely on an equal footing with local employees. From this perspective, "national labour markets" are a thing of the past.

The main impetus for opening national labour markets in Europe was of an economic nature: The free movement of workers was considered, along with the free movement of capital, goods and services, as one of the basic prerequisites for a functioning European internal market. Such a market-centered view was already found in the Treaty of Rome in 1957, which established the European Economic Community (EEC) and enshrined the four freedoms. Mobile workers were allowed to take up employment in another member state and to settle *subsequently* there (Recchi, 2015, p. 22). In other words, the residence permit was tied to an ongoing employment relationship.

Of the six founding members of the EEC, it was primarily Italy that demanded freedom of movement for workers in order to reduce national unemployment and strengthen the position of Italian citizens in other EEC countries, who had been moving north in larger numbers since the beginning of the 1950s (Favell & Hansen, 2002, p. 585). However, the enforcement of the free movement principle was long confronted with significant political and legal hurdles, as the EEC member states in the north were reluctant to give up their national prerogative of controlling immigration. Accordingly, mobile EEC citizens were effectively treated like third-country nationals and were still dependent on residence and employment permits (Recchi, 2006, p. 57). It was not until the Regulation (EEC) 1612/68 on the free movement of workers and Directive 68/360/EEC on the abolition of movement and residence restrictions removed important administrative and bureaucratic hurdles, which restricted freedom of movement. Regulation 1612/68 is particularly noteworthy, as it enshrined the principle of equal treatment and non-discrimination for mobile workers from EEC member states:

> [F]reedom of movement for workers [...] entails *the abolition of any discrimination based on nationality between workers of the Member States as regards employment, remuneration and other conditions of work and employment* as well as the right of such workers to move freely within the Community in order to pursue activities as employed persons subject to any limitations justified on grounds of public policy, public security or public health (Regulation 1612/68, emphasis T.K.).[2]

The anti-discrimination regulation was not given much attention at the time, as national sovereignty over the regulation of migration was considered sacrosanct. In retrospect, however, it had a far-reaching impact, as it institutionalized the principle of equal treatment and decoupled the claim to equal rights in the labour market from the citizenship of the country of residence. In a certain sense, this paved the way for "post-national citizenship" (Soysal, 1994), symbolized by EU citizenship.

While the free movement principle was initially concerned with mobile workers, free movement rights were extended over time from free movement for workers to free movement for persons. On the one hand, an extended concept of workers was used, which also included job seekers and part-time workers. On the other hand, the group of people granted free movement rights was gradually expanded (Bernhard & Bernhard, 2014, p. 174). The European Court of Justice (ECJ) played a central role in this context, strengthening the position of mobile EEC citizens in the 1970s and 1980s through a series of judgments ("case law"), concerning self-employed persons, seasonal workers and trainees (Recchi, 2015, p. 24). The 1992 Maastricht Treaty, which established the European Union, marked the culmination of this development. With the introduction of EU citizenship, which complements

[2]Regulation 1612/68 was replaced in 2011 by Regulation (EU) No. 492/2011 on the freedom of movement for workers, which henceforth enshrines equal treatment in access to and exercise of employment in another Member State.

national citizenship and came into force on 1 November 1993, a general right to freedom of movement was created, which was spelled out in the 2004 free movement directive as follows:

> Citizenship of the Union confers on every citizen of the Union a primary and individual right to move and reside freely within the territory of the Member States, subject to the limitations and conditions laid down in the Treaty (establishing the European Community, T.K.) and to the measures adopted to give it effect (Directive 2004/38/EC).

By making freedom of movement no longer dependent on an employment relationship, its meaning was transformed: Labour migrants became "EU movers" (Favell & Recchi, 2009, p. 9). Although the right to free movement can be restricted if mobile EU/EEA citizens do not have sufficient means of subsistence or health insurance coverage for a stay of more than three months, the scope for such restrictions is constrained by the case law of the ECJ (Joppke, 2010, pp. 165–166).

With the introduction of EU citizenship, various forms of cross-border mobility such as "lifestyle migration" or retirement migration were facilitated, which are not primarily economically motivated (King, 2002; O'Reilly, 2007). Despite this development, however, the majority of intra-European migrations are still work-related (Barslund & Busse, 2016). Therefore, free movement for workers (and also the free movement of services) still represents the core of the EU free movement regime. What are the special features of intra-European labour mobility?

2.4 Characteristics of Transnational Labour Mobility in Europe

The search for employment continues to be of central importance for European migrants. However, this is no longer the classic labour migration of the 20th century. While in the 1950s and 1960s "guest workers" were recruited as part of state recruitment programs, current mobilities take place in a transnational migration space that offers new opportunity structures. Since EU migrants are no longer dependent on a residence or employment permit, they have more agency in constructing flexible "worklife" biographies. This also applies to migrants from the new member states (NMS), who are overrepresented in the low-wage sector, but often pursue flexible mobility projects (Krings et al., 2013).

In other respects, too, current labour mobility differs from the "guest worker" era of the 1950s and 1960s. While migrants were mainly employed in industry and mining at that time, current labour mobility has a "post-industrial" character (Wallace, 2001). In addition, there is more occupational diversity among EU migrants, including a growing share of highly qualified individuals (European Commission, 2012). While "guest workers" in the past were largely confined to low-skilled segments, current EU migrants can also be found in higher-paid professional jobs, which are often part of a transnational career in "global cities" like Amsterdam, Berlin, Brussels, London or Vienna (Favell, 2008; Verwiebe & Müller, 2006).

One feature of intra-European migration is its more transient nature, characterized by repeated border crossings and circular mobility. This "liquid migration" (Engbersen et al., 2010) is less socially prestructured, shows a high degree of flexibility and targets multiple destinations. Its fluid character often reflects a demand for "flexible" workers in the expanding service sectors of Western Europe (Wallace, 2001). Repeated border crossings are made possible by geographical proximity and EU freedom of movement, but also by new transport options. Especially low-cost airlines like Ryanair and easyJet, which are inextricably linked with migration movements to the UK and Ireland (Burrell, 2011), have transformed the European mobility experience:

> In the 1950s and 1960s, migrants from Southern Italy and Turkey took several days to travel to Germany [...] At the beginning of the twenty-first century, Polish migrants would have travelled to Ireland on flights that would have taken less than three hours (Krings et al., 2013, p. 13).

New transport opportunities and easy online booking of flights facilitate the transnational mobility of individuals, and thus also the accessibility of labour markets. In this respect, the expansion of European air traffic has contributed to the redrawing of the boundaries of regional labour markets in the enlarged EU. This development was reinforced by new flight connections, which have not only emerged between international hubs, but especially between regional airports ("point-to-point") (Williams & Baláž, 2009).

A new experience of distance is created not only by a dramatic increase in European air traffic in the 21st century, but also by new information and communication technologies (Urry, 2007). The increased use of these technologies in the form of cheap (video) telephony, emails and virtual social networks enables the maintenance of social contacts over long distances and is an expression of time-space compression (Castells, 2017).[3] ICTs are not only of crucial importance in the post-migration phase, but also beforehand. This not only concerns the online booking of flights, but above all virtual search of possible destinations, including possible employment opportunities. Personal networks are still an important information resource for EU migrants. However, these networks have a more transient and individualistic character. Due to their legal status as EU citizens, but also due to the possibility of virtual job search, intra-European migrants are less dependent on "strong" ethnic linkages than was the case with previous generations of "guest workers" (Wickham, 2016, p. 194).

The interplay of physical and virtual "technologies of mobility" opens up the possibility of a transnational life *in-between,* in which the boundaries between "home" and "abroad" become more blurry (Pries, 2010a; Vertovec, 2004). At the same time, the "liquid" character of intra-European migration should not be overstated. Especially with regard to the post-2004 migration movements, settlement processes have become more widespread (OECD, 2012). Thus, it seems more fitting to arrange intra-European migration on a continuum of short-term and longer-term mobility, with different temporalities that reflect the flexibility demands of Western European labour markets (van Ostaijen & Scholten, 2018).

Undoubtedly, the EU free movement regime and the role of new transport and communication opportunities have expanded the options for EU migrants. Since employers and state institutions are less directly involved in the recruitment of workers, intra-European labour mobility is more supply-side driven, especially in the low-skilled segment (Wickham, 2016, p. 192). At the same time, East-West migration is also demand-driven. A demand for labour was the decisive factor that, in conjunction with Eastern enlargement, triggered a previously unknown dynamic of labour mobility in Europe (Dølvik, 2013; Galgóczi et al., 2009). Microsociological studies suggest that a demand for labour cannot be separated from the supply of available labour. Especially in some low-wage sectors, employers have developed a preference for "flexible" workers from the new EU (MacKenzie & Forde, 2009; Wiesböck, 2016). In other words, supply and demand can develop in a dynamic relationship to each other in a cross-border employment context. This also includes forms of irregular employment. Although EU citizens are by definition legal migrants, this does by no means preclude precarious employment. Paradoxically, the fact that EU

[3] At the same time, communication in virtual space also has its limits. Even new ICTs "cannot bridge all the gaps of information and expression endemic to long-distance separation" Vertovec (2004, p. 223).

citizens have a secure residence status can even facilitate taking up informal employment, as unlike third-country migrants, they do not have to fear consequences as regards their residency status (Recchi, 2015, p. 73). Often, precarious employment is associated with "atypical" employment such as self-employment, posted work, and temporary agency work, which are sometimes used as an alternative route into the labour market (Dølvik et al., 2014; Wagner, 2018). In this context, questions about the regulation of employment and compliance with labour and social standards in a transnational mobility space keep arising.

2.5 The Regulation of European Labour and Service Mobility

The completion of the European internal market with its four fundamental freedoms (goods, capital, services, and people) was characterized by a logic of "negative integration" and the tearing down of market barriers (Scharpf, 1998). The creation of supranational regulatory institutions in the sense of "positive integration" proved to be much more difficult against the background of the institutional and social heterogeneity of the EU (Ferrera, 2005; Pernicka, 2015). This is visible especially in the field of cross-border labour mobility, where market opening processes are much further advanced than market-correcting measures (Dølvik et al., 2014).

Despite this bias towards "market making", a transnational regulatory regime has emerged in the EU over time, which aims at "re-embedding" cross-border mobility flows (Evju, 2014). This regime is partially enshrined in the various EU treaties (primary law) and in regulations and directives (secondary law). While regulations apply directly in all EU member states, directives must be transposed into national law. This often leads to tensions between national and European regulatory levels, in particular in the area of labour market and social policy (Ahlberg et al., 2014).

In EU primary law, the Treaty on the Functioning of the European Union (TFEU) is particularly relevant, especially Article 21 on freedom of movement and Article 45 on freedom of movement for workers. In EU secondary law, Regulation (EU) No. 492/2011 is particularly noteworthy, which replaced Regulation 1612/68 and henceforth enshrines the principle of equal treatment in access to employment in another member state. To help implement

Tab. 2.1 Various forms of cross-border labour mobility in the EU. (Source: Own representation)

Regulatory framework	Temporality		Place of work principle (collective agreements and statutory minimum wages)
	Short-term and recurring mobility	Long-term mobility	
Worker mobility	Seasonal work, commuting mobility, circular migration	Individual migration	Yes
Freedom to provide services/ Posting of Workers Directive	Cross-border posting of workers/temporarily mobile self-employed	N/A (Cross-border postings are by definition temporary)	(Yes) (with regard to remuneration); (after 12/18 months (almost) all employment rules of the host country apply)
Freedom of establishment	N/A (it is assumed that there is permanent self-employment in another MS)	Self-employed (with and without employees)	(Yes) (although collective agreements and statutory minimum wages do not apply to the self-employed)

these rights, Directive 2014/54/EU was adopted, which includes measures for improved legal protection for mobile workers, among other things.

Central to the regulation of worker mobility is the extent to which the place of work principle *(lex loci labouris)* applies in terms of remuneration and employment conditions (see Tab. 2.1). The introduction of this principle in the 1960s was intended to prevent the respective labour and social standards in the individual member states from being undermined in the context of cross-border mobility (Dølvik, 2013, p. 36). In this regard, one can certainly speak of a process of "positive" integration, in which mobile EU workers were integrated into the employment system of the host country, not unlike the situation of earlier "guest workers" (Bommes & Geddes, 2000).

In case of the freedom to provide services (Article 56 TFEU) and the freedom of establishment (Article 49 TFEU), which also count among the EU fundamental freedoms, a different regulatory logic applies. Above all, the freedom to provide services stands for a process of market opening through EU competition law, which traditionally is based on the principle of the country of origin (Bosch, 2013, pp. 280–281). In this respect, an "institutional disembedding" (Holst, 2015) of work is observed, which is disconnected from the regulatory environment of the host country. This mobility is controversial, especially when it is used to circumvent employment standards or as an alternative route into the labour market

(Hunger, 2000; Wagner & Hassel, 2016). The mobility of self-employed persons sometimes involves 'bogus' self-employment, where work is carried out under instruction, but labour law does not apply (Molitor, 2015). Thus, the core of the dispute about cross-border mobility is often about the question of whether local collective agreements and minimum standards apply. Table 2.1 provides an overview of various forms of cross-border labour mobility along the dimensions of regulation, temporality, and place of work.

As can be seen from Tab. 2.1, the place of work principle now also applies to the remuneration of posted workers. This is the result of a decades-long, contested process of re-regulating service mobility. When the country of origin principle caused controversy in the case of posted work in the early 1990s, calls for a Europe-wide regulation grew. The driving forces were the European social partners in the construction industry, led by the European Federation of Building and Woodworkers (EFBWW) but also the European Construction Industry Federation (FIEC), who feared increasing competition from cheap companies from abroad. In their efforts to achieve a re-regulation of posted work, the social partners formed a coalition with parts of the European Commission and national governments, supported by the respective national construction unions (Dølvik et al., 2014, pp. 140–142). This political lobbying was ultimately successful when the Posting of Workers Directive 96/71/EC was adopted in 1996. The directive established a regulatory framework for universally binding minimum standards such as minimum wages and maximum working hours based on the host country principle (Eichhorst, 1999).

The implementation of the Posting of Workers Directive revealed significant national variations, reflecting the different traditions of industrial relations in the EU. While Austria and France largely enshrined the equal pay principle, in Germany only a statutory minimum wage for the construction sector was established, in addition to some minimum standards. In Sweden and Denmark, wage determination was completely excluded from the implementation of the directive, as there is a strong tradition of collective bargaining autonomy. This regulatory model came under pressure due to a series of controversial rulings by the ECJ ("Laval Quartet") and had to be adjusted accordingly (Døvik et al., 2014, p. 151–152). Thus, the dispute over posted work does not only reflect difficulty in (re-) regulating cross-border mobility, but also"implementation differences" (Staples et al., 2013) between the national and the supranational level.

In the first decade of the 21st century, with the exception of the 2008 Temporary Agency Work Directive, hardly any measures were adopted at EU level in the field of labour market and social policy. This was also a reflection of changed power relations in the enlarged EU, where liberal states like the UK effectively formed a coalition with the NMS to prevent initiatives that were seen as restrictions of the four fundamental freedoms (Dølvik et al., 2014, p. 146). In addition, the European Commission increasingly set other priorities, which were primarily aimed at completing the internal market for cross-border services. In this context, a deregulation agenda was pursued, which occasionally collided with the re-regulation efforts of nation-states. The dispute over the 2006 Services Directive is a case in point (Ahlberg et al., 2014).

In the second decade of the 21st century, the political wind turned again. Under the impression of the economic and financial crisis of 2008/2009, but also of the 2016 Brexit referendum, which left the European institutions in a collective state of shock, a free-market agenda increasingly came on the defensive. Under Commission President Jean-Claude Juncker, the "social dimension of Europe" was again more strongly emphasised from 2017 onwards. This programme included, among other things, a reform of the Posting of Workers Directive, which was adopted with Directive EU 2018/957. The changes at least partially enshrine the place of work principle for cross-border postings. From now on, the working and employment conditions of the host country, including remuneration "on the basis of equal treatment" (amended Article 3 of the Posting of Workers Directive), apply.[4] What was remarkable about this initiative was that, in contrast to the intergovernmental level, where the governments of Poland and Hungary opposed the reform, there were no East-West divisions to be observed on the trade union side. Workers' representatives from the NMS also supported the principle of equal treatment for cross-border postings (Furåker & Larsson, 2020).

The reform of the Posting of Workers Directive undoubtedly marks an important step towards implementing the principle of "equal pay for equal work in the same place". However, it remains to be seen to what extent these equality norms will also be implemented in transnational workplaces. Research on posted work in the construction industry or in the meat industry indicate a discrepancy between formal standards and actual implementation

[4] The labour law of the host country applies in full if a posting lasts longer than 12 or, in justified cases, 18 months.

(Berntsen & Lillie, 2016; Staples et al., 2013; Wagner & Hassel, 2015). In this context, an "enforcement gap" (Wagner & Berntsen, 2016) can be observed, particularly in connection with cross-border subcontracting chains. This is indicative of a structural problem in the EU, where a transnational mobility space has emerged, but no corresponding enforcement institutions beyond the nation-state (Bosch, 2013, p. 283).

The Posting of Workers Directive now largely enshrines the principle of equal treatment, but does not specify how the rights of posted workers can be enforced in a cross-border context (Ahlberg et al., 2014). Against this background, Directive 2014/67/EU ("Enforcement Directive") was adopted, which aims to "guarantee respect for an appropriate level of protection of the rights of posted workers for the cross-border provision of services, in particular the enforcement of the terms and conditions of employment that apply in the Member State where the service is to be provided (Article 1, paragraph 1). This directive provides, among other things, for improved access to information on working and employment conditions for posted workers "in a clear, transparent, comprehensive and easily accessible way" (Article 5, paragraph 1), as well as improved cooperation between national authorities through the Internal Market Information System. This is particularly important in view of the role of service providers, which are often "mailbox companies" established solely for the purpose of posting (Cremers, 2020).

The Enforcement Directive has been criticised for not specifying which rights apply in case of de facto posting disguised as (bogus)self-employment. It also remains unclear how administrative penalties can be enforced in a cross-border context (Seeliger & Wagner, 2020, p. 1120). To address this imbalance, the establishment of a European Labour Authority (ELA) was announced in 2017 as part of the "European Pillar of Social Rights". The ELA's task is to provide information for mobile EU citizens and businesses, to bring together existing instruments in the field of cross-border mobility, and to carry out joint enforcement activities in cooperation with national authorities (Cremers, 2020). It remains to be seen to what extent the ELA, which started its work in 2019, can help to close an "enforcement gap" with regard to cross-border mobility. The decisive factor will be to what extent the ELA will actually be able to exercise its powers in a transnational context (ibid., p. 29).

While postings now at least pro forma adhere to the host country principle as regards remuneration, social security continues to be based on the country of origin. According to EU Regulation 883/2004 on the coordination of social security systems, posted workers remain insured in their country of origin as long as the posting lasts less than 24 months (which is the norm). This fact is particularly important in the "Bismarckian" welfare states (Austria, Belgium, Germany, France, Netherlands), which not only have comparatively high wages, but also high social contributions (Krings, 2019, p. 26).

Posted work and the mobility of the self-employed are somewhat exceptional in the field of social security. In most other forms of cross-border mobility, the principle of equal treatment applies and EU migrants are included in the social security systems of the host country. This was by no means a tension-free process, as the individual member states showed little enthusiasm for giving up the previous coupling of welfare rights to *national* citizenship. Thus, the initiatives of the EU were limited to coordination (rather than standardization) of national social security systems (Recchi, 2015, p. 22). As early as 1971, Regulation (EEC) 1408/71 on the application of social security systems was adopted, which sets out the four essential principles for receiving social security benefits in the EU: 1) The principle of non-discrimination: EU citizens residing in another Member State must not be disadvantaged in access to social security benefits; 2) Portability of social security benefits: Benefits can be claimed in a Member State other than the one in which they were acquired; 3) Accumulation of qualifying periods: Insurance and waiting periods from other Member States must be taken into account when examining eligibility; 4) Principle of place of work: The regulations of the Member State in which the person was or is employed apply in relation to social security benefits (Ferrera, 2005, pp. 103–104).

Thus, the EU has created a regulatory framework for the portability of social benefits with the explicit aim of promoting cross-border labour mobility. However, there are still implementation problems when transferring social security benefits. A major reason for this is that mobile EU citizens often move between very different social security systems in Europe (Wickham, 2016, p. 185). In addition, supplementary pensions from occupational and private old-age provision, which have gained in importance in recent years, do not fall within the scope of Regulation (EC) 883/2004 on the coordination of social security systems (which replaced Regulation (EEC) 1408/71).[5] Moreover, especially in the context of recent EU enlargement, there is an increasing conditionality in access to welfare benefits, which challenges the EU's principle of equal treatment (Rooss & Westerveen, 2020).

[5] Regulation 883/2004 came into force on 1 May 2010 following the adoption of Implementing Regulation 987/2009.

The difficulty of cross-border transfer of social benefits is presumably one reason why the proportion of mobile EU citizens who live and work in another member state remains relatively low at four percent. Other reasons cited include language and cultural barriers, and difficulties in the recognition of professional qualifications (Dobson, 2009, p. 123). To simplify procedures in this area, a regulatory framework for the recognition of professional qualifications was created with EU directives 2005/36/EC and 2013/55/EU. However, these only apply to regulated professions that require a national certificate. As the most recent initiative in this area, the introduction of a European Professional Card (EPC) was implemented under Directive 2013/55/EU, which is intended to facilitate the recognition of qualifications in regulated professions via electronic means.[6] Despite this framework, there are still limits to the recognition of qualifications. Professional degrees are produced by different *national* education systems and sometimes not only involve different levels of qualification, but also different societal prestige (Bosch & Charest, 2008). Also, parts of the public sector that deal with the "exercise of public authority and duties designed to safeguard the general interest of the state" are still reserved for citizens only. Other jobs in the public service not only require appropriate language skills, but also contextual knowledge, for instance in the field of social services, which is less available to newly arrived migrants (Wickham, 2016, p. 184).

Other pro-mobility initiatives include the introduction of the European Health Insurance Card and the EURES employment platform, which brings together mobile workers and employers (https://ec.europa.eu/eures). The creation of a European Higher Education Area ("Bologna") is also significant as it enables the comparability of university degrees and can thus have a mobility-promoting effect. Indirectly, the introduction of the euro at the end of the 1990s can also be seen as a mobility-promoting initiative, as it has reduced the costs of cross-border transactions and allows for direct comparability of prices and wages (Recchi, 2008, pp. 205–207).

To summarize, in the EU a regulatory regime has emerged, which aims both at promoting and re-embedding transnational mobility. The debate on cross-border service mobility has shown that these two goals can be in a conflicting relationship with each other. Undoubtedly, this regime contains elements of "positive integration", but remains incomplete against the background of social and institutional diversity in the enlarged EU.

---

[6] So far, however, the EPC can only be used in the following five professions: pharmacist, mountain guide, real estate agent, nurse, and physiotherapist (European Union, 2021).

2.6 Summary

This chapter examined the transnationalization of work, using the example of labour migration. It argued that processes of transnationalization are not only shaped by company strategies, but also by employee strategies, whose work-related mobility within and outside of organizations has increased. To a significant extent, this mobility is demand-driven, as there is still a need for additional workers with different qualifications in "post-industrial" societies. But it is also supply-driven, as workers and service companies can offer themselves in a cross-border context, which is facilitated by the EU free movement regime. Therefore, for an understanding of EU labour mobility, it is central how economic push and pull factors interact with the free movement regime and the institutional regulation of employment in the European system of multi-level governance. In this context, the mutual actions of employers and employees play an important role in establishing cross-border linkages that are embedded in multiple institutional settings. The EU enlargement process in the 21st century in particular has added a new dynamic of transnational labour mobility, as will be examined in more detail in the next chapter.

References

Afonso, A., & Devitt, C. (2016). State of the Art paper: Comparative political economy and international migration. *Socio-Economic Review, 14*(3), 591–613.

Ahlberg, K., Johansson, C., & Malmberg, J. (2014). Monitoring compliance with labour standards: Restriction of economic freedoms or effective protection of rights? In S. Evju (Ed.), *Regulating transnational labour in Europe: The quandaries of multilevel governance* (pp. 87–216). University of Oslo, Department of Private Law.

Anderson, B., Ruhs, M., Spencer, S., & Rogaly, B. (2006). Fair enough? Central and East European migrants in low wage employment in the UK. Joseph Rowntree Foundation. Retrieved May 20, 2020, from https://www.compas.ox.ac.uk/2006/pr-2006-changing_status_fair_enough

Baccaro, L., & Howell, C. (2017). *Trajectories of neoliberal transformation: European industrial relations since the 1970s*. Cambridge University Press.

Barslund, M., & Busse, M. (2016). Labour mobility in the EU: Addressing challenges and ensuring ‚fair mobility'. CEPS Special Report Nr. 139. Retrieved October 2, 2018, from https://www.ceps.eu/download/publication/?id=9493&pdf=SR139%20MB%20and%20MB%20LabourMobility.pdf

Bernhard, S., & Bernhard, S. (2014). Arbeitnehmerfreizügigkeit und Diskriminierung auf dem deutschen Arbeitsmarkt – Europäische Politik, transnationaler Mobilitätsraum und nationales Feld. *Berliner Journal für Soziologie, 24*(2), 169–199.

Berntsen, L., & Lillie, N. (2016). Hyper-mobile migrant workers and Dutch trade union representation strategies at the Eemshaven construction sites. *Economic and Industrial Democracy, 37*(1), 171–187.

Bommes, M., & Geddes, A. (Eds.). (2000). *Immigration and welfare: Challenging the boundaries of the welfare state*. Routledge.

Bosch, G. (2013). Grenzüberschreitende Arbeitsmärkte und nationale Beschäftigungssysteme in der EU. In H. G. Soeffner (Ed.), *Transnationale Vergesellschaftungen* (pp. 279–292). Springer VS.

Bosch, G., & Charest, J. (2008). Vocational training and the labour market in liberal and coordinated economies. *Industrial Relations Journal, 39*(5), 428–447.

Bosch, G., & Kalina, T. (2008). Low-wage in Germany: An overview. In G. Bosch & C. Weinkopf (Eds.), *Low-wage in Germany* (pp. 19–112). Russell Sage Foundation.

Bosch, G., Lehndorff, S., & Rubery, J. (Eds.). (2009). *European employment models in flux. A comparison of institutional change in nine European Countries*. Palgrave Macmillan.

Burrell, K. (2011). Going steerage on Ryanair: cultures of migrant air travel between Poland and the UK. *Journal of Transport Geography, 19*(5) 1023–1030. https://doi.org/10.1016/j.jtrangeo.2010.09.004

Castells, M. (2017). *Der Aufstieg der Netzwerkgesellschaft. Das Informationszeitalter: Wirtschaft. Gesellschaft. Kultur* (2nd ed.). Springer.

Castles, S. (2006). Guestworkers in Europe: A resurrection? *International Migration Review, 40*(4), 741–766.

Castles, S., & Miller, M. J. (2009). *The age of migration: International population movements in the modern world* (4th ed.). Palgrave Macmillan.

Conradson, D., & Latham, A. (2005). Transnational urbanism: Attending to everyday practices and mobilities. *Journal of Ethnic and Migration Studies, 31*(2), 227–233.

Cremers, J. (2020). The European Labour Authority and rights-based labour mobility. *ERA-Forum, 21*, 21–34.

de Gijsel, P., Janssen, M., Wenzel, H. J., & Woltering, M. (Eds.). (1999). *Understanding European cross-border labour markets: Issues in economic cross-border relations*. Metropolis.

De Wispelaere, F., De Smedt, L., & Pacolet, J. (2020). *Posting of workers: Report on A1 portable documents issued in 2018*. European Kommission, Generaldirektion Beschäftigung, Soziales und Integration.

den Adel, M., Blauw, W. Dobson, J., Hoesch, K., & Salt, J. (2004). Recruitment and the migration of foreign workers in health and social care. In M. Bommes, K. Hoesch, U. Hunger, & H. Kolb (Eds.), *Organisational recruitment and patterns of migration: Interdependencies in an integrating Europe*. IMIS-Beiträge 25 (pp. 201–230). Institut für Migrationsforschung und Interkulturelle Studien.

Dobson, J. R. (2009). Labour mobility and migration within the EU following the 2004 Central and East European enlargement. *Employee Relations, 31*(2), 121–138.

Dølvik, J. E. (2013). European movements of labour: Challenges for European social models. In E. Jurado & G. Brochmann (Eds.), *Europe's immigration challenge: Reconciling work, welfare and migration* (pp. 33–58). Tauris.

Dølvik, J. E., Eldring, L., & Visser, J. (2014). Setting wage floors in open labour markets: The role of the social partners in Europe's multilevel governance. In S. Evju (Ed.), *Reg-*

ulating transnational labour in Europe: The quandaries of multilevel governance (pp. 131–186). University of Oslo, Department of Private Law.

Eichhorst, W. (1999). Europäische marktgestaltende Politik zwischen Supranationalität und nationaler Autonomie: Das Beispiel der Entsenderichtlinie. *Industrielle Beziehungen, 6*(3), 340–359.

Emmenegger, P., & Careja, R. (2012). From dilemma to dualization: Social and migration policies in the 'reluctant countries of immigration'. In P. Emmenegger, S. Häusermann, B. Palier, & M. Seeleib-Kaiser (Eds.), *The age of dualization: The changing face of inequality in deindustrializing societies* (pp. 124–148). Oxford University Press.

Engbersen, G., Snel, E., & Boom, J. D. (2010) 'A van full of Poles': Liquid migration from Central and Eastern Europe. In R. Black, G. Engbersen, M. Okólski, & C. Pantîru (Eds.), *A continent moving West? EU enlargement and labour migration from Central and Eastern Europe* (pp. 115–140). Amsterdam University Press.

Europäische Kommission. (2010). Special Eurobarometer 337: Geographical and labour market mobility. Retrieved May 12, 2020, from https://data.europa.eu/data/datasets/s803_72_5_ebs337?locale=en

Europäische Kommission, Generaldirektion Beschäftigung, Soziales und Integration. (2012). *Employment and social developments in Europe 2011*. Amt für amtliche Veröffentlichungen der Europäischen Gemeinschaften.

Europäische Union. (2021). Europäischer Berufsausweis. Retrieved September 13, 2021, from https://europa.eu/youreurope/citizens/work/professional-qualifications/european-professional-card/index_de.htm

Evju, S. (Ed.). (2014). *Regulating transnational labour in Europe: The quandaries of multilevel governance*. University of Oslo, Department of Private Law.

Favell, A. (2008). *Eurostars and Eurocities: Free movement and mobility in an integrating Europe*. Blackwell.

Favell, A., & Hansen, R. (2002). Markets against politics: Migration, EU enlargement and the idea of Europe. *Journal of Ethnic and Migration Studies, 28*(4), 581–601.

Favell, A., & Recchi, E. (2009). Pioneers of European integration: An introduction. In: E. Recchi & A. Favell (Eds.), *Pioneers of European integration. Citizenship and mobility in the EU* (pp. 1–25). Edward Elgar.

Felbo-Kolding, J., Leschke, J., & Spreckelsen, T. F. (2019). A division of labour? Labour market segmentation by region of origin: The case of intra-EU migrants in the UK, Germany and Denmark. *Journal of Ethnic and Migration Studies, 45*(15), 2820–2843.

Ferrera, M. (2005). *The boundaries of welfare: European integration and the new spatial politics of social protection*. Oxford University Press.

Flecker, J. (2017). *Arbeit und Beschäftigung. Eine soziologische Einführung*. Facultas/UTB.

Fóti, K. (2013). Mobility and migration of healthcare workers in central and eastern Europe. European Foundation for the Improvement of Living and Working Conditions. Retrieved June 20, 2020, fromhttps://www.eurofound.europa.eu/sites/default/files/ef_publication/field_ef_document/ef1335en.pdf

Furåker, B., & Larsson, B. (2020) *Trade union cooperation in Europe: Patterns, conditions, issues*. Springer.

Galgóczi, B., Leschke, J., & Watt, A. (Eds.). (2009). *EU labour migration since enlargement: Trends, impacts and policies*. Ashgate.

Harvey, M. (2003). Privatization, fragmentation and inflexible flexibilization in the UK construction industry. In G. Bosch & P. Philip (Eds.), *Building chaos: An international comparison of deregulation in the construction industry* (pp. 188–209). Routledge.

Hirsch-Kreinsen, H. (2010). Multinationale Unternehmen. In F. Böhle, G. G. Voß, & G. Wachtler (Eds.), *Handbuch Arbeitssoziologie* (pp. 597–617). VS Verlag.

Holst, H. (2015). Europäisierung als institutionelle Entbettung – Finanzialisierte multinationale Konzerneund die Arbeitsbeziehungen im europäischen Paketsektor. In S. Pernicka (Ed.), *Horizontale Europäisierung im Feld derArbeitsbeziehungen* (pp. 151–182). Springer VS.

Hunger, U. (2000). Temporary transnational labour migration in an integrating Europe and the challenge to the German Welfare State. In M. Bommes & A. Geddes (Eds.), *Immigration and welfare: Challenging the boundaries of the welfare state* (pp. 189–208). Routledge.

Joppke, C. (2010). *Citizenship and integration*. Polity Press.

King, R. (2002). Towards a new map of European migration. *International Journal of Population Geography, 8*(2), 89–106.

Koser. K., & Salt, J. (1997). The geography of highly skilled international migration. *International Journal of Population Geography, 3*(4), 285–303.

Köhler, C., & Krause, A. (2010). Betriebliche Beschäftigungspolitik. In F. Böhle, G. G. Voß, & G. Wachtler (Hrsg.), *Handbuch Arbeitssoziologie* (S. 387–412). VS Verlag.

Krings, T. (2019). *Posted Workers* in Österreich: Grenzüberschreitende Entsendearbeit im Spannungsfeld von offenen Märkten und (supra)nationaler Regulierung. *Österreichische Zeitschrift für Soziologie, 44*(1), 21–41.

Krings, T., Moriarty, E., Wickham, J., Bobek, A., & Salamońska, J. (2013). *New mobilities in Europe: Polish migration to Ireland post-2004*. Manchester University Press.

MacKenzie, R., & Forde, C. (2009). The rhetoric of the 'good worker' versus the realities of employers' use and the experiences of migrant workers. *Work, Employment and Society, 23*(1), 142–159.

Maletzky, M., & Pries, L. (2014). Die Transnationalisierung von Arbeitsmobilität. Entwicklungstrends und ausgewählte Herausforderungen ihrer Regulierung. *Arbeits- und Industriesoziologische Studien (AIS-Studien), 7*(2), 56–79.

Martin, P., Abella, M., & Kuptsch, C. (2006). *Managing labour migration in the twenty-first century*. Vail-Ballou.

McCollum, D., & Findlay, A. (2015). 'Flexible' workers for 'flexible' jobs? The labour market function of A8 migrant labour in the UK. *Work, Employment and Society, 29*(3), 427–443.

Mense-Petermann, U. (2020). Theorizing transnational labour markets: A research heuristic based on the new economic sociology. *Global Networks, 20*(3), 410–433.

Molitor, C. (2015). *Geschäftsmodell Ausbeutung: Wenn Europaische Arbeitnehmer_Innen in Deutschland um Ihre Rechte betrogen werden*. Friedrich Ebert Stiftung.

OECD. (2012). *Free movement of workers and labour market adjustment: Recent experiences from OECD countries and the European Union*. OECD.

O'Reilly, K. (2007). Intra-European migration and the mobility – Enclosure dialectic. *Sociology, 41*(2) 277–293. https://doi.org/10.1177/0038038507074974

Pernicka, S. (Ed.). (2015). *Horizontale Europäisierung im Feld der Arbeitsbeziehungen*. Springer VS.

Piore, M. J. (1979). *Birds of passage: Migrant labour and industrial societies*. Cambridge University Press.

Piore, M. J. (1986) The shifting grounds for immigration. *The Annals of the American Academy of Political and Social Science, 485*(1), 23–33.

Platzer, H. W., Klemm, M., & Dengel, U. (2020). Einleitung. In H. W. Platzer, M. Klemm, & U. Dengel (Eds.), *Transnationalisierung der Arbeit und der Arbeitsbeziehungen: Interdisziplinäre Perspektiven* (pp. 9–16). Nomos.

Pries, L. (2010a). *Transnationalisierung: Theorie und Empirie grenzüberschreitender Vergesellschaftung*. VS Verlag.

Pries, L. (2010b). Internationalisierung von Arbeitsmobilität durch Arbeitsmigration. In F. Böhle, G. G. Voß, & G. Wachtler (Eds.), *Handbuch Arbeitssoziologie* (pp. 729–747). VS Verlag.

Pries, L. (2013). Transnationalisierung. In S. Mau & N. M. Schöneck (Eds.), *Handwörterbuch zur Gesellschaft Deutschlands* (pp. 881–894). VS Verlag.

Pries, L. (2017). *Erwerbsregulierung in einer globalisierten Welt: Theoretische Konzepte und empirische Tendenzen der Regulierung von Arbeit und Beschäftigung in der Transnationalisierung* (2nd ed.). VS Verlag.

Pries, L. (2020). *Transnationalisierung von Arbeit – (Inter-)Nationalisierung von Erwerbsregulierung?* In H. W. Platzer, M. Klemm, & U. Dengel (Eds.), *Transnationalisierung der Arbeit und der Arbeitsbeziehungen: Interdisziplinäre Perspektiven* (pp. 19–42). Nomos.

Quack, S., Schulz-Schaeffer, I., Shire, K., & Weiß, A. (Hrsg.). (2018). *Transnationalisierung der Arbeit*. Springer VS.

Recchi, E. (2006). From migrants to movers: Citizenship and mobility in the European Union. In M. Smith & A. Favell (Eds.), *The human face of global mobility* (pp. 53–77). Transaction.

Recchi, E. (2008). Cross–state mobility in the EU: Trends, puzzles and consequences. *European Societies, 10*(2), 197–224.

Recchi, E. (2015). *Mobile Europe: The theory and practice of free movement in the EU*. Palgrave Macmillan.

Rhodes, M. (2010). Employment policy: Between Efficacy and Experimentation. In H. Wallace, M. A. Pollock, & A. R. Young (Hrsg.), *Policy-making in the European Union* (6th ed., pp. 283–306). Oxford University Press.

Rooss, C., & Westerveen, L. (2020). The conditionality of EU freedom of movement: Normative change in the discourse of EU institutions. *Journal of European Social Policy, 30*(1), 63–78.

Ruhs, M. (2013). *The price of rights*. Princeton University Press.

Sassen, S. (1988). *The mobility of labour and capital*. Cambridge University Press.

Sassen, S. (1991). *Global city: New York, London, Tokyo*. Princeton University Press.

Scharpf, F. W. (1998). Balancing positive and negative integration. The regulatory options for Europe. In D. Dettke (Eds.), *The challenge of globalization for Germany's social democracy: A policy agenda for the twenty-first century* (pp. 29–57). Berghahn.

Schulz-Schaeffe, I., & Bottel, M. (2018). Die Herstellung transnationalmobiler Arbeitstätigkeiten in der Softwareentwicklung. In S. Quack, I. Schulz-Schaeffer, K. Shire, & A. Weiß (Eds.), *Transnationalisierung der Arbeit* (pp. 99–128). Springer VS.

Sciortino, G. (2013). Immigration in Italy: Subverting the logic of welfare reform. In E. Jurado & G. Brochmann (Eds.), *Europe's immigration challenge: Reconciling work, welfare and migration* (pp. 77–94). Tauris.
Seeliger, M., & Wagner, I. (2020). A socialization paradox: Trade union policy cooperation in the case of the enforcement directive of the posting of workers directive. *Socio-Economic Review, 18*(4), 1113–1131.
Sengenberger, W. (1987). *Struktur und Funktionsweise von Arbeitsmärkten: Die Bundesrepublik Deutschland im Vergleich*. Campus.
Soysal, Y. N. (1994). *Limits of citizenship: Migrants and postnational membership in Europe*. Chicago University Press.
Spindler, S. (2011). Feminisierung von Migration – Formen und Folgen weiblicher Wanderungsprozesse. In G. Hentges & H. W. Platzer (Eds.), *Europa quo vadis? Ausgewählte Problemfelder der europäischen Integrationspolitik* (pp. 171–186). VS Verlag.
Staples, R., Trinczek, R., & Whittall, M. (2013). Posted Workers: Zwischen Regulierung und Invisibilisierung. *Arbeit, 22*(4), 271–286.
Ulrich, K. (2019). Arbeitsmarkt und Rentenpolitik: Die unendliche Diskussion um den Fachkräftemangel. Retrieved June 22, 2021, from https://www.dw.com/de/die-endlose-diskussion-um-den-fachkr%C3%A4ftemangel/a-48479097
Urry, J. (2007). *Mobilities*. Polity.
Van Ostaijen, M., & Scholten, P. (2018). Between mobility and migration: The consequences and governance of intra-European movement. In M. Van Ostaijen & P. Scholten (Eds.), *Between mobility and migration: The multi-level governance of intra-European movement* (pp. 1–20). Springer.
Verwiebe, R. & Müller, M. C. (2006). Gelungene Integration in den Arbeitsmarkt? Die flexiblen Biografien transnational mobiler Europäer zu Beginn des 21. Jahrhunderts. *Berliner Zeitschrift für Soziologie, 16*(1), 95–114.
Verwiebe, R., Wiesböck, L., & Teitzer, R. (2014). New forms of intra-European migration, labour market dynamics and social inequality in Europe. *Migration Letters, 11*(2), 125–136.
Vertovec, S. (2004). Cheap calls: The social glue of migrant transnationalism. *Global Networks, 4*(2), 219–224.
Wagner, I. (2018). *Workers without borders: Posted work and precarity in the EU*. Cornell University Press.
Wagner, I., & Berntsen, L. (2016). Restricted rights: Obstacles in enforcing the labour rights of mobile EU workers in the German and Dutch construction sector. *Transfer: European Review of Labour and Research, 22*(2), 193–206.
Wagner, B., & Hassel, A. (2015). Labour migration and the German meat processing industry: Fundamental freedoms and the influx of cheap labour. *South Atlantic Quarterly, 114*(1), 204–214.
Wagner, B., & Hassel, A. (2016). Move to work, move to stay? Mapping atypical labour migration into Germany. In J. E. Dølvik & L. Eldring (Eds.), *Labour mobility in the enlarged Single European Market* (Comparative Social Research) (Vol. 32, pp. 125–158). Emerald Group.
Waldinger, R., & Lichter, M. I. (2003). *How the other half works: Immigration and the social organization of labour*. University of California Press.

Wallace, C. (2001). Conceptual challenges from the new migration space. In C. Wallace & D. Stola (Eds.), *Patterns of Migration in Central Europe* (pp. 45–71). Palgrave.

Wickham, J. (2016). *Unequal Europe: Social divisions and social cohesion in an old continent*. Routledge.

Wickham, J., & Bruff, I. (2008). Skills shortages are not always what they seem: Migration and the Irish software industry. *New Technology, Work and Employment, 23*(1–2), 30–43.

Wiesböck, L. (2016). A preferred workforce? Employment practices of East-West cross border labour commuters in the Central European Region. *Österreichische Zeitschrift für Soziologie, 41*(4), 391–407.

Williams, A. M., & Baláž, V. (2009). Low-cost carriers, economies of flows and regional externalities. *Regional Studies, 43*(5), 677–691.

3 Transnational Labour Mobility in Europe in the 21st Century

This chapter examines the extent and impact of intra-European labour mobility in the recent past. It distinguishes between three phases of mobility. The first phase covers the 1990s to the beginning of the 21st century, which was characterized by a moderate increase in cross-border migration. The second phase includes Eastern enlargement since 2004, which triggered a previously unknown dynamic of labour mobility and overlaps with the third phase, which began in 2008/09 with the economic and financial crisis and included a diversion effect towards Northern Europe. It is shown that intra-European migration is driven by economic push and pull factors, but also by the institutional configuration of a transnational mobility space, which structurally promotes mobility. Before going into more detail on the socio-demographic characteristics of mobile EU citizens and the labour market effects, the development of intra-European migration over the last three decades is first traced.

3.1 A New Dynamic of Labour Mobility in Europe

Labour Migration in the 1990s

Intra-European migration in the 1990s was primarily shaped by two trends. On the one hand, there was an increase in employment-related mobility within the "old" EU-15, which was linked to the completion of the single market and the introduction of EU citizenship. On the other hand, there was an increase in East-West migration, which was linked to the transformation processes in Central and Eastern Europe (CEE).

T. Krings, *Labour Mobility in the European Union as an Example of the Transnationalization of Employment*,
https://doi.org/10.1007/978-3-658-43977-4_3

With regard to intra-EU mobility, the European Commission has increasingly emphasized the importance of freedom of movement for the creation of a pan-European employment area since the 1990s. This was not only intended to help mobile workers improve their job prospects, but also to make the European labour markets more "flexible", to balance labour market inequalities and to contribute to a more effective distribution of the "production factor" labour (Boswell & Geddes, 2011, pp. 181–182).

So far, however, there is only limited evidence that worker mobility *within the EU-15* is an economic adjustment mechanism that balances labour market inequalities between countries and regions. Even though intra-European mobility has been on the rise since the 1990s, it only involves mobility between regions with weak and strong growth to a comparatively small extent. This was succinctly formulated by James Wickham (2016, p. 195): "Within the old EU the least important migrations are those most desired by traditional policy: movement from areas of high unemployment to areas of labour shortage". In addition to institutional and cultural barriers, social security systems in the countries of origin are likely to play a role in why particularly low-skilled EU-15 citizens show a comparatively low willingness to move (ibid.).

It is therefore not surprising that the increase in employment-related migration in the EU in the 1990s was mostly about the mobility of higher qualified individuals who found employment in the expanding knowledge-intensive service industries in 'global cities' such as Berlin, Brussels or London (Favell, 2008; Verwiebe & Müller, 2006). Overall, however, the scope of this mobility was relatively modest, even compared to a new wave of employment-related East-West mobility.

An increase in East-West mobility did not start with Eastern enlargement in the 21st century, but already in the 1990s against the backdrop of a "new map of migration in Europe" (King, 2002). At that time, the transformation of state socialist economies into capitalist market economies virtually overnight led to processes of social disintegration, and generated new migration movements mainly towards Germany and Austria, the traditional destinations for mobile citizens from CEE. These were mostly short-term forms of labour migration, which either took place as part of new "guest worker programs" (Rudolph, 1996), or outside of official programs in the growing informal economies of Western Europe (Wallace, 2001).

Due to geographical proximity and the possibility of repeated border crossings, most of these migration movements were short-term and circular. In particular in regions like Berlin - Western Poland and Vienna - Bratislava, "transnational social spaces" (Pries, 2010) emerged, characterized by frequent border crossings and circular migration. The often temporary and transient character of this migration reflected the increasing demand for flexible workers in the rapidly growing service industries of Western Europe (Wallace, 2001, pp. 54–56). The labour market opportunities of these commuters were limited. They were, not unlike the "guest workers" of the 1960s and 1970s, employed in low-skilled jobs, mainly in agriculture, construction and the expanding service sector. They often worked under precarious conditions in a grey zone of regular and irregular employment (Wilpert, 1998). With the EU enlargements in 2004, 2007 and 2013, the residence status of many migrants from CEE changed. Virtually overnight, former "guest workers" with precarious residence status became EU citizens with all the rights of Union citizenship. The following quote from a Polish architect, who had already moved to Ireland in 2002 in advance of enlargement, shows how much their new status has transformed the migration experience:

> I remember the stress on the Austrian border when we were crossing it, when we were going there with the labourers … It was an illegal job, a completely different, different world, the one from before the EU … There was this division, this definite division between East and West (Krings et al., 2013, p. 1).

EU Enlargement in 2004 and 2007

Even before enlargement, there was a noticeable increase in East-West mobility in Europe, which was made possible, among other things, by the abolition of the visa requirement in 2001 for the countries that joined in 2004. Many of the migrants took advantage of eased border crossing and then retrospectively legalized their residence status (Recchi, 2015, p. 55). However, the decisive event that triggered a new dynamic of employment-related migration in Europe was the EU accession of eight former state socialist countries from Central and Eastern Europe (CEE) along with Malta and Cyprus in 2004 (EU-10), followed by the accession of Bulgaria and Romania in 2007 (EU-2).[1]

The 2004 and 2007 enlargement significantly increased the interstate income differences in the EU. Out of concern for wage and social dumping, most EU-15 countries restricted access to the labour market for the new EU citizens for up to seven years. Such transitional arrangements were also introduced during Southern enlargement in the 1980s when Greece, Portugal, and Spain acceded to the then European Community. After the expiry of the deadlines, there was no significant increase in labour migration from these countries (Kvist, 2004). However,

[1] The eight former state socialist countries were Estonia, Latvia, Lithuania, Poland, Slovakia, Slovenia, the Czech Republic, and Hungary.

the income differences at the time of Eastern enlargement were much higher than during Southern enlargement: While the per capita income of Greece, Portugal, and Spain in the 1980s was about two-thirds of the then EU-9 average, the per capita income in the eight 2004 accession countries from CEE at the beginning of the 21st century was just under 50% of the EU-15 average, and in the case of Bulgaria and Romania, even just under 30% (in purchasing power standards) (European Commission, 2020). Against the backdrop of these substantial wage differences and already existing migration paths, higher migration in the context of Eastern enlargement was to be expected. In fact, post-2004 migration turned out to be significantly higher than predicted in research reports for the European Commission (Dølvik, 2013, p. 39).

Especially in the first years after the 2004 and 2007 enlargement, there was a rapid increase in East-West mobility. The main destination countries for citizens from the 2004 accession countries were the UK and Ireland, which, along with Sweden, were the only old member states (OMS) that immediately opened their labour market. Between 2003 and 2008, the number of EU-10 citizens in the UK increased by over 700,000, making the island the most important destination in absolute numbers for the 2004 accession countries. In relative terms, migration to Ireland was even higher. Between 2003 and 2008, the number of EU-10 migrants there increased by almost 200,000 (European Commission, 2012, p. 250). In both English-speaking countries, Polish migrants made up two-thirds of these movements. Interestingly, an increase in migration was already evident in both countries before the official EU accession, so the post-2004 development cannot be explained solely by the opening of the labour market.

Germany and Austria, the traditional destinations for migrants from CEE, also experienced an increase in labour mobility post-2004 despite transitional periods. This suggests that economic push and pull factors have a greater impact on the labour market than restrictions. One result of the transitional arrangements was an increase in "atypical" mobility such as self-employment and cross-border posting in both countries, which provided an alternative route into the labour market (SVR, 2013, p. 78; Tamas & Münz, 2006; Wagner & Hassel, 2016). While self-employment, which was mostly "bogus" self-employment, was particularly noticeable in the construction industry, where the freedom to provide services was restricted, postings were primarily an entry point into the labour market in the meat industry in Germany. As a result of this atypical mobility, a trend towards dualization has intensified, which continues to have an impact even after the transitional periods have expired (Dølvik, 2013; Wagner & Hassel, 2016).

There can be little doubt that the transitional arrangements in Germany and Austria have led to a diversion effect of migration to the UK and Ireland. However, this effect should not be overestimated. The main dynamic for the high migration to the UK and Ireland was an open and booming labour market in combination with new transport opportunities (especially low-cost airlines) and the English language. These pull factors increased the willingness to migrate among parts of the population in the accession countries, who were previously not among the traditional migrant groups. These included younger and better-qualified individuals, for whom relatively high youth unemployment at the beginning of the 21st century was an important push factor (Fihel et al., 2015; Pollard et al., 2008).

The main destinations for the 2007 accession countries were Italy and Spain, with this process beginning several years before the official EU accession. For many Romanian and Bulgarian migrants, accession represented the opportunity to regularize their residency status. In Italy and Spain, the number of EU-2 citizens increased by 700,000 each between 2003 and 2008 (European Commission, 2012, p. 250). This sharp increase occurred even though both countries imposed transitional arrangements for Romania and Bulgaria.[2] This suggests that intra-European labour mobility is largely demand and network-driven, and the transitional arrangements had only a limited impact on the scale and direction of this mobility. The booming Spanish construction sector, as well as a chronic shortage of labour in the Italian care sector, were pull factors for EU-2 migrants, who often found employment in an informal work environment (Recchi, 2015, pp. 96–98). In the case of Romanian migrants, the Romance language tradition in Italy and Spain was another pull factor, underlining the importance of cultural and linguistic factors for migration. As in the case of EU-10 migrants, it was mainly younger and relatively well-qualified individuals,[3] who made use of their new free movement rights. High youth unemployment (20% in Romania and 15% in Bulgaria) was likely a significant push factor, as was a comparatively low employment rate in both countries (European Commission, 2012, p. 262).

In terms of intra-European migration post-2004, it is noticeable that the target countries have increasingly diversified. Countries such as Ireland, Belgium,

---

[2] Spain had lifted the transitional arrangements for Romania in 2009, but reintroduced them in 2011 due to a difficult labour market situation as a result of the economic and financial crisis.

[3] However, among EU-2 migrants, those with low qualifications were also relatively strongly represented, (see Fig. 3.5 below).

Denmark, the Netherlands, and Norway, which were not significant destinations for East-West mobility before enlargement, have now become important destinations (Fihel et al., 2015, p. 23). This suggests that intra-European mobility is a more individualistic form of migration, which is less socially prestructured than previous migration movements. The EU free movement regime, new transport options (especially low-cost airlines), and virtual information sources enable mobility projects that vary significantly in terms of motivation, duration, and destination (Krings et al., 2013).

The first decade of the 21st century was characterized by a new dynamic of East-West migration. At the same time, there was also an increase in transnational mobility within the old EU-15, albeit to a much lesser extent. The main destinations in the first decade were Spain and, to a lesser extent, Austria, Switzerland, and the UK. In the second decade, the UK became the most important destination for EU-15 migrants from Southern Europe, who moved north as a result of the economic crisis, closely followed by Germany, which increasingly positioned itself as a destination for qualified migration (SVR, 2013, p. 103).

Intra-European Mobility since the Economic Crisis of 2008/2009

In the wake of the economic and financial crisis of 2008/2009, a redirection of intra-European migration was observed. Countries like Ireland and Spain, which were heavily hit by the crisis due to a bursting property bubble, experienced a sharp decline in new immigration and an increase in emigration. The outflows from both countries differed. While in Ireland, both native and foreign citizens emigrated since the outbreak of the crisis, in Spain it was primarily the latter (both third-country nationals and EU-2 migrants) (Fries-Tersch & Mabilia, 2015, p. 30). At the same time, countries like Poland, where the labour market remained relatively robust even in the crisis, saw an increase in return migration, with highly qualified individuals showing less inclination to return than those with a medium level of qualification (Anacka & Fihel, 2012). One possible reason for this could be that for highly qualified individuals, a stay abroad opens up new career opportunities that do not necessarily disappear even in an economic crisis (Krings et al., 2013).

Return migration was by no means the only migration pattern in the crisis years, as intra-European mobility continued to be characterized by new migration, circular mobility, and repeated border crossings (Fihel et al., 2015). As already mentioned, there was noticeable emigration of Romanian and Bulgarian migrants, especially from Spain, where the crisis in the construction sector led to dramatic job losses. Since return migration to the countries of origin did not increase to the same extent, it can be assumed that some of these mobile EU citizens moved north, especially to Germany (Fries-Tersch & Mabilia, 2015, pp. 27–28).

Also native citizens in Southern Europe moved north in larger numbers, but not to the extent as suggested by reports of the "new guest workers" (Becker et al., 2013). When migrants, who return or move on, are left out, it becomes clear that the extent of South-North mobility is nowhere near as high as would be suggested by an "optimal currency area" (Mundell, 1961) in the face of dramatically increased unemployment in Southern Europe. In fact, the majority of intra-European labour mobility both before and after the economic crisis took place from countries *outside* the Eurozone to countries *within* the Eurozone.[4]

A significant reason why the propensity to migrate was much higher in the new member states (NMS) than in Southern Europe is the larger income disparities. In social science migration research, there is broad agreement that migration movements are not just a response to wage differences, but represent a combination of push and pull factors, often triggered by political decisions such as recruitment programmes and maintained at the micro level by networks (Martin et al., 2006, pp. 6–13; Massey et al., 1993, pp. 432–433). However, income differences are *one* important factor for migration movements, as is visible in the context of intra-European mobility.

As can be seen from Fig. 3.1, there is a statistical correlation between income differences and emigration. The greater the differences from the average per capita income in the old EU-15 is, the higher the willingness to migrate is. By far the highest emigration rate can be observed in Romania, the second poorest EU member state, where between 2008 and 2018 almost 12% of the adult population of working age migrated to another member state. Bulgaria, Latvia and Lithuania also have a high emigration rate of almost eight percent. Interestingly, Poland, the most important country of emigration in the EU in absolute terms after Romania, has only a medium-high emigration rate (and measured by per capita income only a below-average high rate).[5]

Compared to the NMS, the propensity to migrate is much less pronounced in the OMS. For example, less than one percent of the working-age population

[4] An exception is the UK, which is not part of the Eurozone, but in the 21st century was one of the most important destinations for migrants from both the old and the new member states.

[5] However, it should be taken into account that short-term transnational mobility is relatively widespread among Polish migrants, which are only partially captured by Eurostat's emigration data (Kaczmarczyk, 2010).

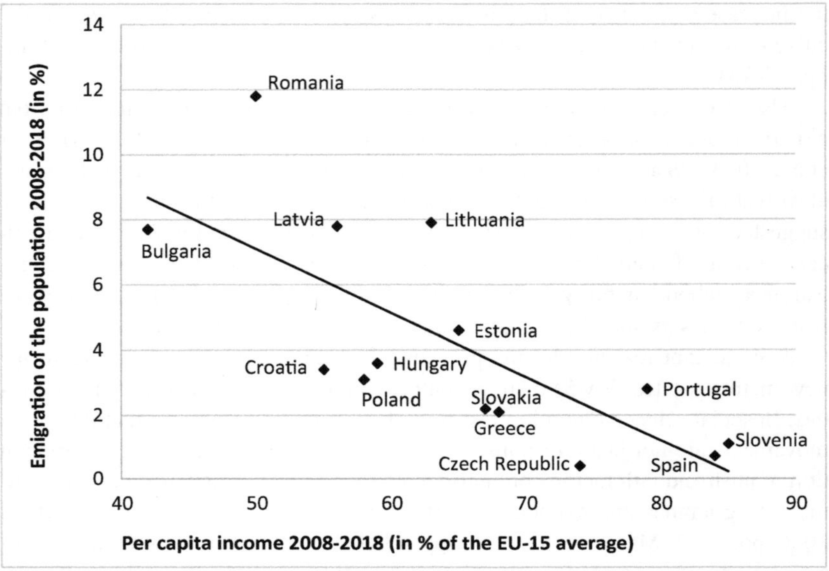

Fig. 3.1 Proportion of the population who have moved to another EU member state in relation to per capita income in the EU-15 (2008–2018) (age 20–64 years). (Source: Eurostat, 2020a, 2020b)

in Spain moved to another member state between 2008 and 2018, despite dramatically increased unemployment, especially among younger cohorts. The social security systems, which cushion the effects of unemployment in Southern Europe more than in the NMS, are likely to have played a role here (Wickham, 2016, p. 195). Portugal is the only Southern European country with an above-average willingness to emigrate, which can be traced back to a decades-long migration tradition that has made the Iberian country the most important country of emigration in Western Europe (Recchi, 2015, p. 167).

Where did intra-European migrants move to since the economic and financial crisis of 2008/2009? As shown in Fig. 3.2, Germany and the UK became the most important destinations for intra-European mobility. The context for this was a booming German labour market and the expiry of the transition periods in 2011 and 2014, which provided the context for a sharp increase in new migration from Poland, Romania, and Bulgaria. This once again underlines that the transitional arrangements had an impact on the scope of migration flows, but only in a complex interplay with other push and pull factors, especially a demand for labour in the destination countries (Galgóczi et al., 2009). This was also evident in the case

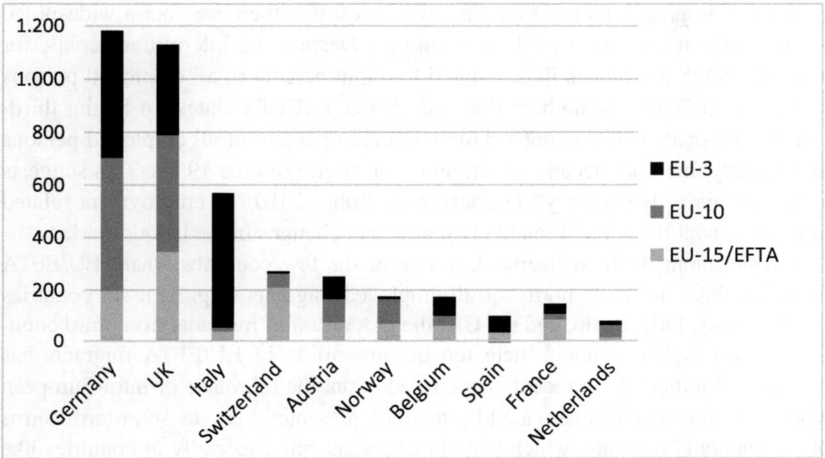

Fig. 3.2 Increase in the EU/EFTA migrant population (18–67 years) in ten EU/EFTA member states (2008–2018) (in 000)[6]. (Source: EU-LFS, own calculations)

of Italy, which, despite transitional arrangements, became the most important destination for Romanian migrants, which in turn points to the importance of cultural factors and networks for migration movements (Recchi, 2015, p. 76). After the three major countries, the most important destination country for NMS migrants was Austria, which, like Germany, only opened its labour market in 2011 and 2014. The main countries of origin here were Hungary and Romania. With regard to migration from the old EU-15, Switzerland stands out as an important destination alongside the UK. The latter experienced an increase in migration mainly from Germany, but also from France and Italy.

Spain and Ireland, which were among the most important destinations for intra-European migration in the first decade, experienced a significant decline in new migrant inflows against the backdrop of the crisis. At the same time, data

[6] All the following information on intra-European migration is based on the European Labour Force Survey (LFS) (unless otherwise stated). "Migrant" is defined partly by country of birth and partly by citizenship for data-pragmatic reasons. To avoid defining domestically born persons with foreign citizenship as "migrants", the criterion "migration" is also used. EU-3 includes Bulgaria and Romania as well as Croatia, which joined the EU in 2013. The data provided by the LFS are weighted (annual weight).

from the European Labour Force Survey suggest that there was not a widespread wave of return migration in these countries. Despite the intermittent crisis, for example, NMS migrants still accounted for eight percent of all employed persons in Ireland in 2018, the highest share of all EU-15/EFTA states. In Spain, third-country nationals still account for more than ten percent of all employed persons in a country that only became an immigration country in the 1990s. This suggests a certain "path dependency" (Anderson & Ruhs, 2010) of employment-related migration, which results in relatively permanent changes in the labour market.

Ireland, along with Switzerland, is one of the few countries where EU/EFTA migrants make up the majority of all employed migrants (Fig. 3.3).[7] In countries like Germany, Italy, Spain, and the UK, the proportion of migrants from third countries is still higher, although here too the proportion of EU/EFTA migrants has increased significantly in recent years. In addition, the relevance of intra-European labour mobility is underestimated by the data presented here, as short-term forms of transnational mobility, which play an important role especially in countries like Germany and Austria, are usually not captured by the EU-LFS. This of course also applies to irregular employment, which is relatively widespread in Southern Europe (Reyneri, 2013). Thus, it seems plausible to assume that about one in five workers in the EU-15 was born abroad, if we take a broad understanding of employment. If we were to also include people with a migration background who were born domestically, this proportion would be even higher. Thus, the increasing diversity of European societies shows itself especially in the labour market.

In the "age of migration" (Castles & Miller, 2009), intra-European labour mobility is just one facet of international migration movements. However, it has gained more importance in most EU/EFTA member countries over the last 15 years than employment-related migration from third countries. This can be largely explained by the EU free movement regime, which structurally promotes mobility, but also by enlargement since 2004, which has triggered a new dynamic of cross-border labour mobility.

Who are the mobile EU citizens who make use of their right to freedom of movement? What qualifications do they have and in which areas of the labour market can they be found? To explore this further, the socio-demographic profile of intra-EU migrants in the six main destination countries since 2004 (Germany, Ireland, Italy, Austria, Spain, and the UK) is now examined in more detail.

[7] In Ireland, not only is the proportion of migrants from the NMS above average, but also that from the OMS. The latter is mainly due to inflows from the UK, often by people of Irish descent who were born there.

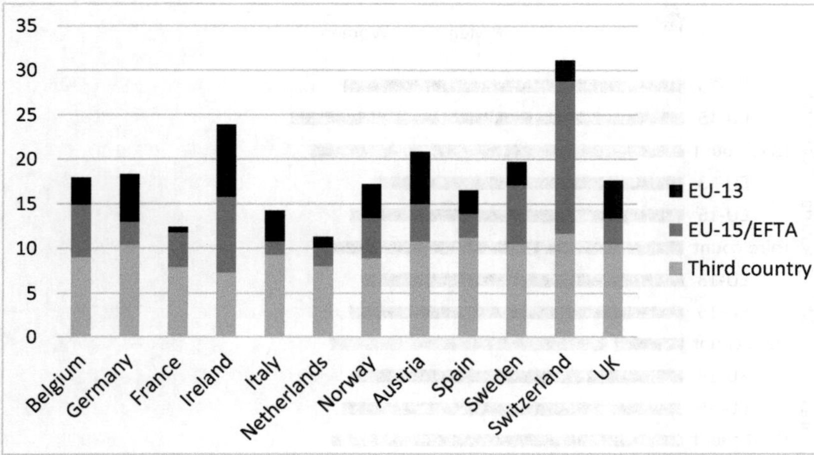

Fig. 3.3 Proportion of all foreign-born employed persons in twelve EU/EFTA member states (2018) (in %)[8]. (Source: EU-LFS, own calculations)

3.2 Socio-Demographic Characteristics of Intra-European Labour Mobility

When the socio-economic profile of the mobile EU working population is examined more closely, it is noticeable that migrants from the NMS have a comparatively young age profile.[9] This is not only in comparison with migrants from the OMS, but also in comparison with the respective domestic population (and people from third countries). This is particularly noticeable in the UK and Ireland, where 70% of all workers from the EU-13 countries are under 37 years old. On the one hand, this may be due to the fact that in both countries the large migrant inflows from the NMS is still a relatively recent phenomenom. On the other hand,

[8] EU-13 includes all new member states that have joined the EU since 2004, thus also Malta and Cyprus, which, however, are insignificant as countries of origin of intra-European mobility.

[9] The working population is the economically active part of the population, which includes both employed and unemployed people.

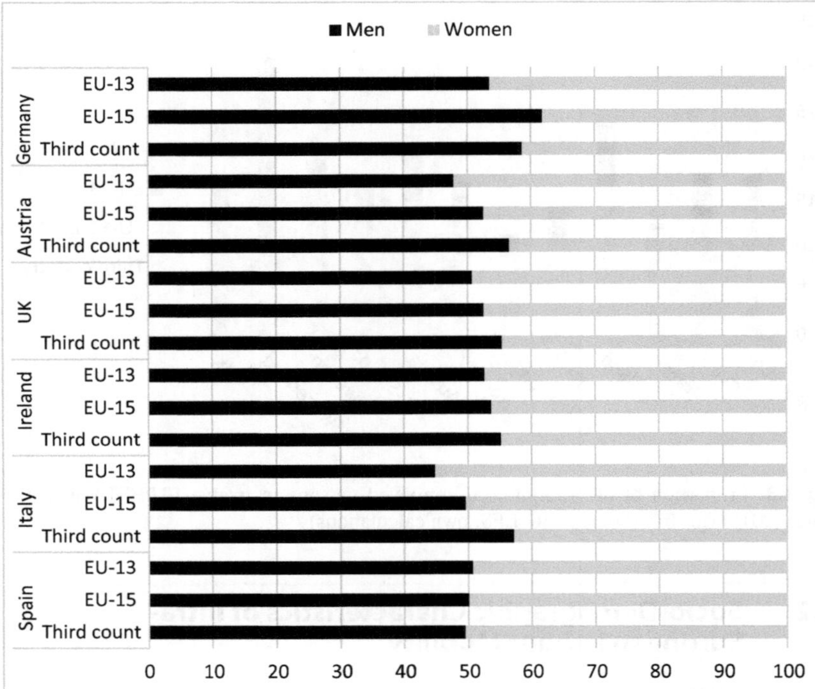

Fig. 3.4 Migrant working population by country of birth and gender (2018) (in %)[10]. (Source: EU-LFS, own calculations)

the two English-speaking countries were a particularly attractive destination for younger and better-qualified migrants post-2004 (see below).

A push factor for young EU-13 migrants may have been the relatively high youth unemployment in some NMS at the time of EU accession. Another reason may be that the enlargement process opened up new mobility opportunities ("work and travel"). Even though the search for employment and a higher income played an important role in this mobility, this mobility went beyond purely economic rationality, with motives such as the desire to travel and adventure also playing a significant role (Krings et al., 2013).

[10] For reasons of clarity, EU-10 and EU-3 migrants are combined into one category (EU-13).

In terms of gender and country of birth, the proportion of female workers from the EU-13 was higher than in other migrant groups (except in Spain) (Fig. 3.4). Since their proportion was also higher than in their countries of origin, we can speak of a "feminization of migration" (Spindler, 2011). This is related to structural change in "post-industrial" societies, where growth in knowledge-intensive services goes hand in hand with an increasing demand for low-paid services (e.g., cleaning or domestic services), which are traditionally associated with female labour. This is particularly noticeable in Italy, where the high proportion of Romanian migrant women cannot be considered separately from an underdeveloped public care sector (Sciortino, 2013).

In addition to age and gender, mobile workers from the old and new member states differ primarily in terms of educational background and occupational status. As can be seen from Fig. 3.5, EU-15 migrants have the highest qualifications in all six destination countries, even compared to the domestic population (and third-country nationals).[11] Thus, this group represents a case of "positive selection", as they also have on average higher educational qualifications than the population in the countries of origin.

Regarding NMS migrants, it is noticeable that they are strongly represented in the middle segment, so they are not necessarily a particularly highly qualified group (see also Recchi, 2015). There are some significant differences between the destination countries. For example, EU-13 migrants (similar to third-country nationals) in Italy and Spain have an above-average proportion of people with a low level of education, reflecting a corresponding demand for low-skilled workers in the local labour markets (Meardi et al., 2012; Sciortino, 2013). The UK and Ireland, on the other hand, attract an above-average proportion of highly qualified citizens from the NMS. This confirms previous research that shows that the liberal, English-speaking countries are an attractive destination for better-qualified migrants (Kaczmarczyk, 2010; Kogan, 2007).

In terms of labour market participation, it is noticeable that male migrants from the old and new EU have higher employment rates everywhere than the domestic population and third-country nationals (age group 17–62 years, own calculations). In addition to relatively high qualifications especially among EU-15 migrants, this suggests that the institutional framework of the free movement regime promotes a

[11] The only exception is Ireland, where the proportion of highly qualified workers from third countries is slightly higher at two-thirds.

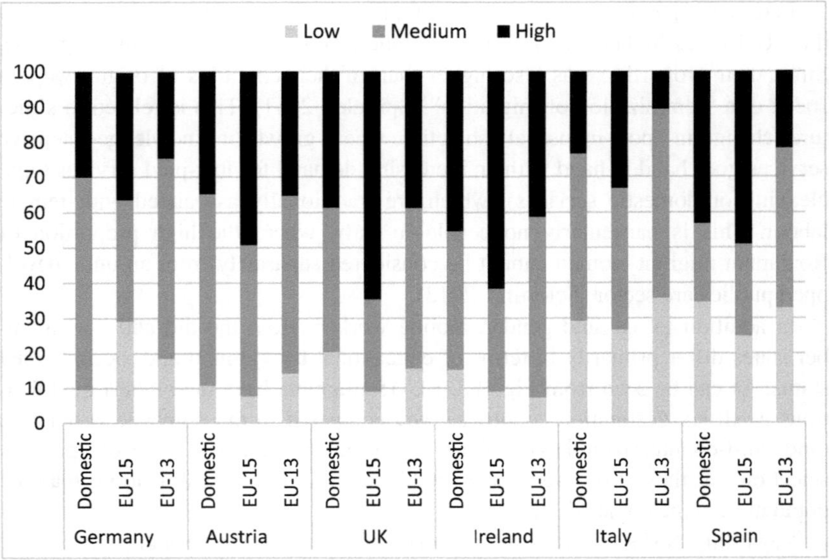

Fig. 3.5 Qualification level of the working population by country of birth (2018) (in %)[12]. (Source: EU-LFS, own calculations)

comparatively easy entry into the labour market. Especially in Germany, Ireland, and the UK, male EU migrants have employment rates of well above 80%. The picture is less clear for female migrants. In Austria (and to some extent in Germany), female domestic and EU workers have similar employment rates, while in the UK and Ireland, female EU-13 migrants have above-average rates. It is noticeable that third-country migrant woment have significantly lower employment rates everywhere (except in Southern European countries, which generally have a low female labour force participation). This applies particularly to female migrants from poorer third countries (Wickham, 2016, p. 189).

In terms of unemployment, it is noticeable that third-country migrants usually have higher rates than EU migrants and natives. An exception are Spain and Italy

[12]In all case countries (with the exception of the UK), EU-10 migrants have a slightly higher qualification structure than EU-3 migrants (see also European Commission, 2012).

where EU-3 migrants and third-country migrants have similar high unemployment rates. However, these figures, which come from the European Labour Force Survey, are likely to be distorted. As already mentioned, irregular employment in the context of migration is relatively widespread in Southern Europe, with extra-European migrants being most affected (Reyneri, 2013). A significant proportion of these migrants are likely not to appear in official labour market data such as the EU-LFS. As with employment rates, differences in unemployment rates are likely not only due to human capital endowment, but also to different regulation of cross-border mobility. In this respect, EU migrants have significantly better opportunities to find new employment in the event of unemployment, not least because the loss of their job cannot be a reason for expulsion.

What is the situation in terms of occupational status? Do mobile EU citizens find employment that matches their qualifications, or can we observe a process of de-skilling? As shown in Fig. 3.6, EU-15 migrants are strongly represented in high-skilled positions, reflecting their comparatively high level of education. With the exception of Germany, they have a higher proportion in high-skilled jobs than the domestic population everywhere. They are part of a transnational service class, for whom a stay abroad is increasingly part of a European career (Wickham, 2016, p. 195). For this highly qualified group, a job-related move to another EU country represents a "top notch option" (Recchi, 2015, p. 72).

The situation is different for mobile citizens from the NMS, where a process of de-skilling can be observed to some extent. Despite a predominantly medium level of qualification, they are overrepresented in low-skilled jobs. For example, only about 7-8% of mobile workers from the NMS in Ireland and Austria are considered low-skilled, but 23% are employed as unskilled workers, a development that can also be observed to some extent in the UK. At the other end of the occupational hierarchy, NMS migrants are underrepresented in higher-skilled jobs, especially in the UK and Ireland. There, about 40% of them have a tertiary education, but only about 25% and 22% respectively work in a corresponding job with high qualification requirements.

Thus, East-West migration often involves *skilled* workers in *low-skilled* jobs. This is explained by the willingness of NMS migrants to "trade off" a higher wage at the destination for a job that is below their qualifications, especially when the job is seen as temporary (Anderson et al., 2006). But the demand side is also relevant: companies are occasionally less willing to hire NMS migrants in skilled positions, partly because of a lack of information about foreign qualifications or difficulties in recognizing them (European Commission, 2012, p. 272).

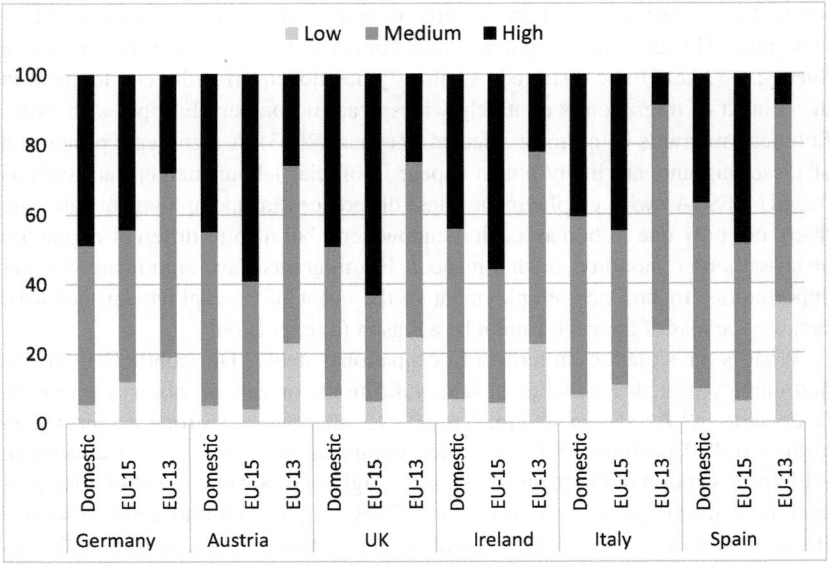

Fig. 3.6 Occupational position by country of birth (2018) (in %)[13]. (Source: EU-LFS, own calculations)

Further, in some low-wage segments, employers have developed a preference for qualified migrants, not least because of assumed *soft skills* such as manners, attitude to work, teamwork and flexibility. Employers are well aware that NMS

[13] The job requirement profile is based on the main groups of the International Standard Classification of Occupations (ISCO-08).

High (1–3) = executives, academic professions, technicians & equivalent non-technical professions.

Medium (4–8) = office workers & related professions, service professions & salespeople, skilled workers in agriculture & fishing, craft & related professions, plant & machine operators & assembly professions.

Low (9) = unskilled workers.

In all case countries (with the exception of Ireland), EU-10 migrants have a slightly higher qualification level than EU-3 migrants (see also European Commission, 2012).

migrants sometimes take a job that is far below their qualifications, especially at the beginning, as this English construction employer freely admitted: "I've had doctors…bit of a sad waste really, doing laddering work because it paid better than being a doctor back home" (in Anderson et al., 2006, p. 74).

From a human capital theory perspective, it is assumed that the longer the stay in the destination country, the better the prospects for occupational mobility (e.g., Chiswick et al., 2005). To what extent does a process of "structural integration" (Esser, 2009) take place with regard to the employment situation of NMS migrants? Figure 3.7 compares the occupational position of domestic workers and EU-13 migrants who have been in the host country for ten years or longer. As can be seen, there is hardly a general convergence of the occupational status. To some extent, it can be observed in Germany and the UK. In contrast, in the more segmented labour markets of Southern Europe, there is hardly any process of upward occupational mobility of NMS migrants (see also Recchi, 2015). What is noticeable is that the various typologies of "varieties of capitalism" (Hall & Soskice, 2001) or "welfare regimes" (Esping-Andersen, 1999) only have limited explanatory value to capture interstate differences in labour market integration. If we

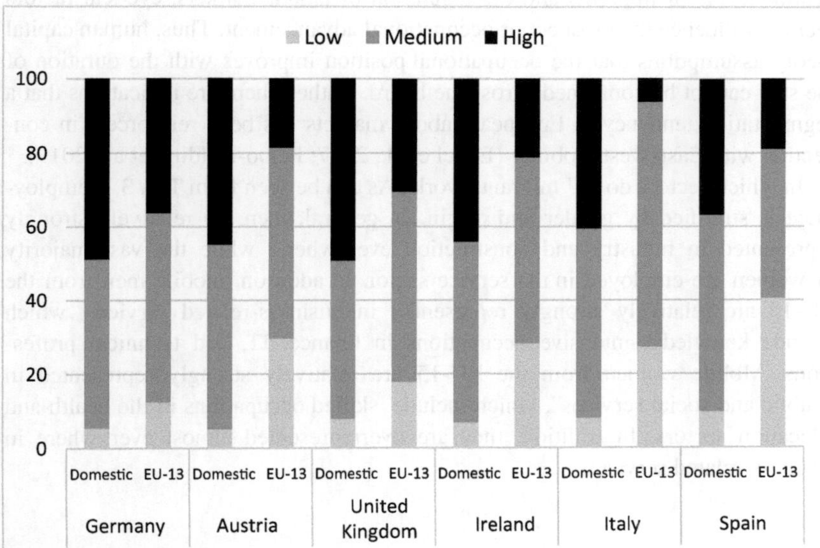

Fig. 3.7 Occupational position of NMS migrants (≥ 10 years in the host country) compared to the domestic population (2018) (in %). (Source: EU-LFS, own calculations)

look at Fig. 3.7, it is noticeable that intra-regime differences weigh more heavily than inter-regime differences. This is particularly noticeable in the comparison of Germany and Austria, both of which are considered coordinated market economies and conservative welfare states. Although EU-13 migrants who have been resident in Austria for a longer period have on average higher qualifications than in Germany, there is less occupational convergence than in the neighboring country. This confirms previous research (e.g., Fassmann et al., 1997), which has shown that the Austrian labour market is less permeable than the German one. This is explained, among other things, by the importance of large companies in Germany, which enable greater permeability in internal labour markets, and by a larger protected segment in Austria, including formerly nationalized companies (ibid., p. 743). Even in liberal market economies, there are interstate differences with regard to the occupational position of long-term resident NMS migrants. Although they have similar qualifications in the UK and Ireland, a process of convergence is more observable in the former than in the latter. This seems to be due, among other things, to the fact that the Irish labour market was more affected by the economic crisis of 2008/09, so that occupational mobility of NMS migrants came to a standstill to some extent (Mühlau, 2012). Thus, in addition to qualification levels of migrants and the regulation of labour markets, cyclical factors seem to influence the prospect of occupational advancement. Thus, human capital theory assumptions that the occupational position improves with the duration of the stay cannot be confirmed across the board. Rather, there are indications that a segmentation tendency in European labour markets has been reinforced in connection with East-West mobility (Eppel et al., 2017; Felbo-Kolding et al., 2019).

In which sectors do EU migrants work? As can be seen from Tab. 3.1, employment is stratified by gender and origin. In general, men are relatively strongly represented in industry and construction everywhere, while the vast majority of women are employed in the service sector. In addition, mobile men from the EU-15 are relatively strongly represented in business-related services, which include knowledge-intensive occupations in finance, IT, and technical professions. Mobile women from the EU-15 are relatively strongly represented in "public and social services", which include skilled occupations in the health and education sectors. In addition, they are overrepresented almost everywhere in business-related services.

Tab. 3.1 Employment of EU migrants by sector and gender (2018) (in %)[14]. (Source: EU-LFS, own calculations)

Men

	Germany		Austria		UK		Ireland		Italy		Spain	
	EU15	EU13	EU15	EU13	EU15	EU13	EU15	EU13	EU15	EU13	EU15	EU13
Industry	25	**30.8**	22.7	21.8	10	**21.2**	17	**25**	21	**29.9**	15.6	19.3
Construction	8.7	**20.2**	6.8	**23.5**	9.8	**19**	8	**12.5**	**10.5**	**32.9**	9.8	**29.6**
Distributive Services	20.3	**23.4**	14.8	**21.8**	14	**32.6**	17	**25**	21	19.8	16.8	**26.2**
Personal & other services	**17.6**	6	**14.8**	**10.9**	**12.1**	7.7	**12**	**16.7**	**10.5**	9.5	**23.8**	**(13.7)**
Business-related services	**18.6**	12.2	**22.7**	15.1	**35.7**	13.4	**33.1**	16.7	6.2	6.1	**23.4**	(9)
Public & social services	9.8	7.4	**18.2**	6.7	**18.5**	6.1	13	4.2	(21)	(1.8)	10.7	(XXX)
Total	*100*	*100*	*100*	*100*	*100*	*100*	*100*	*100*	*100*	*100*	*100*	*100*

(continued)

[14]Excluding agriculture. Bold numbers mean that the proportion of EU-15 or 13-EU migrants is above average compared to all workers in the respective sector. Seasonal employment in sectors such as construction and tourism is underestimated. Figures in brackets indicate case numbers of less than 30. Case numbers of less than 10 are not shown (XXX). Deviations from the sum of 100% are due to rounding.

Tab. 3.1 (continued)

Women

	Germany		Austria		UK		Ireland		Italy		Spain	
	EU15	EU13	EU15	EU13	EU15	EU13	EU15	EU13	EU15	EU13	EU15	EU13
Industry	**13.1**	**15.5**	**11.3**	**10.8**	**7.7**	**11.9**	6.9	**15.3**	11.9	10.9	(7.4)	**(8.8)**
Construction industry	1.5	1.5	1.3	1.5	(XXX)	(2.6)	(XXX)	(1.2)	(XXX)	(0.5)	(XXX)	(XXX)
Distributive Services	15.4	17.4	**21.3**	19.2	11.1	**22.7**	13.8	**22.4**	**(17.4)**	8.5	16	16.4
Personal & other services	**19.7**	**16.9**	**15**	**23.8**	**13.5**	**16.1**	14.9	**27.1**	(16.5)	**58.6**	**27**	**58**
Business-related services	**21.5**	16.9	16.3	**20.8**	**22.5**	**20.9**	**24.1**	**20**	**(22)**	9.9	**29.1**	11.5
Public & social services	28.8	31.6	35	23.8	43.4	26	**40.2**	**14.1**	**31.2**	11.6	20.1	(4.9)
Total	*100*	*100*	*100*	*100*	*100*	*100*	*100*	*100*	*100*	*100*	*100*	*100*

EU-13 migrants are relatively strongly represented in retail (distributive services), where there are many low-wage jobs, especially in the UK and Ireland, in addition to positions with medium qualification requirements (MAC, 2014). Men from the EU-13 are strongly represented in construction and industry everywhere, which have a high proportion of manual, labour-intensive occupations. The fact that almost 30% of all EU-13 migrants in Spain are still employed in construction is an interesting finding. Despite a dramatic collapse of the industry during the 2008/09 crisis, many migrants from the EU (and third countries) obviously still find employment there. This suggests a certain path dependency of cross-border labour mobility.

Mobile women from the EU-13 are overrepresented everywhere in "personal and other services". This is especially visible in Southern Europe, where almost 60% of all migrant women, who mainly come from Romania, work in this sector, which mainly covers household-related services. As already mentioned, the strong presence of migrant workers in this sector is an expression of an underdeveloped care sector in the local familialistic welfare model. Against the background of increased employment of domestic women, which was particularly noticeable in Spain before the 2008 economic crisis, this generates a demand for additional foreign workers (Esping-Andersen, 1999; Sciortino, 2013).

So far, it has been argued that a segmentation tendency in the labour market has become entrenched, particularly in connection with East-West migration. To what extent can this be confirmed with data from the European Labour Force Survey? An approximation for labour market divisions is the extent of unstable "atypical" employment, especially temporary contracts and temporary agency work (Eichhorst et al., 2017). In principle, not every atypical employment is to be classified as precarious, as it also occurs in higher qualified segments or is often chosen in relation to part-time work. However, atypical employment has a higher "potential for precariousness" (Keller & Seifert, 2013) than the standard employment relationship (SER). A decisive factor is whether temporary contracts or part-time employment are entered on a voluntary basis. If this is not the case, it can be assumed that these jobs can rather be attributed to the secondary segment (Nicolaisen et al., 2019). Based on these considerations, a precariousness indicator was built that combines involuntary temporary and involuntary part-time employment. Temporary agency work was also included, as it is a paradigmatic example of insecure employment (Holst, 2012).

Tab. 3.2 Proportion of precarious employment by origin (2018) (in %)[15]. (Source: EU-LFS, own calculations)

	Germany	Austria	UK	Ireland	Italy	Spain
Domestic	5.4	4.4	3.8	6.2	19.4	24.7
EU-15	9	6.3	3.4	6.6	22.9	25.4
EU-13	12.1	14.1	6.8	8.9	36.9	47.1
Third country	12	11.2	6.5	10.8	33.8	43.6

As can be seen from Tab. 3.2, migrants from the EU-13 are everywhere more affected by precarious employment than the domestic population or EU-15 migrants (and in this respect resemble migrants from third countries). In Spain and Italy, this proportion is particularly high and reflects an overall high proportion of temporary contracts and insecure employment. In contrast, the proportion in the "liberal" countries is rather low, suggesting that there, given relatively low dismissal protection, "numerical flexibility" is also achieved as part of the SER (Kalleberg, 2018). Overall, the results in Tab. 3.2 suggest that NMS and third-country migrants disproportionately bear the employment risks in Western European labour markets.

The data presented so far from the EU-LFS relate to long-term forms of mobility, as they only capture migrants who have been living in their country of residence for at least one year (or intend to do so). What about other forms of circular and temporary mobility, where place of residence and place of work often diverge and which play an important role in the context of intra-European migration (Verwiebe et al., 2014)?

Short-term and recurring forms of work-related mobility

As discussed in Chap. 2, forms of intra-European work mobility differ not only in terms of temporality, but also in terms of the regulation of the employment relationship. Of central importance is the extent to which the place

[15] Due to very low case numbers for temporary agency work, the indicator for the UK only includes involuntary fixed-term and part-time employment.

of work principle *(lex loci labouris)* applies. This is the case for commuting mobility (in EU parlance "frontier workers") and seasonal work within the framework of the free movement of workers, so that collective agreements, minimum wages and general employment conditions at the place of work apply.[16] In contrast, the place of work principle only applies to a limited extent in the case of cross-border posting (even though the equal pay principle was established as part of the revised EU Posting of Workers Directive in 2018). Posted workers are to be distinguished from mobile self-employed persons, for whom labour law such as collective agreements and working time regulations do not apply.

Commuter migration and seasonal work

Cross-border workers can be defined as individuals who reside in an EU/EFTA member state and are gainfully employed in another EU/EFTA member state (Fries-Tersch & Mabilia, 2015, p. 56).[17] This transnational mobility encompassed almost two million workers in 2018 who regularly cross national borders, usually daily or weekly, or who work abroad for a limited period of time as seasonal workers. This transnational mobility mainly takes place between neighboring countries. The main destination countries for cross-border workers in absolute numbers are Germany and Switzerland, and in relative terms Luxembourg, followed by Switzerland and Austria. The main countries of origin in absolute numbers are France, Germany and Poland, and in relative terms Slovakia, Estonia and Hungary (Fries-Tersch & Mabilia, 2015, pp. 58–59). This once again underlines the increasing importance of East-West migration since 2004, but also shows that cross-border mobility occurs to a considerable extent within the EU-15/EFTA. The following Fig. 3.8 provides an overview of some of the most important mobility flows. It also shows that this migration is not gender-neutral. While male workers are mainly employed in manufacturing and in construction, female workers are mainly employed in the health sector.

Like other forms of work-related migration, commuter migration and seasonal work are socially stratified. While mobile workers from the EU-15/EFTA are

[16] In countries like Germany, there are exceptions to the social insurance obligation for seasonal workers who engage in gainful employment for less than 70 days. This regulation was extended to 102 days during the COVID-19 crisis in 2021. However, this regulation does not contradict the place of work principle, as it also applies to domestic workers.

[17] The NMS as destination countries of short-term labour mobility are marginal compared to the OMS, which is why they are not considered further in the discussion (Fries-Tersch & Mabila, 2015, p. 61).

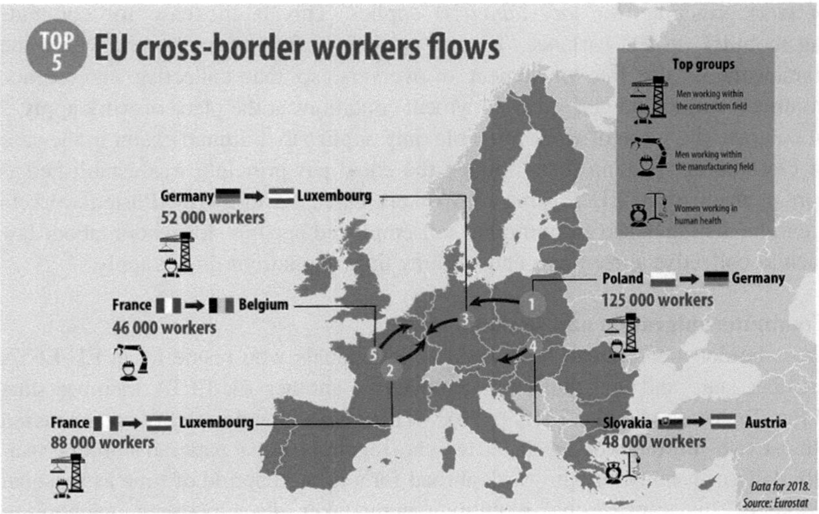

Fig. 3.8 The five most important flows of cross-border work in the EU (2018). (Source: Eurostat, 2020c, © Eurostat)

mainly employed in higher-skilled occupations, mobile workers from the EU-13 are overrepresented in labour-intensive low-wage segments (Fries-Tersch & Mabila, 2015, pp. 63–64). The latter, in turn, have an above-average proportion of temporary employment, reinforcing the impression that they are a group of "hyperflexible" workers (Meardi et al., 2012). This "hyperflexible" character of transnational mobility is also evident in cross-border postings.

Cross-border posting of workers

One of the main features of transnational labour mobility in Europe is not only an increase in individual migration, but also in posted and temporary agency work (Shire et al., 2018). In the case of posted work, employees of a company are sent to another member state for a limited period to perform an economic activity under the guidance of their company. In the case of cross-border temporary agency work, labour is provided to a foreign company.[18] In 2018, nearly two million postings

[18]Cross-border agency work is numerically much smaller than posted work, so it is not separately reported.

(including temporary agency work) were reported in the EU-28/EFTA in the form of A1 social security forms, the vast majority for dependent employees.[19] Almost 60% of all postings take place within the EU-15/EFTA and another third from EU-13 states to EU-15/EFTA states (De Wispelaere et al., 2020, p. 26). Thus, postings are by no means only a phenomenon between low-wage and high-wage countries. However, it is the latter that is usually associated with incidents of wage dumping (Krings, 2019; Staples et al., 2013; Wagner, 2018).

The most important sector for posting is the construction industry, which accounts for approximately 40% of all postings. In construction, the outsourcing of services is not a new phenomenon, as the industry has always been based on the division of work. Already in the 1990s the transnationalization of the construction labour market as part of bilateral work agreements and the EU freedom of services was observed in some countries, especially in Germany (Hunger, 2000). In general, a distinction can be made between a form of "cooperative subcontracting" and a form of "competitive subcontracting" (Harvey, 2003) when outsourcing services. While the former aims at a specialization of work tasks and professional knowledge, the latter primarily aims at cost reduction. Especially the competitive variant of subcontracting has gained a new dynamic in the course of EU enlargement. Transnational subcontracting chains in particular prove to be susceptible to dumping practices, as they are often characterized by an informal work environment in which minimum standards are not adhered to (Krings, 2019; Wagner, 2018).

As can be seen from Fig. 3.9, the most important destination countries for postings are all continental European "Bismarckian" welfare states (Austria, Belgium, Germany, France, Netherlands). This regime type is characterized not only by comparatively high wages, but also by high social contributions. Therefore, there is a "specific incentive structure" (Staples et al.,

[19] Seven percent of all postings were for self-employed individuals. A1 social security forms (formerly E101) are issued by the respective insurance carrier in the country of origin for the posted person and indicate, among other things, the destination of the posting. These certificates do not necessarily provide information about the total number of posted persons, as multiple A1 forms can be issued for one person in the case of repeated postings. At the same time, these certificates can underestimate the extent of posting, as companies occasionally try to avoid the administrative procedures, especially in the case of short-term postings (Fihel et al., 2015, p. 74). A1 forms are also issued for employees who work in two or more member states. Even though it is likely that postings are also included here, these are not further considered in the analysis, as no information is known about the destination countries of this mobility (De Wispelaere et al., 2020).

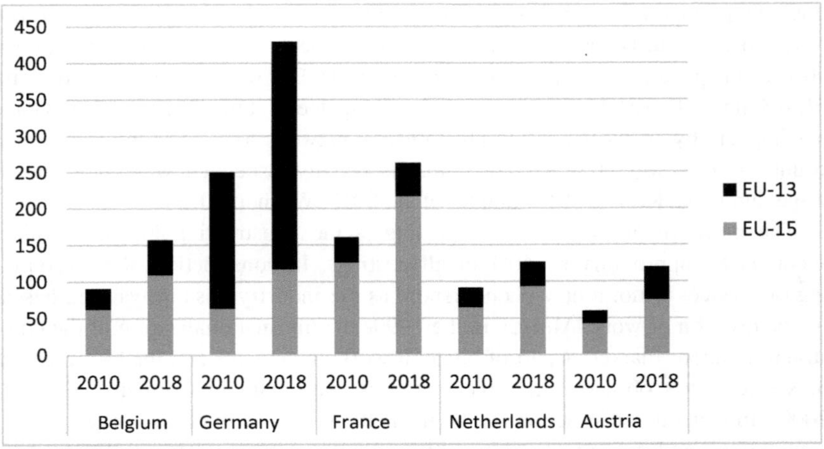

Fig. 3.9 Postings to the top five recipient countries (A1 forms) (2010 and 2018) (in 000). (Source: European Commission, 2012; De Wispelaere et al., 2020, own calculations)

2013) for companies in these countries to resort to cross-border posting. In this respect, the increasing usage of transnational service mobility not only stands for a flexibilization of labour relations, but also for "regime shopping" (Dølvik et al., 2014) between different social security systems in Europe.

As already mentioned, the principle of equal treatment was largely enshrined with the reform of the Posting of Workers Directive. It remains to be seen to what extent this reform as well as the 2014 EU Enforcement Directive will lead to improved compliance with labour standards in the European mobility area. Since reports of dumping practices in connection with cross-border subcontracting continue, an improved "subcontractor management" (Bosch et al., 2019) seems to be necessary. It remains questionable whether the Enforcement Directive goes far enough, as only the direct subcontractor can be held liable, instead of starting at the top of the subcontracting chain in the sense of a general contractor liability (as by now exists in the German construction industry).

From an analytical perspective, it seems sensible to distinguish between on the one hand atypical mobility as part of the freedom to provide services and freedom of establishment and on the other hand individual migration as part of the free movement of workers, because they are based on a different "logic of integration" into the labour market. However, in the social reality of cross-border mobility, this distinction cannot always be maintained. In the enlarged EU, a transnational market for low-wage work has emerged, in which the boundaries between posted work, bogus self-employment, and dependent employment are often fluid (Dølvik et al., 2014, p. 156). What these different forms of work-related mobility have in common is a temporary character, low pay, and a high degree of precarity. This leads us to the question of the labour market effects of intra-European mobility.

3.3 Labour Market Effects

How has the increase in intra-EU mobility affected European labour markets? Have fears of wage dumping and displacement in connection with Eastern enlargement been confirmed? Theoretically, there is no compelling link between an expansion of the labour supply and an increase in unemployment. The number of available jobs is not fixed ("lump of labour fallacy"). Rather, labour markets develop in a dynamic relationship of supply and demand, which is significantly shaped by overall economic development. Thus, employment-related migration not only expands the supply of labour but also the demand. On the one hand, additional jobs can be created by the availability of new labour, and on the other hand, migrants as consumers increase the demand for goods and services, which also results in more jobs. There is little evidence to suggest that in a globalized economy an expansion of the labour supply is associated with a decline in employment opportunities for the domestic population (Crouch, 2019, pp. 33–37; MAC, 2018). Rather, especially qualified migration can fill skill shortages in the labour market (Geis, 2012).

Most studies conclude that, from an overall economic perspective, an increase in intra-European labour mobility has not led to major disruptions in European labour markets (e.g., Holland et al., 2011; Kahanec et al., 2010). There is no clear evidence of a rise in unemployment or generally falling wages. However, these studies, which often rely on highly aggregated data, only capture the impact on specific occupational groups, sectors, or local labour markets to a limited extent. There is some indication that low-skilled groups are more affected by new migration. In some low-wage segments of the labour market, probably some displacement has occured, particularly affecting long-term migrants from third countries who are in direct competion with EU newcomers (Baas et al., 2010; Dølvik, 2013).

With regard to the effects of intra-European migration on wages, different groups in the labour market are affected to varying degrees. As regards higher-skilled migration, which comes disproportionately from the old EU-15, it can be assumed that it has a moderating influence on wages in the upper income distribution (Kahanec et al., 2010, pp. 7–8). Overall, however, this effect is likely to be relatively small. Also in connection with migration from the new EU-13, which has increased more strongly in the last two decades, most studies find no significant effects on general wages. However, some negative wage effects are likely to have occured in some low-skilled segments where EU-13 migrants are over-represented. In contrast, higher occupational groups are likely to have benefited from this migration (Baas et al., 2010; Dustmann et al., 2013; MAC, 2018).

The above-mentioned studies focus almost exclusively on individual migration as part of the free movement of workers. In doing so, other forms of employment-related mobility such as transnational posting or the mobility of the self-employed are overlooked (Bosch, 2013; Dølvik, 2013). While macroeconomic research finds no major effects of intra-European migration, sociological studies on worker posting at the micro-level found displacement effects in construction and the meat-processing industry (Dølvik et al., 2014; Hunger, 2000; Wagner & Hassel, 2015). Thus, different research approaches and levels of analysis can produce contrasting results. Overall, precarious forms of labour mobility such as posting, cross-border agency work, and bogus self-employment, which often occur in a grey area between regular and irregular employment, seem to lead to greater disruptions in the labour market. Thus, there is a need to "look beyond the horizon of traditional migration research" (Bosch, 2013, p. 280) and to include other forms of transnational labour mobility in the analysis.

Another question is whether an increase in precarious labour mobility intensifies a trend towards flexibility in the labour market (Raess & Burgoon, 2015). In principle, there is no simple cause-and-effect relationship between migration and flexibility. A decisive factor is the institutional context. Comparative research on low-wage work has shown that inclusive labour market institutions are capable of absorbing larger numbers of migrants without leading to a significant increase in precarious employment (Appelbaum et al., 2010). There is also some evidence that a deregulation dynamic in Western European labour markets preceded a new wave of labour migration in the 1990s (Schierup et al., 2006; Wilpert, 1998). At the same time, there is little doubt that in less regulated labour markets, an increase in cross-border mobility can boost a trend towards flexibility. This is particularly noticeable when "exit options" (Appelbaum et al., 2010) exist for employers to leave the regulatory context. The presence of a new pool of "flexible" workers can also impact on the recruitment strategies of employers (MacKenzie & Forde, 2009). Thus, supply and demand in a transnational mobility space can mutually reinforce each other and exacerbate a trend towards flexibility and dualization in labour markets.

So far, the labour market impact has been discussed with a view on the destination countries. What about the countries of origin? Can we observe a "brain drain" there, i.e., the emigration of (highly) qualified workers whose human capital is difficult to replace? In principle, it can be assumed that the labour market impact in the countries of origin is greater than in the destination countries. As shown in Fig. 3.1 (see above), the emigration of the working-age population was particularly significant in Romania, but also in Bulgaria, Latvia, and Lithuania. This emigration has helped to reduce unemployment in the short term and to increase average wages, especially at a time when many of these states were still in a transition phase (OECD, 2012).

At the same time, this emigration has not only exacerbated a demographic crisis, particularly in Southeast Europe, but long-term negative effects on overall economic development are also suspected (Atoyan et al., 2016). Estimates suggest that due to emigration in Bulgaria, Romania, and Lithuania, the potential total economic output will be reduced by five to ten percent over a longer period. Remittances from migrants in the destination countries can only partially compensate for this negative effect (Hollande et al., 2011).

The fact that in most Central and (South) Eastern European countries the proportion of academics among emigrants was above average (Ayotan et al., 2016, pp. 12–13) has fueled the debate about a "brain drain". A loss of (highly) qualified personnel, especially in the medical field, poses a significant problem. However, it is too simplistic to reduce the problem of emigration of skilled workers from the NMS to a "brain drain". On the one hand, the mobility of highly qualified scientists and academics is one of the characteristics of a globalized world (Koser and Salt, 1997). On the other hand, there is the possibility of knowledge circulation, especially when highly qualified migrants return to their home countries (Kaczmarczyk, 2010). Until then, there is some evidence that the human capital of NMS migrants in the destination countries is not being utilized in an optimal way, so it often seems more appropriate to speak of a "brain waste" rather than a "brain drain" (Galgóczi et al., 2012).

3.4 Summary

This chapter examined the scale and labour market impact of intra-European migration. Already in the 1990s, there was an increase in this mobility, facilitated by the completion of the single market and the introduction of EU citizenship. However, this increase was relatively moderate and largely limited to higher qualified individuals within the EU-15. In the 21st century, Eastern enlargement triggered an unprecedented dynamic of labour mobility, as hundreds of thousands of mostly young and relatively well-educated new EU citizens moved westward.

This migration served a demand for skilled and less skilled labour, but was also accompanied by concerns about wage dumping. Against this background, the majority of the EU-15 states restricted access to the labour market for up to seven years. However, it turned out that these transitional arrangements had only a limited impact on the direction and scope of intra-European mobility flows, which were driven by relatively strong economic push and pull factors. These factors interacted with the EU free movement regime and new transport and communication opportunities, thus enabling the largest voluntary migration movement in recent European history.

In the course of the economic and financial crisis of 2008/2009, a diversion effect of intra-European mobility towards the north was observed, a development that intensified in the second decade of the 21st century. In addition, some of the main countries of origin, such as Poland, experienced a certain degree of return migration. However, these were by no means the only mobility patterns in the crisis years, as intra-European migration continued to be characterized by circular migration, repeated border crossings, and onward movements. At the same time, the dictum of "liquid migration" should not be overstretched, as this mobility is also accompanied by settlement processes that result in relatively permanent changes in European labour markets.

References

Anacka, M., & Fihel, A. (2012). Return migration to Poland in the post-accession period. In B. Galgóczi, J. Leschke, & A. Watt (Eds.), *EU migration and labour markets in troubled times: Skills mismatch, return and policy responses* (pp. 143–168). Ashgate.

Anderson, B., & Ruhs, M. (2010). Migrant workers: Who needs them? A framework for the analysis of staff shortages, immigration, and public policy. In M. Ruhs & B. Anderson (Eds.), *Who needs migrant workers? Labour shortages, immigration, and public policy* (pp. 15–52). Oxford University Press.

Anderson, B., Ruhs, M., Spencer, S., & Rogaly, B. (2006). Fair enough? Central and East European migrants in low wage employment in the UK. Joseph Rowntree Foundation. Retrieved May 20, 2020, from https://www.compas.ox.ac.uk/2006/pr-2006-changing_status_fair_enough

Appelbaum, E., Bosch, G., Gautié, J., Mason, G., Mayhew, K., Salverda, W., Schmitt, J., & Westergaard-Nielsen, N. (2010). Introduction and overview. In J. Gautié & J. Schmitt (Eds.), *Low-wage work in the wealthy world* (pp. 1–32). Sage.

Atoyan, R., Christiansen, L., Dizioli, A., Ebeke, C. Ilahi, N., & Ilyina, A. (2016). Emigration and its economic impact on Eastern Europe. IMF Staff Discussion Note 16/07. Retrieved May 25, 2020, from https://www.imf.org/external/pubs/ft/sdn/2016/sdn1607.pdf

Baas, T., Brücker, H., & Hauptmann, A. (2010). Labour mobility in the enlarged EU: Who wins, who loses? In M. Kahanec & K. F. Zimmermann (Eds.), *EU labour markets after post-enlargement migration* (pp. 46–70). Springer.

Becker, S., et al. (2013, February 25). Der deutsche Traum. *Der Spiegel*.

Bosch, G. (2013). Grenzüberschreitende Arbeitsmärkte und nationale Beschäftigungssysteme in der EU. In H. G. Soeffner (Ed.), *Transnationale Vergesellschaftungen* (pp. 279–292). Springer VS.

Bosch, G., Hüttenhoff, F., & Weinkopf, C. (2019). *Kontrolle von Mindestlöhnen*. Springer VS.

Boswell, C., & Geddes, A. (2011). *Migration and mobility in the European Union*. Palgrave Macmillan.

Castles, S., & Miller, M. J. (2009). *The age of migration: International population movements in the modern world* (4th ed.). Palgrave Macmillan.

Chiswick, B. R., Lee, Y. L., & Miller, P. W. (2005). A longitudinal analysis of immigrant occupational mobility: A test of the immigrant assimilation hypothesis. *International Migration Review, 39*(2), 332–353.

Crouch, C. (2019). *The globalization backlash*. Polity.

De Wispelaere, F., De Smedt, L., & Pacolet, J. (2020). *Posting of workers: Report on A1 portable documents issued in 2018*. European Kommission, Generaldirektion Beschäftigung, Soziales und Integration.

Dølvik, J. E. (2013). European movements of labour: Challenges for European social models. In E. Jurado & G. Brochmann (Eds.), *Europe's immigration challenge: Reconciling work, welfare and migration* (pp. 33–58). Tauris.

Dølvik, J. E., Eldring, L., & Visser, J. (2014). Setting wage floors in open labour markets: The partners in Europe's multilevel governance. In S. Evju (Ed.), *Regulating transnational labour in Europe: The quandaries of multilevel governance* (pp. 131–186). University of Oslo, Department of Private Law.

Dustmann, C., Frattini, T., & Preston, I. (2013). The effect of immigration along the distribution of wages. *Review of Economic Studies, 80*(1), 145–173.

Eichhorst, W., Marx, P., & Wehner, C. (2017). Labour market reforms in Europe: Towards more flexicure labour markets? *Journal for Labour Market Research, 51*(3), 1–17.

Eppel, R., Leoni, T., & Mahringer, H. (2017). *Österreich 2025: Segmentierung des Arbeitsmarktes und schwache Lohnentwicklung in Österreich*. Österreichisches Institut für Wirtschaftsforschung.

Esping-Andersen, G. (1999). *Social foundations of postindustrial economies*. Oxford University Press.

Esser, H. (2009). Pluralisierung oder Assimilation? Effekte der multiplen Inklusion auf die Integration von Migranten. *Zeitschrift für Soziologie, 38*(5), 358–378.

Europäische Kommission, Generaldirektion Beschäftigung, Soziales und Integration. (2012). *Employment and social developments in Europe 2011*. Amt für amtliche Veröffentlichungen der Europäischen Gemeinschaften.

Europäische Kommission, Generaldirektion Wirtschaft und Finanzen. (2020). Annual Macro-Economic Database (AMECO). Retrieved October 20, 2020, from https://ec.europa.eu/economy_finance/ameco/user/serie/SelectSerie.cfm

Eurostat. (2020a). EU citizens living in another member state – Statistical overview. Retrieved April 15, 2020a, from https://ec.europa.eu/eurostat/statistics-explained/index.php?title=EU_citizens_living_in_another_Member_State_statistical_overview

Eurostat. (2020b). GDP per Capita in PPS. Retrieved April 15, 2020b, from http://ec.europa.eu/eurostat/tgm/table.do?tab=table&init=1&language=en&pcode=tec00114&plugin=1

Eurostat. (2020c). 2.3 Crossing borders. Retrieved March 15, 2020, from https://ec.europa.eu/eurostat/cache/digpub/eumove/bloc-2c.html?lang=en

Fassmann, H., Münz, R., & Seifert, W. (1997). Die Arbeitsmarktposition ausländischer Arbeitskräfte in Deutschland (West) und Österreich. *Mitteilungen aus der Arbeitsmarkt- und Berufsforschung, 30*(4), 732–745.

Favell, A. (2008). *Eurostars and Eurocities: Free movement and mobility in an integrating Europe*. Blackwell.

Felbo-Kolding, J., Leschke, J., & Spreckelsen, T. F. (2019). A division of labour? Labour market segmentation by region of origin: The case of intra-EU migrants in the UK, Germany and Denmark. *Journal of Ethnic and Migration Studies, 45*(15), 2820–2843.

Fihel, A., Janicka, A., Kaczmarczyk, P., & Nestorowicz, J. (2015). *Free movement of workers and transitional arrangements: Lessons from the 2004 and 2007 enlargements*. Centre of Migration Research.

Fries-Tersch, E., & Mabilia, V. (2015). *Annual report on Intra-EU labour mobility*. European Kommission, Generaldirektion Beschäftigung, Soziales und Integration.

Galgóczi, B., Leschke, J., & Watt, A. (Eds.). (2009). *EU labour migration since enlargement: Trends, impacts and policies*. Ashgate.

Galgóczi, B., Leschke, J., & Watt, A. (Eds.). (2012). *EU migration and labour markets in troubled times: Skills mismatch, return and policy responses*. Ashgate.

Geis, W. (2012). Der Beitrag der Zuwanderung zur Fachkräftesicherung. *IW-Trends – Vierteljahresschrift zur empirischen Wirtschaftsforschung, 39*(2), 85–98.

Hall, P. A., & Soskice, D. (Eds.). (2001). *Varieties of capitalism: The institutional foundations of comparative advantage*. Oxford University Press.

Harvey, M. (2003). Privatization, fragmentation and inflexible flexibilization in the UK construction industry. In G. Bosch & P. Philip (Eds.), *Building chaos: An international comparison of deregulation in the construction industry* (pp. 188–209). Routledge.

Holland, D., Fic, T., Rincon-Aznar, A., Stokes, L., & Paluchowski, P. (2011). *Labour mobility within the EU: The impact of enlargement and the functioning of the transitional arrangements*. National Institute of Economic and Social Research.

Holst, H. (2012). Gefahrenzone Absatzmarkt? Leiharbeit und die Temporalstrukturen der Flexibilisierung. In A. Krause & C. Köhler (Eds.), *Arbeit als Ware: Zur Theorie flexibler Arbeitsmärkte* (pp. 141–161). Transcript.

Hunger, U. (2000). Temporary transnational labour migration in an integrating Europe and the challenge to the German Welfare State. In M. Bommes & A. Geddes (Eds.), *Immigration and welfare: Challenging the boundaries of the welfare state* (pp. 189–208). Routledge.

Kaczmarczyk, P. (2010). Brains on the move? Recent migration of the highly skilled from Poland and its consequences. In R. Black, G. Engbersen, M. Okólski, M., & C. Pantîru (Eds.), *A Continent moving West? EU enlargement and labour migration from Central and Eastern Europe* (pp. 165–186). Amsterdam University Press.

Kahanec, M., Zaiceva, A., & Zimmermann, K. F. (2010). Lessons from migration after EU enlargement. In M. Kahanec & K. F. Zimmermann (Eds.), *EU labour markets after post-enlargement migration* (pp. 3–45). Springer

Kalleberg, A. L. (2018). *Precarious lives: Job insecurity and well-being in rich democracies*. Polity.

Keller, B., & Seifert, H. (2013). *Atypische Beschäftigung zwischen Prekarität und Normalität*. Edition sigma.

King, R. (2002). Towards a new map of European migration. *International Journal of Population Geography, 8*(2), 89–106.

Krings, T. (2019). *Posted Workers* in Österreich: Grenzüberschreitende Entsendearbeit im Spannungsfeld von offenen Märkten und (supra)nationaler Regulierung. *Österreich Zeitschrift für Soziologie, 44*(1), 21–41.

Krings, T., Moriarty, E., Wickham, J., Bobek, A., & Salamońska, J. (2013). *New mobilities in Europe: Polish migration to Ireland post-2004*. Manchester University Press.

Kogan, I. (2007). *Working through barriers: Host country institutions and immigrant labour market performance in Europe*. Springer.

Koser. K.,& Salt, J. (1997). The geography of highly skilled international migration. *International Journal of Population Geography, 3*(4), 285–303.

Kvist, J. (2004). Does EU enlargement start a race to the bottom? Strategic interaction among EU member states in social policy. *Journal of European Social Policy, 14*(3) 301–318. https://doi.org/10.1177/0958928704044625

MacKenzie, R., & Forde, C. (2009). The rhetoric of the 'good worker' versus the realities of employers' use and the experiences of migrant workers. *Work, Employment and Society, 23*(1), 142–59.

Martin, P., Abella, M., & Kuptsch, C. (2006). *Managing labour migration in the twenty-first century*. Vail-Ballou.

Massey, D., Arango, J., Hugo, G., Kouaouci, A., Pellegrino, A., & Taylor, J. E. (1993). Theories of international migration: A review and appraisal. *Population and Development Review, 19*(3), 431–466.

Meardi, G., Martin, A., & Riera, M. L. (2012). Constructing uncertainty: Unions und migrant labour in construction in Spain and the UK. *Journal of Industrial Relations, 54*(1), 5–21.

Migration Advisory Committee (MAC). (2014). Migrants in low-skilled work: The growth of EU and non-EU labour in low-skilled jobs and its impact on the UK. Full Report July 2014. Retrieved October 13, 2020, from https://www.gov.uk/government/publications/migrants-in-low-skilled-work

Migration Advisory Committee (MAC). (2018). EEA migration in the UK: Final report. September 2018. Retrieved October 17, 2020, from https://www.gov.uk/government/publications/migration-advisory-committee-mac-report-eea-migration

Mühlau, P. (2012). *Occupational and earnings mobility of Polish migrants in Ireland in the Recession*. IIIS Discussion Paper No. 413. Trinity College Dublin.
Mundell, R. A. (1961). A theory of optimal currency areas. *The American Economic Review, 51*(4), 657–665.
Nicolaisen, H., Kavli, H. C., & Jensen, R. S. (Eds.). (2019). *Dualisation of part-time work: The development of labour market insiders and outsiders*. Policy.
OECD. (2012). *Free movement of workers and labour market adjustment: Recent experiences from OECD countries and the European Union*. OECD.
Pollard, N., Latorre, M., & Sriskandarajah, D. (2008). *Floodgates or turnstiles? Post-EU enlargement migration flows to (and from) the UK*. Institute for Public Policy Research.
Pries, L. (2010). *Transnationalisierung: Theorie und Empirie grenzüberschreitender Vergesellschaftung*. VS Verlag.
Raess, D., & Burgoon, B. (2015). Flexible work and immigration in Europe. *British Journal of Industrial Relations, 53*(1), 94–111.
Recchi, E. (2015). *Mobile Europe: The theory and practice of free movement in the EU*. Palgrave Macmillan.
Reyneri, E. (2013). Irregular immigration and the underground economy in Southern Europe: Breaking the vicious cycle. In E. Jurado & G. Brochmann (Eds.), *Europe's immigration challenge: Reconciling work, welfare and migration* (pp. 143–158). Tauris.
Rudolph, H. (1996). The new Gastarbeiter system in Germany. *New Community, 22*(2), 287–300.
Sachverständigenrat deutscher Stiftungen für Integration und Migration (SVR). (2013). *Erfolgsfall Europa? Folgen und Herausforderungen der EU-Freizügigkeit für Deutschland Jahresgutachten 2013 mit Migrationsbarometer*. SVR.
Schierup, C. U., Hansen, P., & Castles, S. (2006). *Migration, citizenship and the European welfare state: A European dilemma*. Oxford University Press.
Sciortino, G. (2013). Immigration in Italy: Subverting the logic of welfare reform. In E. Jurado & G. Brochmann (Eds.), *Europe's immigration challenge: Reconciling work, welfare and migration* (pp. 77–94). Tauris.
Shire, K., Heinrich, S., Imai, J., Mottweiler, H., Tünte, M., & Wang, C. C. (2018). Private labour market intermediaries in cross-border labour markets in Europe and Asia: International norms, regional actors and patterns of cross-border labour mobility. In S. Quack, I. Schulz-Schaeffer, K. Shire, & A. Weiß (Eds.), *Transnationalisierung der Arbeit* (pp. 155–184). Springer VS.
Spindler, S. (2011). Feminisierung von Migration – Formen und Folgen weiblicher Wanderungsprozesse. In G. Hentges & H. W. Platzer (Eds.), *Europa quo vadis? Ausgewählte Problemfelder der europäischen Integrationspolitik* (pp. 171–186). VS Verlag.
Staples, R., Trinczek, R. & Whittall, M. (2013). Posted Workers: Zwischen Regulierung und Invisibilisierung. *Arbeit, 22*(4), 271–286.
Tamas, K., & Münz, R. (2006). *Labour migrants unbound? EU enlargement, transitional measures and labour market effects*. Institute for Future Studies.
Verwiebe, R., & Müller, M. C. (2006). Gelungene Integration in den Arbeitsmarkt? Die flexiblen Biografien transnational mobiler Europäer zu Beginn des 21. Jahrhunderts. *Berliner Zeitschrift für Soziologie, 16*(1), 95–114.
Verwiebe, R., Wiesböck, L., & Teitzer, R. (2014). New forms of intra-European migration, labour market dynamics and social inequality in Europe. *Migration Letters, 11*(2), 125–136.

Wagner, I. (2018). *Workers without borders: Posted work and precarity in the EU*. Cornell University Press.

Wagner, B., & Hassel, A. (2015). Labour migration and the German meat processing industry: Fundamental freedoms and the influx of cheap labour. *South Atlantic Quarterly, 114*(1), 204–214.

Wagner, B., & Hassel, A. (2016). Move to work, move to stay? Mapping atypical labour migration into Germany. In J. E. Dølvik & L. Eldring (Eds.), *Labour mobility in the enlarged Single European Market* (Comparative Social Research) (Vol. 32, S. 125–158). Emerald Group.

Wallace, C. (2001). Conceptual challenges from the new migration space. In C. Wallace & D. Stola (Eds.), *Patterns of migration in Central Europe* (pp. 45–71). Palgrave.

Wickham, J. (2016). *Unequal Europe: Social divisions and social cohesion in an old continent*. Routledge.

Wilpert, C. (1998). Migration and informal work in the new Berlin: New forms of work or new sources of labour? *Journal of Ethnic and Migration Studies, 24*(2), 269–294.

Germany: New Migration In-Between a Job Market Boom, Skill Shortages and Precarious Mobility

4

Germany emerged in the 21st century as one of the most important destinations for international labour migration (OECD, 2014). This mobility came to a significant extent from the enlarged EU, especially since the opening of the labour market in 2011 and 2014. This chapter examines the scale and impact of this mobility, as well as efforts to 're-embed' it into national labour relations. It is shown that a booming labour market was an important pull factor for migrants from the old and new EU, who found employment relatively quickly. However, this mobility was also accompanied by incidents of wage dumping and non-compliance with labour standards, partly in connection with "atypical" employment such as posted work and (bogus)self-employment. Not least with regard to the opening of the labour market, the debate about a (re-)regulation of standards gained importance. Thus, concerns about a "race to the bottom" were an important factor influencing the introduction of a statutory minimum wage and the extension of sector-wide wage floors. Before going into more detail on these regulatory efforts, the development of intra-European migration since the 1990s is first traced with a particular focus on the labour market integration of mobile Europeans.

4.1 Intra-European Migration, Eastern Enlargement, and a Restricted Labour Market

Since the 1990s, Germany has experienced a significant increase in migration from the then EU-15. Compared to earlier immigration cohorts, these migrants were increasingly found in high-skilled occupations. This was explained not only

T. Krings, *Labour Mobility in the European Union as an Example of the Transnationalization of Employment*, https://doi.org/10.1007/978-3-658-43977-4_4

by their higher qualifications, but also by the introduction of EU citizenship, which opened up new mobility and employment opportunities (Verwiebe & Eder, 2006; Verwiebe & Müller, 2006).

Even more significant in numbers was an increase in mostly temporary and circular mobility from Central and Eastern Europe (CEE) in the 1990s (Nowicka, 2007; Rudolph, 1996). In the wake of societal upheavals in CEE, but also due to visa facilitations, this mobility gained importance especially in regions close to the border. Although there had been an official recruitment stop since the 1970s, there was still a demand for "flexible" workers in labour-intensive sectors such as construction, hospitality, agriculture, care, and the food industry (Tamas & Münz, 2006). This mobility developed in-between regular and irregular employment. On the one hand, there was a revival of bilateral recruitment agreements between Germany and a number of CEE countries. The aim of these new "guest worker" programs was to channel the growing migration flows into legal channels, to cover the labour needs in certain sectors, and to politically stabilize the transformation countries (Rudolph, 1996). On the other hand, a significant part of this cross-border mobility took place in the growing informal economy of "global cities" like Berlin. It is important to emphasize that a partial deregulation of the labour market preceded this migration (Wilpert, 1998). At the same time, there can be little doubt that this mobility reinforced a trend towards precarious employment in some low-wage segments (Krings, 2021).

Especially in the construction industry, precarious employment emerged in connection with transnational mobility in the 1990s (Hunger, 2000). At that time, domestic companies increasingly relied on foreign subcontractors from CEE, whose workers were mostly employed under rather precarious contracts for work. This form of employment was controversial, as it was repeatedly accompanied by allegations of wage dumping. Against this background, the bilateral quotas for contract workers from CEE were restricted, mainly due to pressure from the construction union IG BAU (Treichler, 1998). However, by this point, construction companies had come to appreciate the benefits of cross-border subcontracting. To continue to utilize it in the future, they "discovered" the EU freedom of services, which now allowed them to rely on foreign subcontractors and low-wage labour, especially from Southern Europe (Menz, 2005, pp. 107–109).

The increased use of posted work marked a "paradigm shift" (Hunger, 2000) in the employment of foreign workers in Germany. The incorporation of earlier waves of "guest workers", despite contrary political rhetoric, was characterized by a logic of inclusion in national welfare states (Bommes & Geddes,

2000). In contrast, posted workers were at best only partially integrated into the national labour market institutions of the host country. Since the Posted Workers Act of 1996, sector-specific minimum wages were set for the construction industry (and now also to 12 other sectors such as the meat-processing industry or the care sector). However, this re-regulation no longer took place (unlike in Austria [see Chap. 5]) on the basis of the entire wage grid. Unions no longer possessed the organizational and institutional power resources to enforce such re-regulation on the basis of equal pay (Menz, 2005).

In Germany (and in Austria), there was widespread skepticism towards EU enlargement in general and the freedom of movement for workers in particular. When then-Chancellor Gerhard Schröder announced at the end of 2000 that Germany would impose transition periods of up to seven years for workers from the 2004 accession countries, he had the support of almost all political parties and most social partners. In particular, the IG Bau, which had already been confronted with posted work in the 1990s, vehemently demanded transition periods not only for workers, but also for service companies from the new member states (NMS) (Krings, 2009).

The Confederation of German Employers' Associations (BDA) and the Federation of German Industries (BDI) also supported "flexible" transition periods, but saw the period of seven years as too long and demanded sector-specific opening clauses in view of a shortage of skilled workers in some parts of the labour market (Jeřábek, 2011, pp. 233–234). Unlike trade unions, there was some disagreement on the employer side. The BDI rejected an additional restriction of the freedom to provide services. This suggests that especially larger industrial companies have an interest in using cost-effective service providers (Thelen, 2014, p. 53). In contrast, the Main Association of the German Construction Industry (HDB) and the Central Association of the German Construction Industry (ZDB), together with IG BAU, demanded multi-year transition periods, out of concerns for smaller domestic companies (and their workers), which are in competition with foreign service providers.

In the end, Germany restricted both the free movement for workers and service providers in sectors such as construction and industrial cleaning for up to seven years in 2004 ("2+3+2 rule"). The same transition rules were imposed for the 2007 accession countries. In the case of Croatia, which joined the EU in 2013, the two-year transition period was not extended, so Croatian workers and service providers had free access to the German labour market since July 1, 2015.

The transitional arrangements probably had the effect of redirecting some migration flows to the UK and Ireland (see Chap. 3). Nevertheless,

Germany still experienced an increase in employment-related migration from the NMS. This can be explained on the one hand by new mobility opportunities for citizens from the accession countries, who could benefit from their newly gained status as EU citizens despite restricted freedom of movement for workers, and on the other hand by an increasing demand for additional labour. This particularly affected the low-wage sector, which increasingly became a pull factor for migrant workers from the NMS (Krings, 2021). The latter were still employed during the transition phase as part of the bilateral recruitment programs from the 1990s, which remained largely unaffected by the enlargement process (Tamas & Münz, 2006, p. 138). Overall, a slight increase in work permits for citizens from the NMS has been observed since 2004. A more significant increase was seen in atypical employment such as self-employment and posted work, which represented an alternative way into the German labour market (Wagner & Hassel, 2016).

In construction and in industrial cleaning, an increase in (bogus)self-employment was observed in connection with the transition phase. In these sectors, not only the free movement for workers but also the freedom to provide services was restricted, but not the freedom of establishment. Thus, workers from the NMS could register as one-person companies (OPCs), the number of which increased sharply since 2004 (Wagner & Hassel, 2016, p. 148). Since work was often carried out under supervision, it can be assumed that these were often cases of bogus self-employment, where labour law provisions such as collective agreements or working time regulations were circumvented.

In other sectors such as the meat industry, where the freedom to provide services was not restricted, there has been a strong increase in cross-border postings since Eastern enlargement. According to estimates by the Food, Beverages and Catering Union (NGG), the number of posted workers from CEE increased fivefold between 2003 and 2014 to about 25,000 (Bosch et al., 2019, p. 205). Since there was no wage floor in the industry until 2014, posted workers could be employed under the conditions of the country of origin. In the case of the main sending countries Romania, Poland and Bulgaria, this meant significantly lower labour costs and generally precarious employment. As a result of an increased use of "service contracts as a business model" (Bosch et al., 2019), a displacement process could be observed in the industry, as especially larger companies began to replace parts of the core workforce with contract workers (ibid., p. 197).Thus, the meat industry served as a paradigmatic example of the regulatory challenges in a transnational mobility space characterized by significant income differences. However, it also stands as an example of re-regulation of labour relations with strong support from state institutions (see below).

In the 21st century, intra-European labour migration was predominantly not mobility of the low-skilled. Especially in the case of mobility from the EU-15/ EFTA, it was qualified migration. The proportion of skilled migrants from the NMS has also increased since EU accession. However, as the proportion of low-skilled individuals also increased at the same time, there was a certain degree of polarization in their qualifications (SVR, 2013, p. 102). Politically, measures were taken to enable qualified migration from the NMS. As early as 2007, access to the German labour market was facilitated for certain occupational groups such as mechanical or electrical engineers. Thus, despite a restricted labour market, Germany pursued a flexible policy, which was also the result of skill shortages in some sectors (ibid.). Overall, even before the transition periods expired, there was a rise in labour-related migration from the NMS. However, only the opening of the labour market in conjunction with a sharply rising demand for labour would trigger a hitherto unknown dynamic of cross-border mobility.

4.2 Post-2011: Migration into an Open and Booming Labour Market

The transition periods had only a limited impact on intra-European migration, which were primarily demand- and network-driven (see Chap. 3). However, they were not without effect, as the German experience showed. Since the labour market opening in 2011, there has been a rapid increase in work-related migration from the NMS. At the beginning of 2011, 349,000 people from the accession countries were in employment subject to social insurance or marginal employment. Five years later it was almost three times as many (1,023,900). During the same period, the number of employees from the Southern European "crisis states" Greece, Italy, Portugal and Spain only increased by almost 150,000 people to 550,900 (Bundesagentur für Arbeit, 2016, p. 20). Thus, the "new guest workers", if one wants to use this term, came less from Southern Europe (Becker et al., 2013), but rather from Poland, Romania and Bulgaria.

In terms of regional distribution, EU migrants predominantly settled in West Germany, especially in the economically prosperous states of Baden-Württemberg, Bavaria and Hesse, but also in Berlin. In these states, they represented between seven and eight percent of the total population at the end of 2017. In most other old states in the West, their share was over five percent, while it was less than two percent in the new states in the East (Dubois, 2019, p. 49).

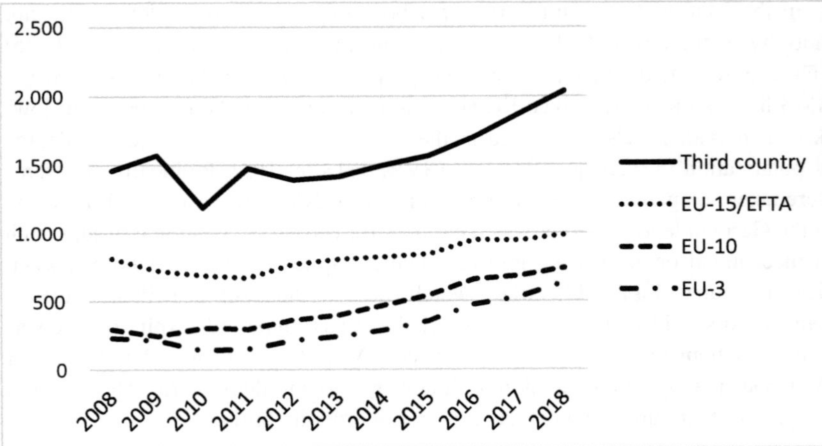

Fig. 4.1 Development of migrant employment in Germany (000s) (2008–2018). (*Source*: EU-LFS, own calculations)

This certainly reflects existing economic inequalities and different employment opportunities between East and West Germany. However, it must also be taken into account that part of the work-related migration to East Germany is not recorded by this registration data. Especially cross-border commuters, whose number in Brandenburg and Saxony has increased significantly since the opening of the labour market in 2011 (Buch et al., 2020, p. 4), are not recorded by the registration register, as the place of residence is still in the country of origin.

Overall, labour market development in Germany in the second decade of the 21st century was significantly shaped by migration. Between 2010 and 2018, the number of migrant workers (2.5 million) increased significantly more than that of non-migrant workers (1.6 million).[1] As Fig. 4.1 shows, this increase was mainly due to migration from the NMS (EU-10 and EU-3). At the same time, however, the figure also documents an increase in migration from third countries, which particularly reflects the growing refugee migration since 2015.

[1]Unless otherwise stated, all data now cited are own calculations based on the European Labour Force Survey (EU-LFS).

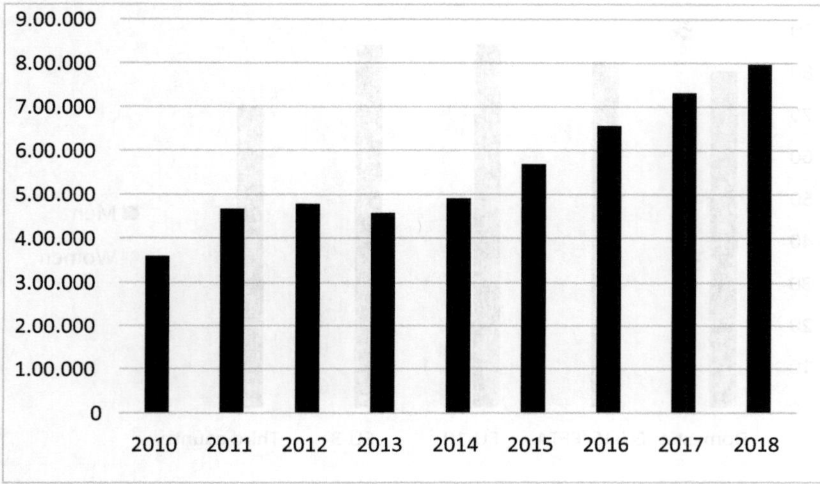

Fig. 4.2 Job vacancies in Germany (annual average (2010–2018). (Source: Bundesagentur für Arbeit, 2021b)

It is plausible that a decline in employment opportunities in crisis countries such as Ireland and Spain, which were important destinations for NMS migrants, has partially led to a redirection of migration flows (Fries-Tersch & Mabilia, 2015). However, the decisive factor that promoted a new dynamic of labour mobility was the opening of the German labour market in conjunction with an extraordinary labour market boom. As shown in Fig. 4.2, the number of job vacancies has continuously increased in the second decade. New jobs were created primarily in those sectors where NMS employees are disproportionately represented (industry, nursing and health, business services, hospitality and construction (Bundesagentur für Arbeit, 2021a). This suggests that supply and demand mutually reinforced each other in the context of a booming and open labour market, leading to a significant increase in employment.

One of the central arguments of this book is that the institutional architecture of the EU free movement regime not only structurally promotes mobility but also facilitates entry into the labour market. This is reflected in the high labour market participation of mobile EU citizens, which is also enabled by their relatively high qualifications. As the following Fig. 4.3 shows, male migrants from the NMS (EU-10 and EU-3) have the highest employment rates of all groups in the labour market, something that can also be observed in

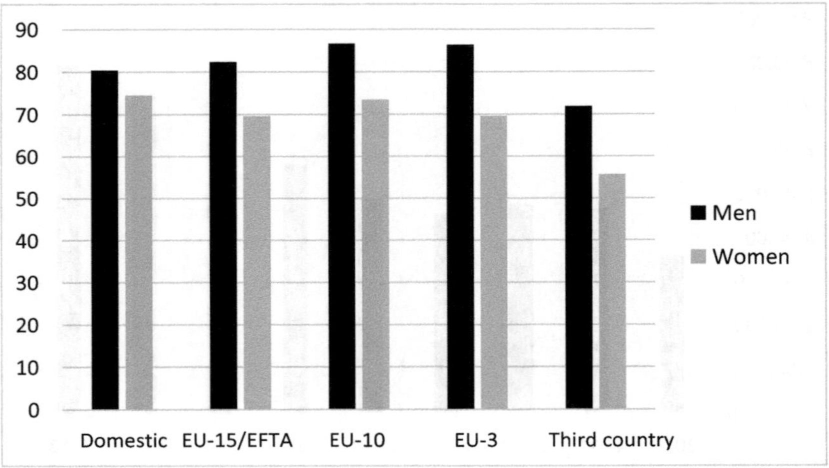

Fig. 4.3 Employment rates by country of birth and gender (17–62 years, in %) (2018). (Source: EU-LFS, own calculations)

other countries (see Chap. 3). Overall, it is noticeable that third-country migrants have significantly lower employment rates, both among men and especially among women.

What about unemployment? In 2018, both men and women from third countries had the highest unemployment rates at just under nine percent and six percent respectively, although it should be noted that these rates (like those of other groups in the labour market) have been declining since the beginning of the decade. This suggests that the booming German labour market also offered employment opportunities for third-country migrants. Among the other groups, the unemployment rate was between three and four percent, with female EU-3 migrants being somewhat of an outlier at nearly five percent.[2]

What about the qualifications of migrants? Especially among EU-15 migrants who migrated in recent years, a significant increase in the level of education can be observed. The cohort that migrated since 2010 has a graduate rate of 50%. The situation is different for NMS migrants who migrated

[2]Age group 17 to 62 years, EU-LFS, own calculations.

Tab. 4.1 Occupational status by country of birth (main occupational groups ISCO-08) (2018) (in %)[4]. (Source: EU-LFS, own calculations)

	Domestic	EU-15/EFTA	EU-10	EU-3	Third country
1. Managers	5.1	7.1	2.8	2.3	3.3
2. Professionals	19.1	24.7	10.9	10	13.4
3. Technicians & Associate Professionals	24.6	13.5	16.3	12.2	13.5
4. Clerical Support Workers	13.9	8.7	10	8.3	7.9
5. Services & Sales Workers	13.7	16.2	13.7	13.4	16.2
6. Skilled Agricultural, Forestry & Fishery Workers	1.4	0.7	1.1	0.5	0.7
7. Craft & Related Trades Workers	11.6	10.6	19.5	16.6	14.7
8. Plant & Machine Operators & Assemblers	5.2	6.8	9.1	12.9	10.9
9. Elementary Occupations	5.5	11.6	16.7	23.7	19.4
Total	*100*	*100*	*100*	*100*	*100*

since 2010. In this cohort, a certain polarization of qualifications can be observed. Especially among EU-3 migrants, the proportion of low-skilled workers has grown relatively strongly and now comprises over 35%. This suggests that the German labour market is also attractive for unskilled and semi-skilled workers from abroad (SVR, 2013, p. 102).

In which professional positions do mobile EU citizens find themselves in the German labour market? As shown in Tab. 4.1, EU-15/EFTA migrants are overrepresented among managers and professionals. This is likely due to their above-average education, but also to institutional regulations in the European mobility space, which facilitate the recognition of academic and professional qualifications. Particularly noteworthy is the high proportion of STEM graduates in this group, whose skills and abilities are in high demand in the German labour market (SVR, 2013, p. 102).[3] NMS migrants are overrepresented at nearly 50% (EU-

[3] STEM stands for Science, Technology, Engineering, and Mathematics.

[4] Deviations from the sum of 100% are due to rounding.

10) and above 50% (EU-3) in manual occupations (ISCO groups 6-9). EU-10 migrants are also well represented at 30% in higher skilled positions (especially in ISCO group 3), while EU-3 migrants are overrepresented in elementary occupations (24%), reflecting their comparatively high share of low-skilled workers. Overall, there can be little doubt that mobile individuals from the old and new EU (and also from third countries) make a significant contribution to filling skill shortages in the German labour market (see also Geis-Thöne, 2021).

There is some evidence to suggest that the booming German labour market has become more permeable for migrant workers. Both EU-10 migrants and EU-3 migrants are more likely to be in high-skilled positions (ISCO groups 1-3) than their formal education levels would suggest. It is therefore plausible that missing qualifications are partially compensated by work experience in the host country and "learning on the job", as has been observed, for example, with refugees in Germany (Brücker et al., 2020). However, it must be taken into account that working in a skilled job does not automatically lead to better pay. For instance, more than 50% of NMS migrants employed as skilled workers in Germany are low-wage earners (Burkert, 2015). Reasons for this high low-wage share include a lack of language skills, which can result in qualifications acquired abroad not being fully utilized. A lower "reservation wage" (Waldemar, in Burkert, 2015) is also suspected, which is likely to weigh more heavily on transnational commuters. It also seems plausible that there are at least occasional practices of wage discrimination (Krings, 2021). How has the increase in migrant employment affected the German labour market? Have concerns about dumping practices and displacement been confirmed?

4.3 Labour Market Effects and "Atypical" Employment

In general, there is so far little empirical evidence that intra-European migration has led to major disruptions in the German labour market. Between 2010 and 2018, when migration increased particularly strongly, the general employment rate rose from 71.2% to just under 76% (age group 17–62 years). A favorable development was also observed with regard to the unemployment rate, which, after a temporary high under the impact of the economic and financial crisis of 2008/09, fell to below four percent by the end of the second decade. Unlike in Austria (see Chap. 5), a decrease in unemployment was also observed among third-country nationals. Apparently, the booming German labour market was able to incorporate a large number of migrants without this leading to a decrease in employment opportunities for locals. There were also no major negative effects on wage development, as real wages grew more strongly in the second decade than in the two previous decades (Seils & Emmler, 2020). Overall, the negative employment and wage effects of migration from the enlarged EU were likely to be small (Bonin et al., 2020).

However, it must be taken into account that this mobility has different effects on different sectors and occupations. In some low-wage segments, which have a high proportion of mobile workers from the NMS, wage growth probably has been moderated. These effects mainly affected low-skilled migrants who have been living in Germany for a longer period of time and who are in direct competition with the new migrants (Baas et al., 2010; Brenke et al., 2010).

There is also some evidence to suggest that cross-border posting of workers led to a displacement effect in some sectors that is not necessarily reflected in macroeconomic studies on the impact of labour mobility (Bosch, 2013; Wagner & Hassel, 2015). As discussed above, posted work and (bogus)self-employment represented an alternative strategy into the German labour market. However, the experiences since the opening of the labour market suggest that these "atypical" employment relationships have not only been an entry option into the labour market, but have now become a structural part of employment. Thus, between 2010 and 2018, the number of postings (A1 forms) increased from 250,000 to 428,900. The main countries of origin were Poland, Slovakia, Slovenia, Croatia, and Hungary (Pacolet & De Wispelaere, 2020, pp. 24, 28). The high proportion of postings from the NMS (almost 75%) suggests that a preference for a strategy of "cost-driven outsourcing" (Bosch et al., 2019) has become entrenched especially in the construction industry, which accounts for about half of all postings.

Experience in the meat industry suggest that a "cost-driven outsourcing" does not necessarily require transnational posting of workers. In the industry, the number of postings has significantly decreased since the introduction of a sector-specific minimum wage in 2014. The new minimum wage resulted in a significant wage increase for contract workers. In addition, the large meat companies voluntarily committed themselves to directly employing foreign contract workers and extend the main provisions of German labour and social law to them (Erol & Schulten, 2021, p. 10). However, since the system of contract work remained largely untouched, there was no sustainable improvement in the conditions of the workers, who mainly came from Romania, Poland, and Bulgaria. Ironically, it required an "exogenous shock" in the form of the COVID-19

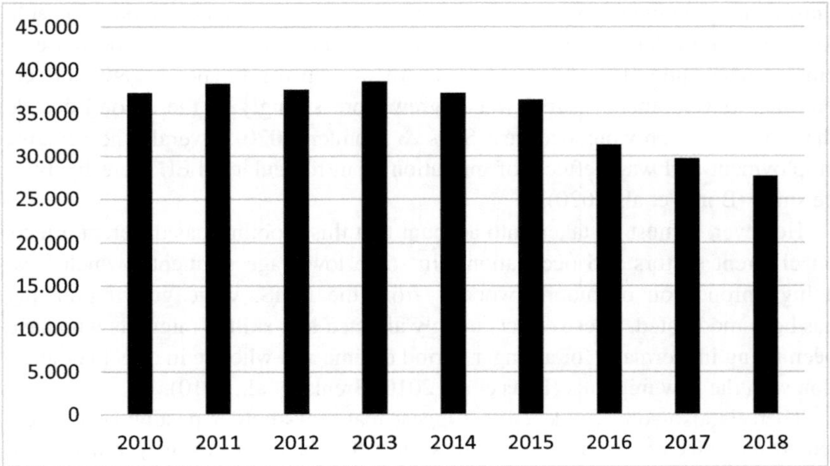

Fig. 4.4 Registrations of Polish one-person companies in Germany (2010–2018). (Source: Federal Statistical Office, 2021)

pandemic, which shattered the system of contract work and enabled a "reorganization of labour relations" (Erol & Schulten, 2021) in the industry (see below).

Another example of atypical employment in the context of cross-border mobility is self-employment. As already mentioned, the number of one-person companies (OPCs) increased relatively strongly during the transition periods, especially in the construction industry. Initially, this was primarily an alternative strategy for entering the labour market. However, Fig. 4.4 shows, using the example of Polish OPCs which registered in Germany, that their number has remained at relatively high level even after the opening of the labour market in 2011 and has only been declining slightly in recent years. This suggests that (bogus)self-employment and posting of workers were not only an entry option into the labour market, but were also used as "cost-driven outsourcing" (Bosch et al., 2019). Bogus self-employment exists when formally self-employed individuals work under instructions and are integrated into operational work processes. It is noticeable that, unlike posting or temporary agency work (see Chap. 2), there have so far been hardly any initiatives at the European level to regulate this form of "dependent self-employment" (Mühlberger, 2007) in a transnational context.

Atypical labour mobility such as posting of workers, cross-border temporary agency work, and (bogus)self-employment often takes place in an informal work environment where minimum standards are only partially adhered to. Often, these employment relationships are characterized by a stark power imbalance between both sides, with posted workers or self-employed migrants often confronted with exploitative working conditions (Staples et al., 2013; Wagner, 2018). However, it would be an oversimplification to portray such forms of employment only as a story of exploitation and oppression. Migrants are active agents who make their own decisions, albeit often under constrained circumstances (Datta et al., 2007). Given significant income differences in the enlarged EU, a salary below the minimum standards in the host country may still appear acceptable if the "frame of reference" (Waldinger & Lichter, 2003) remains the country of origin. Such reasoning becomes apparent in the following quote from a Polish posted worker on a German construction site:

> Our firm told us what to say when controls takes place. We agreed to that before we came to Germany. That's normal. If someone offers you to work in Germany for €8.50 and you earn €3 in Poland than you agree, even if €8.50 is less than the German minimum wage *(at the time of the interview (2012), the industry minimum wage in the construction industry was €11.05, T.K.)*. Management also told us that we would not receive vacation pay even though the contract states that we do (in Wagner, 2015a, p. 700)

A certain degree of informality can therefore be in the interest of both sides of the employment relationship. However, it must be considered that the conditions can be unilaterally changed or even terminated by the employer side at any time. The result is often unpaid wages, cheated employees and subcontractors who quite literally make off (Molitor, 2015, p. 17).

In principle, it is difficult to quantify the extent of non-compliance with labour standards. However, a number of industry studies suggest that the circumvention of minimum standards and collective agreements are not isolated cases (e.g. Mense-Petermann, 2018; Staples et al., 2013; Wagner & Berntsen, 2016; Wagner & Hassel, 2015). In particular, practices of wage dumping seem to be relatively widespread in the case of postings, cross-border temporary agency work, contracts for work, and bogus self-employment, with the construction and the meat industry, as well as industrial cleaning, the care sector, and transport and logistics (including the parcel industry) being mentioned as "vulnerable sectors" (Molitor, 2015, p. 4). Cross-border subcontracting chains prove to be particularly vulnerable. Although general contractors are obliged under Article 14 of the Posted Workers Act to ensure compliance with minimum standards among their subcontractors, monitoring the standards in a transnational work environment, which is characterized by language barriers, informal norms and asymmetric power relations between employers and employees, is often difficult. Although minimum wages now exist in many sectors, which have been complemented by a statutory national minimum wage in 2015, effective monitoring especially in smaller companies is quite limited. In fact, in industries such as construction, the number of inspections has been declining for several years (Bosch et al., 2019, p. 167).

Precarious labour mobility is not a causal factor for labour market flexibility, but rather a consequence of politically induced deregulation and a decline in trade union density and interest representation at company level (works councils) (Appelbaum et al., 2010; Wilpert, 1998). However, there is little doubt that the use of posted work, temporary agency work, contracts for work, and (bogus) self-employment reinforces a dualization trend in some low-wage segments of the German labour market (Palier & Thelen, 2010). While the core workforces in many companies are still relatively well protected (even if trends towards precarization have an impact on them (Bosch, 2018), a secondary segment with unstable jobs is growing at the margins of the German labour market. Precarious and mobile workers from the NMS represent a flexible "reserve army" *par excellence* in this respect, as they are often only employed in Germany for a short period of time and often work in a grey area between regular and irregular employment. What efforts have been made to re-embed this transnational mobility in national labour relations?

4.4 An "Ratchet Effect Downwards"? The (Re-) Regulation of the Labour Market

Efforts to regulate cross-border mobility with the help of the legislator date back to the adoption of the Posted Workers Act in 1996. Until this point, the setting of wage floors was the exclusive task of the social partners (Bosch et al., 2019, p. 3). However, as a "compliance gap" (Wagner & Berntsen, 2016) became more widespread especially in connection with transnational posting, political actors became more influential in the regulation of labour standards. The Posted Workers Act was preceded by several years of lobbying by employer associations and trade unions in the construction industry, which was affected by cross-border mobility like no other. In this campaign for industry-wide binding standards, cracks appeared in the German model of social partnership. While the sectoral employer associations HDB and ZDB, who were concerned about foreign cheap competition, supported minimum standards in the construction industry, the cross-sector employers' association BDA rejected them with reference to collective bargaining autonomy (Eichhorst, 1999, p. 351).

The construction union IG BAU, for its part, had to make fundamental concessions in its demands for the re-regulation of labour standards. After it had originally demanded statutory extension of all collective agreements in connection with posted work, it had to grudgingly accept that only the two lowest minimum wage levels were declared generally binding (Menz, 2005, pp. 110–115). It seemed as if the times were over in Germany when trade unions possessed the political and institutional power resources to enforce "equal pay for equal work".

In the 21st century, further industry-wide minimum wages were negotiated between the social partners in industrial cleaning and temporary agency work. In these sectors, employers saw the need to create universally binding wage floors, especially with a view to the labour market opening in 2011, in order to prevent a "race to the bottom" through transnational service mobility (Bosch & Weinkopf, 2008, p. 309). However, such industry-wide agreements remained the exception, as the employer camp largely rejected universally binding minimum wages.

In view of the growing "bargaining-free zones" (Bispinck, 2012), but also with regard to the opening of the labour market to workers and service providers from the NMS, the debate on the introduction of a statutory minimum wage gained importance. Although a legal determination of wages is traditionally alien to the German system of industrial relations with its primacy of "collective bargaining autonomy", it takes into account new realities, especially the fact that the "formative power of the collective agreement as a central instrument for regulating income and working conditions is in decline" (Bispinck, 2012, p. 506). In 2018, only 49% of all workers in West Germany and 35% of all workers in East Germany worked in a company covered by an industry-wide collective agreement (Kohaut, 2019). These average values also conceal the fact that the collective bargaining coverage rate is significantly lower in private service industries.

The demand for the introduction of a statutory minimum wage was already raised at the end of the 1990s by the Food, Beverages and Catering Union (NGG), which was affected by the expansion of the low-wage sector like no other union (Schroeder, 2020, p. 27). The demand for a national minimum wage initially met with rejection from other unions such as IG Metall and IG Bergbau, the mining and chemicals union, which were still able to negotiate relatively high wage agreements and hence feared an erosion of the latter. Overall, it took more than ten years "until the German trade unions had fully realized their loss of power and the impossibility of curbing the low-wage sector through collectively agreed industry-wide minimum wages" (Bosch et al., 2019, p. 37).

Finally, the trade unions agreed on a uniform policy position in favor of the introduction of a national minimum wage at the DGB Congress 2006, which was accompanied by demands for an easier statutory extension of collective agreements and the extension of the Posting of Workers Act to all economic sectors. With these demands, the German trade union movement recognized that statutory minimum wages do not necessarily contradict collective bargaining autonomy, but rather can stabilize the whole bargaining system in view of a decline in membership of interest groups (Bispinck, 2012, pp. 504–505).

The trade union movement was unable to reach an agreement with the employer side on the introduction of a statutory minimum wage, which was rejected by the BDA as an intervention in collective bargaining autonomy. From then on, the DGB intensified its political lobbying work towards the SPD, after there had been a falling out between the two sides in the course of the Hartz labour market reforms (Hassel & Schiller, 2010). This lobbying work was relatively successful, as the SPD made the introduction of a minimum wage a precondition for a grand coalition with the CDU/CSU after the 2013 federal election. The 2014 Act on the Strengthening of the Bargaining Autonomy not only included the introduction of a national minimum wage of €8.50, but also an extension of the scope of the Posted Workers Act. Further, it provided for the statutory extension of collective agreements by removing the condition that at least 50% of the workers in the respective sector must be covered by such an agreement. In the future, "public interest" will be sufficient as a prerequisite for the statutory extension of collective agreements (Bispinck, 2015, p. 23). However, this reform has not yet been able to stop the decline in generally binding collective agreements since the 1990s, as the BDA in particular blocked an extension of this bargaining policy tool (Schulten, 2021).

With the introduction of the minimum wage, a "ratchet effect" (Traxler, 1996, p. 24) downwards was created, especially in those sectors that pretty much had lost their capacity for collective bargaining. This "ratchet effect" was of particular importance with regard to the opening of the German labour market in 2011 and 2014. As experiences from the meat industry showed, without universally binding minimum standards, a cut throat competition can occur in a transnational mobility space characterized by huge income differences. The introduction of a wage floor not only has a labour policy dimension, but also a socio-political one. Especially with regard to "social integration at enterprise-level" (Schmidt, 2006), universally valid (minimum) standards are of considerable importance in order to minimize conflicts among diverse workforces.

While the trade union movement has been relatively successful at the political level in its demand for the re-regulation of labour standards, worker representative bodies are increasingly unable to monitor compliance with these standards. Traditionally, works councils in Germany played an important role in this context, not least because the level of union density was always relatively low compared to, for example, Scandinavia (Streeck, 2009, p. 64).[5] However, the monitoring role of works councils is undermined by the fact that there are increasingly "works council-free zones" (Artus et al., 2006) especially in private service industries. Against this background, worker representative bodies are increasingly dependent on state enforcement, as the regulatory role of the social partners is in decline in many sectors. Accordingly, the DGB demanded an increase in the staff of the Financial Control of Undeclared Work Unit (FKS) (DGB, 2014) even before the official introduction of the national minimum wage.

The FKS is a sub-department of the Federal Customs Administration, whose official mandate is the "fight against undeclared work and illegal employment". As part of its control activities, the FKS is responsible, among other things, for monitoring industry-wide and statutory minimum wages, with the main focus on whether taxes and social security contributions are properly paid (Bosch et al., 2019, p. 104). Accordingly, the FKS has no mandate to collect withheld wages; this is, in the words of a staff member, "a civil or labour law matter" (in Molitor, 2015, p. 22). In Germany, the problem arises that, unlike in some European neighboring countries, there is no central state authority to which employees can turn in the event of violations of workers' rights (Bosch et al., 2019, p. 309). The conventional route via the labour court is in most cases not a realistic option for mobile workers, as a representative of the DGB advice center "Fair Mobility" explains:

> You need time and money to go to court... But people have neither money for a lawyer, nor time to attend the many appointments. They are usually no longer in Germany when the proceedings begin. Moreover, it is difficult for many workers to provide evidence for their claims because contracts, statements or records of hours worked are missing. *The legal system simply does not ask about the lives of mobile workers* (in Molitor, 2015, p. 24, emphasis T.K.).

Overall, institutional support in Germany for enforcing wage claims appears to be "completely inadequate" (Bosch et al., 2019, p. 308). To help mobile workers assert their rights, the introduction of a right for trade unions to initiate

[5] Currently, only about 17% of all dependent employees in Germany are organized in trade unions (OECD/AIAS, 2021).

proceedings, as already exists in France or the Netherlands (ibid., p. 309), could help, as does improved information policy. With regard to the latter, it is difficult to understand why the official website of the Customs Administration, which is supposed to implement Article 5 of the EU Enforcement Directive (better access to information), does not inform about minimum working conditions in the main languages of the sending countries (Polish, Slovenian/Serbo-Croatian, Slovak and Hungarian).[6]

In addition to their political campaign for universally binding minimum standards, some trade unions have intensified their efforts to reach out to mobile workers from the NMS. While worker representative bodies were relatively successful in the past in organizing the "guest workers" who settled in the host country (Penninx & Roosblad, 2000), the organization of mobile workers who regularly cross national borders is much more difficult. This is not made easier by the above-average proportion of these workers in atypical employment such as posting and temporary agency work, which pose particular challenges for trade unions (Tapia & Holgate, 2018).

Against this background, worker representative bodies began to develop new organizing initiatives in the 21st century that went beyond traditional forms of workplace-based organizing. The IG BAU, which like no other German trade union was affected by transnational posting in the 1990s, founded the European Migrant Workers Union (EMWU) as early as 2004. The EMWU was intended as a professional association for posted workers from Central and Eastern Europe who are repeatedly sent abroad. The background was that the IG BAU was unable to reach these workers within its national organizational structures. The EMWU did not develop as planned, both in terms of membership and responses from other European construction unions, some of which perceived the union as a competitor (Greer et al., 2013). Eventually, the EMWU was reorganized in 2008 as an information center and lobbying organization and has since functioned as part of a network of advice centers within the DGB project "Fair Mobility".

The advisory centers of "Fair Mobility" have developed into important contact points for mobile workers from the NMS. The centers, which are based in 13 German cities and regions, provide information in the workers' native languages about German labour and social law (www.faire-mobilitaet.de). The setting up of these centers was thus an example of union

[6]The website is available in German and English at the following address: https://www.zoll.de/DE/Fachthemen/Arbeit/arbeit_node.html.

learning processes, recognizing that "hypermobile migrant workers" (Berntsen & Lillie, 2016) cannot be reached with the traditional means of membership recruitment. Instead, the focus was placed on information and campaigning in cooperation with civil society actors to enable mobile workers to assert their rights in a transnational mobility space (Wagner, 2015b). While German unions pursued a more protectionist policy in the past regarding the transition rules (Krings, 2009), projects like Faire Mobilität stand for a more inclusive approach towards migrant workers.

The DGB also increasingly focuses on expanding transnational contacts and cooperation in connection with cross-border mobility. The European network "Fair Working Conditions" was established with trade union cooperation partners in Croatia, Poland, Hungary, Romania, and Slovenia to strengthen the rights of posted workers (www.fair-labour-mobility.eu/). Overall, trade unions are increasingly realizing that exclusively nationally oriented strategies fall short in a transnational mobility space. The aim of cross-border collabouration is less about creating European-wide collective agreements, but rather about enforcing national wage and social standards for all workers.

Above all, the introduction of the national minimum wage in 2015, but also the extension of the Posted Workers Act, marked a political success for the trade union movement. However, these developments could hardly stop the further erosion of collective bargaining. It became apparent that a social partnership approach towards the enforcement of universally binding minimum standards (with the exception of a few industries) became less and less an option, chiefly because the employers' association BDA was no longer interested in it (Schulten, 2021). Thus, worker representative bodies increasingly have to rely on forming political alliances especially with a social democratic party in government. This development has to be seen against the background of the growing role of the state in enforcing labour standards in a transnational mobility space (Dølvik et al., 2014).

The importance of state regulatory policy was visible most recently in the COVID-19 pandemic. When hundreds of foreign contract workers became infected with the coronavirus in German slaughterhouses in the spring of 2020, the abusive conditions in the industry became known to a wider public. The broad scandalization of these conditions opened up the political space for state regulatory measures, which culminated in the Occupational Safety and Health Control Act at the end of 2020. This new law includes, among other things, a ban on contracts for work and temporary agency work for the core areas of the meat industry, as well as an increase in inspections in the industry (Erol & Schulten, 2021, p. 16). It remains to be seen to what extent this new legislation, which was complemented in the summer of 2021 by the conclusion of a minimum wage collective agreement (Specht & Schulten, 2021), will improve working conditions in the industry.

The Occupational Safety and Health Control Act is indicative of the importance of national labour market institutions in the European system of multi-level governance. Despite constraints of EU law, there is room for national regulatory measures (Dølvik et al., 2014). However, adapting to the requirements of EU law can certainly pose a challenge, as the "Rüffert" judgment of the ECJ in 2008 showed. At that time, the collective bargaining regulations of the federal state of Lower Saxony was deemed in violation of the EU freedom to provide services due to lack of a statutory extension provision. But here too, adjustments were made by most federal states in adapting their minimum wage regulations to the requirements of public procurement rules as enshrined in EU law (Nassibi et al., 2016). Thus, there seems to be no inherent contradiction between the four freedoms of the European internal market and universally binding minimum standards.

4.5 Summary

Germany experienced a strong increase in employment-related migration from the enlarged EU in the second decade of the 21st century. The context for this was a booming and open labour market, in which supply and demand mutually reinforced each other and led to a rapid increase in employment. EU migrants are predominantly employed in qualified positions, which helped to alleviate skill shortages in some sectors. People from the new EU, especially from Romania and Bulgaria, also have a relatively high proportion of unskilled workers, which indicates a certain polarization in their qualification level. Given the boom, a large number of new migrants found employment relatively easily, without leading to major displacement. Although wage growth in some parts of the labour market was probably moderated, overall the labour market effects of intra-European migration was likely to have been rather small (Bonin et al., 2020). However, this mobility was accompanied by incidents of wage and social dumping, especially when atypical employment such as posted work, (bogus)self-employment and contract work was involved. These represented an alternative strategy into the German labour market during the transition period for NMS migrants, but continue to be used even after the labour market opening in the sense of "cost-driven outsourcing" (Bosch et al., 2019). This suggests that a secondary segment has become entrenched in the German labour market, characterized by unstable employment and low wages (Krings, 2021).

Concerns about a "race to the bottom" in the context of the labour market opening was a key factor influencing the trade union campaign for universally binding minimum wages. This campaign was relatively successful, with the trade union movement forming an alliance with the SPD, the latter being able to push through the introduction of a national minimum wage and the extension of the Posted Workers Act when in government. At the same time, it became clear that unions can less rely on their own power resources to accomplish labour market regulation. Against this backdrop, the DGB has embarked on new ways of mobilizing migrant workers, focusing on "compliance through empowerment". The emphasis is on multilingual advice and support, which should enable workers to assert their rights in a transnational mobility space.

As regards labour market regulation, worker representative bodies are increasingly dependent on state support, as was most recently evident during the COVID-19 crisis. Faced with dramatically rising infections in German slaughterhouses, the legislator intervened and banned the use of contract work in the core areas of the meat industry as part of the Occupational Safety and Health Control Act 2020. This initiative curtailed a low-wage model, which was primarily based on the exploitation of foreign contract workers. It remains to be seen to what extent these and similar measures will lead to improved working and living conditions of mobile workers. What has become apparent is that state authorities alongside the social partners will continue to play an important role in enforcing universally binding minimum standards in open labour markets.

There are indications that intra-European migration will continue to be the most significant labour mobility to Germany. This was observed in the context of the COVID-19 pandemic in 2020, when net migration from the enlarged EU did not collapse unlike migration from third countries (Geis-Thöne, 2021). A key reason for this was likely to be the EU free movement regime, which enabled repeated border crossings despite temporary restrictions (see Chap. 6) even in a pandemic. The German labour market will continue to depend on labour migration in the future. There can be little doubt that a significant part of this will continue to come from the EU/EFTA.

References

Appelbaum, E., Bosch, G., Gautié, J., Mason, G., Mayhew, K., Salverda, W., Schmitt, J., & Westergaard-Nielsen, N. (2010). Introduction and overview. In J. Gautié & J. Schmitt (Eds.), *Low-wage work in the wealthy world* (pp. 1–32). Russell Sage Foundation.

Artus, I., Böhm, S., Lücking, S., & Trinczek, R. (Eds.). (2006). *Betriebe ohne Betriebsrat: Informelle Interessenvertretung in Unternehmen*. Campus.

Baas, T., Brücker, H., & Hauptmann, A. (2010). Labour mobility in the enlarged EU: Who wins, who loses? In M. Kahanec & K. F. Zimmermann (Eds.), *EU labour markets after post-enlargement migration* (pp. 46–70). Springer.

Becker, S., et al. (2013, February 25). Der deutsche Traum. *Der Spiegel.*

Berntsen, L., & Lillie, N. (2016). Hyper-mobile migrant workers and Dutch trade union representation strategies at the Eemshaven construction sites. *Economic and Industrial Democracy, 37*(1), 171–187.

Bispinck, R. (2012). Allgemeinverbindlicherklärung von Tarifverträgen – Vom Niedergang zur Reform? *WSI-Mitteilungen, 9*(2012), 496–507.

Bispinck, R. (2015). *Zwischen Mindestlohn und Tarifeinheit. Informationen zur Tarifpolitik.* Wirtschafts- und Sozialwissenschaftliches Institut (WSI) der Hans-Böckler-Stiftung.

Bommes, M., & Geddes, A. (Eds.). (2000). *Immigration and welfare: Challenging the boundaries of the welfare state*. Routledge.

Bonin, H., Krause-Pilatus, A., Rinne, U., & Brücker, H. (2020). *Wirtschaftliche Effekte der EU-Arbeitskräftemobilität in den Ziel- und Herkunftsländern. Ein kurzer Literaturüberblick.* IZA Research Report Nr. 102. Forschungsinstitut zur Zukunft der Arbeit.

Bosch, G. (2013). Grenzüberschreitende Arbeitsmärkte und nationale Beschäftigungssysteme in der EU. In H. G. Soeffner (Ed.), *Transnationale Vergesellschaftungen* (pp. 279–292). Springer VS.

Bosch, G. (2018). The making of the German minimum wage: A case study of institutional change. *Industrial Relations Journal, 49*(1), 19–33.

Bosch, G., Hüttenhoff, F., & Weinkopf, C. (2019). *Kontrolle von Mindestlöhnen.* Springer VS.

Bosch, G., & Weinkopf, C. (2008). Summary and conclusions. In G. Bosch & C. Weinkopf (Eds.), *Low-wage work in Germany* (pp. 288–314). Sage.

Brenke, K., Yuksel, M., & Zimmermann, K. F. (2010). EU enlargement under continued mobility restrictions: Consequences for the German labour market. In M. Kahanec & K. F. Zimmermann (Eds.), *EU labour markets after post-enlargement migration* (pp. 111–129). Springer.

Brücker, H., Kosyakova, Y., & Schuß, E. (2020). Fünf Jahre seit der Fluchtmigration 2015: Integration in Arbeitsmarkt und Bildungssystem macht weitere Fortschritte. IAB-Kurzbericht 4/2020. Retrieved November 13, 2020, from https://doku.iab.de/kurzber/2020/kb0420.pdf

Buch, T., Carstensen, J., Hamann, S., Otto, A., Seibert, H., & Sieglen, G. (2020). Grenzpendler aus dem Ausland: Immer mehr Beschäftige in Deutschland mit ausländischem Wohnort. IAB-Kurzbericht 9/2020. Retrieved November 9, 2021, from https://doku.iab.de/kurzber/2020/kb0920.pdf

Bundesagentur für Arbeit. (2016). Auswirkungen der Migration auf den deutschen Arbeitsmarkt. Retrieved November 13, 2021, from https://www.landkreis-prignitz.de/globalcontent/documents/asyl/Auswirk_Migration_Arbeitsmarkt.pdf

Bundesagentur für Arbeit. (2021a). Monatsbericht zum Arbeits- und Ausbildungsmarkt. Retrieved November 13, 2021a, from https://statistik.arbeitsagentur.de/DE/Navigation/Statistiken/Statistiken-aktuell/Statistiken-aktuell-Nav.html

Bundesagentur für Arbeit. (2021b). Gemeldete Arbeitsstellen – Deutschland und Westdeutschland (Zeitreihe Monats-/Jahreszahlen ab 1950). Retrieved November 13, 2021b, from https://statistik.arbeitsagentur.de/SiteGlobals/Forms/Suche/Einzelheftsuche_Formular.html?nn=1459936&topic_f=zr-stea-1950

Burkert, C. (2015). EU-Migranten landen oft im Niedriglohnsektor. Retrieved November 17, 2021, from https://www.iab.de/389/section.aspx/Publikation/k150626901

Datta, K., McIlwaine, C., Evans, Y., Herbert, J., May, J., & Wills, J. (2007). From coping strategies to tactics: London's low-pay economy and migrant labour. *British Journal of Industrial Relations, 45*(2), 404–432.

Deutscher Gewerkschaftsbund (DGB). (2014). Gemeinsam gegen Schwarzarbeit und Scheinselbstständigkeit. Retrieved November 13, 2021, from https://www.dgb.de/themen/++co++3b77dd10-8788-11e4-a4e5-52540023ef1a

De Wispelaere, F., De Smedt, L., & Pacolet, J. (2020). *Posting of workers: Report on A1 portable documents issued in 2018*. European Kommission, Generaldirektion Beschäftigung, Soziales und Integration.

Dølvik, J. E., Eldring, L., & Visser, J. (2014). Setting wage floors in open labour markets: The role of the social partners in Europe's multilevel governance. In S. Evju (Ed.), *Regulating transnational labour in Europe: The quandaries of multilevel governance* (pp. 131–186). University of Oslo, Department of Private Law.

Dubois, M. (2019). Demografie der EU-Zugewanderten in Deutschland. In C. Pfeffer-Hoffmann (Ed.), *EU-Zuwanderung nach Deutschland: Analysen zur Diversität von EU-Zugewanderten in Deutschland* (pp. 35–52). Mensch und Buch.

Eichhorst, W. (1999). Europäische marktgestaltende Politik zwischen Supranationalität und nationaler Autonomie: Das Beispiel der Entsenderichtlinie. *Industrielle Beziehungen, 6*(3), 340–359.

Erol, S., & Schulten, T. (2021). *Neuordnung der Arbeitsbeziehungen in der Fleischindustrie*. WSI-Report Nr. 61. Wirtschafts- und Sozialwissenschaftliches Institut (WSI) der Hans-Böckler-Stiftung.

Fries-Tersch, E., & Mabilia, V. (2015). *Annual report on Intra-EU labour mobility*. European Kommission, Generaldirektion Beschäftigung, Soziales und Integration.

Geis-Thöne, W. (2021). *Corona hat die Zuwanderung nach Deutschland gedämpft*. IW-Kurzbericht Nr. 30. Institut der Deutschen Wirtschaft.

Greer, I., Ciupijus, Z., & Lillie, N. (2013). The European migrant workers union and the barriers to transnational industrial citizenship. *European Journal of Industrial Relations, 19*(1), 5–20.

Hassel, A., & Schiller, C. (2010). *Der Fall Hartz IV: Wie es zur Agenda 2010 kam und wie es weitergeht*. Campus.

Hunger, U. (2000). Temporary transnational labour migration in an integrating Europe and the challenge to the German Welfare State. In M. Bommes & A. Geddes (Eds.), *Immigration and welfare: Challenging the boundaries of the welfare state* (pp. 189–208). Routledge.

Jeřábek, M. (2011). *Deutschland und die Osterweiterung der Europäischen Union*. VS Verlag.

Kohaut, S. (2019). Tarifbindung: Weiterhin erhebliche Unterschiede zwischen Ost- und Westdeutschland. IAB-Forum 22. Mai 2019. Retrieved November 13, 2021, from https://www.iab-forum.de/tarifbindung-weiterhin-deutliche-unterschiede-zwischen-ost-und-westdeutschland/

Krings, T. (2009). A race to the bottom? Trade unions, EU enlargement and the free movement of labour. *European Journal of Industrial Relations, 15*(1), 49–69.

Krings, T. (2021). "Good" bad jobs? The evolution of migrant low-wage employment in Germany (1985–2015). *Work, Employment and Society, 35*(3), 527–544.

Menz, G. (2005). *Varieties of capitalism and Europeanization: National response strategies to the Single European Market*. Oxford University Press.
Mense-Petermann, U. (2018). Eastern European service contract workers in the German meat industry: A case study in market making of a transnational labour market. *ZiF-Mitteilungen, 2*, 23–31.
Molitor, C. (2015). *Geschäftsmodell Ausbeutung: Wenn Europaische Arbeitnehmer_Innen in Deutschland um Ihre Rechte betrogen werden*. Friedrich Ebert Stiftung.
Mühlberger, U. (2007). *Dependent self-employment: Workers on the border between employment and self-employment*. Palgrave Macmillan.
Nassibi, G., Rödl, F., & Schulten, T. (2016). *Perspektiven vergabespezifischer Mindestlöhne*. WSI Policy Brief Nr. 3. Wirtschafts- und Sozialwissenschaftliches Institut (WSI) der Hans-Böckler-Stiftung.
Nowicka, M. (Ed.). (2007). *Von Polen nach Deutschland und zurück*. Transcript.
OECD. (2014). *International migration outlook 2014*. OECD.
OECD/AIAS. (2021). *Institutional characteristics of trade unions, wage setting, state intervention and social pacts*. OECD. Retrieved July 5, 2021, from http://www.oecd.org/employment/ictwss-database.htm
Palier, B., & Thelen, K. (2010). Institutionalizing dualism: Complementarities and change in France and Germany. *Politics & Society, 38*(1), 119–148.
Penninx, R., & Roosblad, J. (Eds.). (2000). *Trade unions, immigration, and immigrants in Europe, 1960–1993: A comparative study of the attitudes and actions of the trade unions in seven West European countries*. Bergham.
Rudolph, H. (1996). The new *Gastarbeiter* system in Germany. *New Community, 22*(2), 287–300.
Sachverständigenrat deutscher Stiftungen für Integration und Migration (SVR). (2013). *Erfolgsfall Europa? Folgen und Herausforderungen der EU-Freizügigkeit für Deutschland Jahresgutachten 2013 mit Migrationsbarometer*. SVR.
Schmidt, W. (2006). Pragmatische Zusammenarbeit: Kollegialität und Differenz bei Beschäftigten deutscher und ausländischer Herkunft in Industriebetrieben. *Zeitschrift für Soziologie, 35*(6), 465–484.
Schroeder, W. (2020). Mindestlohn, Tarifautonomie und Gewerkschaften. *Aus Politik und Zeitgeschichte, 70*(39/40), 26–31.
Schulten, T. (2021). Reform der AVE: Schlüssel zur Stärkung der Tarifbindung. Retrieved November 13, 2021, from https://www.wsi.de/de/blog-17857-reform-der-ave-schluessel-zur-staerkung-der-tarifbindung-33324.htm
Seils, E., & Emmler, H. (2020). *Die Lohnentwicklung im vergangenen Jahrzehnt*. WSI Policy Brief Nr. 64. Wirtschafts- und Sozialwissenschaftliches Institut (WSI) der Hans-Böckler-Stiftung.
Specht, J., & Schulten, T. (2021). Mindestlohntarifvertrag in der Fleischwirtschaft. Retrieved November 13, 2021, from https://www.wsi.de/de/blog-17857-mindestlohntarifvertrag-in-der-fleischwirtschaft-33525.htm
Staples, R., Trinczek, R., & Whittall, M. (2013). Posted Workers: Zwischen Regulierung und Invisibilisierung. *Arbeit, 22*(4), 271–286.
Statistisches Bundesamt. (2021). Unternehmen und Arbeitsstätten: Gewerbeanzeigen in den Ländern. Retrieved November 9, 2021, from https://www.statistischebibliothek.de/mir/receive/DESerie_mods_00000645

Streeck, W. (2009). *Re-forming capitalism: Institutional change in the German political economy*. Oxford University Press.

Tamas, K., & Münz, R. (2006). *Labour migrants unbound? EU enlargement, transitional measures and labour market effects*. Institute for Future Studies.

Tapia, M., & Holgate, J. (2018). Fighting precariousness: Union strategies towards migrant workers in the UK, France, and Germany. In V. Doellgast, N. Lillie, & V. Pulignano (Eds.), *Reconstructing solidarity: Labour unions, precarious work, and the politics of institutional change in Europe* (pp. 188–206). Oxford University Press.

Thelen, K. (2014). *Varieties of liberalization: The new politics of social solidarity*. Cambridge University Press.

Traxler, F. (1996). Sozialpartnerschaft am Scheideweg: Zwischen korporatistischer Kontinuität und neoliberalem Umbruch. *Wirtschaft und Gesellschaft, 1*(1), 13–33.

Treichler, A. (1998). *Arbeitsmigration und Gewerkschaften*. LIT.

Verwiebe, R., & Eder, K. (2006). The positioning of transnationally mobile Europeans in the German labour market: An analysis of its causes and effects. *European Societies, 8*(1), 141–167.

Verwiebe, R., & Müller, M. C. (2006). Gelungene Integration in den Arbeitsmarkt? Die flexiblen Biografien transnational mobiler Europäer zu Beginn des 21. Jahrhunderts. *Berliner Journal für Soziologie, 16*(1), 101–121.

Wagner, B., & Hassel, A. (2015). Labour migration and the German meat processing industry: Fundamental freedoms and the influx of cheap labour. *South Atlantic Quarterly, 114*(1), 204–214.

Wagner, B., & Hassel, A. (2016). Move to work, move to stay? Mapping atypical labour migration into Germany. In J. E. Dølvik & L. Eldring (Eds.), *Labour mobility in the enlarged Single European Market* (Comparative Social Research) (Vol. 32, pp. 125–158). Emerald Group.

Wagner, I. (2015a). Rule enactment in a Pan-European labour market: Transnational posted work in the German construction sector. *British Journal of Industrial Relations, 53*(4), 692–710.

Wagner, I. (2015b). EU posted work and transnational action in the German meat industry. *Transfer, 21*(2), 201–213.

Wagner, I. (2018). *Workers without borders: Posted work and precarity in the EU*. Cornell384 University Press.

Wagner, I., & Berntsen, L. (2016). Restricted rights: Obstacles in enforcing the labour rights of mobile EU workers in the German and Dutch construction sector. *Transfer: European Review of Labour and Research*, *22*(2), 193–206.

Waldinger, R., & Lichter, M. I. (2003). *How the other half works: Immigration and the social organization of labour*. University of California Press.

Wickham, J. (2016). *Unequal Europe: Social divisions and social cohesion in an old continent*. Routledge.

Wilpert, C. (1998). Migration and informal work in the new Berlin: New forms of work or new sources of labour? *Journal of Ethnic and Migration Studies, 24*(2), 269–294.

5 Austria: Migration Into an Open Labour Market in the "Middle of Europe"

One of the most important developments in the Austrian labour market over the past ten years has been employment-related migration from the enlarged EU. The focus of this chapter is on the labour market consequences of this mobility. It is shown that the main dynamic for an extraordinary increase in migration was an open labour market in conjunction with a demand for labour, huge income differences, and geography. EU migrants can be found in different occupational positions. While migrants from the old EU are overrepresented in highly qualified positions, migrants from the new EU are overrepresented in low-skilled positions. Before going into more detail on their labour market integration, the development of intra-European mobility since the 1990s is first traced. Subsequently, efforts by Austrian social partners to re-embed this mobility into national labour relations are discussed. It is shown that a relatively comprehensive regulatory regime could be negotiated, which largely enshrines the equal pay principle on the basis of collective agreements. However, there is a problem of enforcement, especially with regard to "atypical" employment and transnational subcontracting. Thus, there are indications that the effect of recent migration from the new EU member states has reinforced a trend towards segmentation and precarization in the Austrian labour market.

T. Krings, *Labour Mobility in the European Union as an Example of the Transnationalization of Employment*,
https://doi.org/10.1007/978-3-658-43977-4_5

5.1 Intra-European Mobility, Eastern Enlargement, and a Restricted Labour Market

Austria has traditionally been one of the most important destinations for intra-European migration, not least because of its geographical location in the center of Europe. On the one hand, it borders Germany, which is by now the most important source country for EU migrants in Austria. On the other hand, Austria shares almost 1300 km of common land borders with four of the new member states (NMS): Slovakia, Slovenia, the Czech Republic, and Hungary (Walterskirchen, 2009, p. 149), all with significantly lower wage levels. The capital Vienna is only 65 km west of Bratislava, making daily commuting a realistic option. Other larger cities such as Graz, Linz, and Klagenfurt are also located close to the borders with the NMS.

Austria already experienced an increase in migration from Central and Eastern Europe (CEE) in the 1990s, which included both regular and irregular inflows. Part of this mobility took place within the framework of quotas for agriculture and tourism (seasonal employment). The main countries of origin were Hungary and Slovakia, but also the successor states of the former Yugoslavia, from which traditionally most migrants in Austria originate. At the same time, there was an increase in irregular employment since the 1990s, especially in the area of domestic care and private services (Tamas & Münz, 2006, pp. 118–119).

A decisive turning point was Austria's EU accession in 1995, which heralded the "end of the national labour market" (Perchinig, 2010, p. 22). Since then, the regulation of labour supply from the EU/EFTA has only been possible to a very limited extent. This change in the regulatory framework had significant effects on migration flows to Austria as work-related migration from the EU/EFTA, especially from Germany, has increased significantly, a development that intensified in the 21st century as a result of Eastern enlargement.

In the context of EU enlargement in 2004 (as well as subsequent enlargements), Austria restricted access to the labour market for citizens from the accession countries.[1] Against the background of widespread skepticism regarding Eastern enlargement and out of concern about displacement in the labour market,

[1] Similar to Germany, not only was the freedom of movement for workers restricted, but also the freedom to provide services in sectors such as construction and industrial cleaning.

all major political parties and interest groups (with the exception of the Federation of Austrian Industries) supported the transition period (Tamas & Münz, 2006, pp. 100–102).[2]

For the Austrian Economic Chamber (WKO), the transition period represented a certain dilemma. The employers' association, which is dominated by smaller firms, supported the transitional period, as price-cutting competition by foreign service providers was feared in sectors such as the construction industry. At the same time, the WKO adopted a pro-migration stance with regard to skilled workers. To resolve this dilemma, the WKO reached an agreement with the Austrian Trade Union Federation (ÖGB) and the Federal Chamber of Labour (BAK) within the framework of social partnership to achieve a selective opening of the labour market for skilled workers from the NMS. As a concession to the employee side, it was emphasized that the labour demand should be covered "primarily through the labour potential available domestically" (The Social Partners, 2007, p. 3). In addition, various measures were agreed upon in the field of education and training, such as a basic training guarantee for young people (Krings, 2013, p. 273).

As regards low-skilled occupations, there was no comparable opening of the labour market. However, the rightwing coalition of the Austrian People's Party (ÖVP) and the Freedom Party of Austria (FPÖ) had already decided in 2002 to expand temporary "seasonal" employment. Seasonal workers, who mainly came from the 2004 EU accession countries and the former Yugoslavia, could now be employed in all sectors. In addition, the duration of employment was extended to up to twelve months, effectively creating a new "guest worker program" (Kraler, 2012, p. 36). While employers' associations had long demanded a liberalization of seasonal employment, trade unions rejected this, but could hardly influence the political decision-making process anymore (Krings, 2022).

Thus, Austria pursued a flexible migration policy during the transition period in order to meet a growing demand for labour at different skill levels. At the same time, work-related migration from Germany grew in importance in the first decade of the 21st century (Biffl, 2019, p. 114). This suggests that the EU free movement regime in conjunction with economic factors (and in the case of Germany also cultural and linguistic proximity), structurally promotes mobility. This

[2] The Federation of Austrian Industries (IV), which primarily represents the interests of larger industrial companies and is not part of the official social partnership, had from the outset demanded free access to the labour market for citizens from the NMS. However, this demand was not politically viable in Austria.

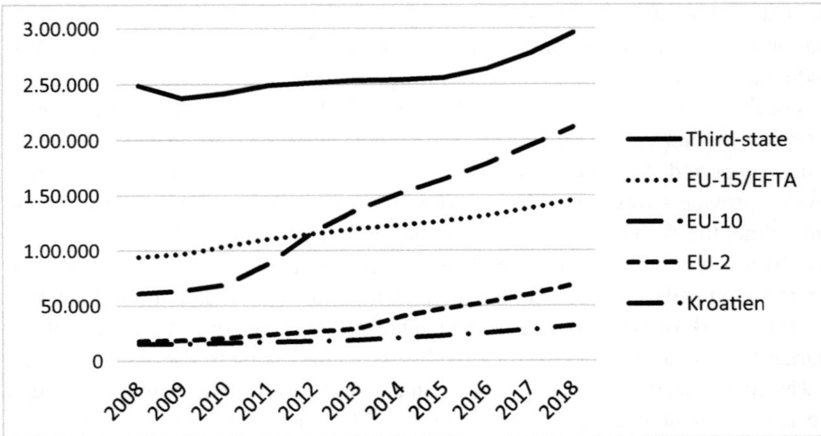

Fig. 5.1 Development of foreign dependent employment (2008–2018). (Source: BMA, 2021)

became also apparent with regard to the opening of the Austrian labour market in 2011 and 2014.

5.2 Post-2011: Migration into an Open Labour Market

The expiry of the transitional period set the context for a previously unknown dynamic of employment-related migration. Between 2010 and 2018, the number of employed persons from the NMS in Austria tripled. As can be seen from Fig. 5.1, the decisive event was the opening of the labour market in 2011 for the EU-10 countries and in 2014 for the EU-2 countries. Since then, there has been a strong increase in migration in all federal states, especially in Vienna and Burgenland (Biffl, 2019, p. 121–123). The main countries of origin were Hungary, Romania, Slovakia, and Poland, which together accounted for three-quarters of all new inflows. In comparison, migration from the EU-15/EFTA only increased incrementally in the second decade. The employment of third-country nationals, traditionally the most important migrant group in the Austrian labour market, remained relatively unchanged for a long time and only increased since 2015

(Fig. 5.1). This reflects growing inflows from Bosnia-Herzegovina and Serbia, but also the refugee migration of 2015 and afterwards, when Austria, along with Germany and Sweden, was the most important destination in Europe.

Especially in a longer-term perspective, the growing importance of intra-European mobility becomes apparent. While the share of EU citizens in foreign employment in Austria was still below ten percent in 1998, twenty years later it was already over 60% (Biffl, 2019, p. 114; BMA, 2021). This development was hugely facilitated by Eastern enlargement in the 21st century, which redrew the borders of the EU ("variable geometry"). Almost overnight, mobile individuals from CEE, who often took up irregular employment, became EU citizens with (almost) all the rights of EU citizenship.

The largest group of NMS migrants who are employed in Austria originate from Hungary (92,300), followed by mobile workers from Romania (56,000) and Poland (40,000) (annual averages for 2018) (BMA, 2021). Interestingly, in recent years the proportion of Hungarian workers in the Austrian labour market has increased significantly more than that of Romanians, even though the latter have immigrated to a greater extent. This suggests that many Hungarians (and also Slovaks) are transnational commuters who still reside in their country of origin (Höller, 2015). In fact, after Luxembourg and Switzerland, Austria has the third highest proportion of cross-border commuters in the EU/EFTA (Fries-Tersch & Mabilia, 2015, p. 59). While in the former two countries the commuters mainly come from the EU-15, in Austria they mainly come from the NMS. Especially in some regions close to the border, their proportion is exceptionally high. For example, in Burgenland alone, it is estimated that almost 20% of all workers are transnational commuters, mainly from Western Hungary (Der Standard, 2016).

As can be seen from Fig. 5.2, male migrants from the old and new EU have a higher employment rate than locals. Among female workers, there are no such differences. Both men and especially women from third countries have significantly lower rates. In addition to the on average higher qualification of EU migrants, this once again underlines that the EU free movement regime promotes cross-border mobility and an easier entry into the labour market.

As already discussed in Chap. 3, EU-15 migrants have an above-average share of highly qualified professionals. On the other hand, workers from the NMS (EU-10 and EU-3) predominantly have a medium level of qualification. As Table 5.1 shows, these differences are also reflected in their occupational status in Austria. While approximately 60% of all workers from the EU-15/EFTA are found in higher qualified occupations (ISCO groups 1–3), NMS migrants are mainly

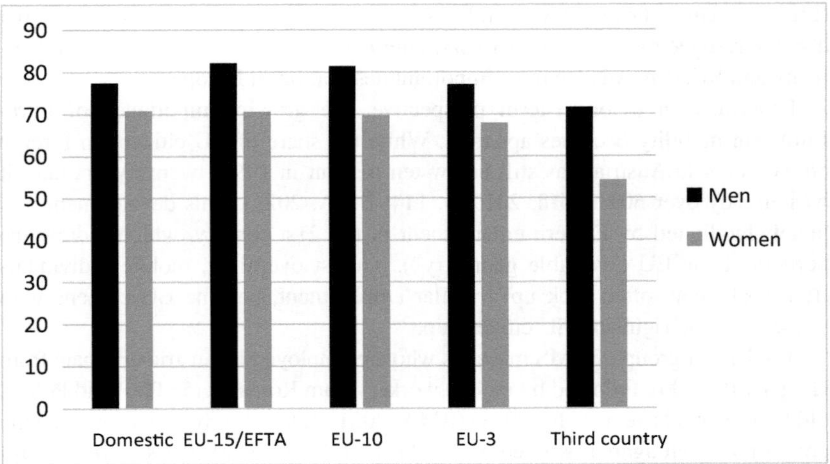

Fig. 5.2 Employment rates by country of birth and gender (17–62 years, in %) (2018). (Source: European Labour Force Survey (EU-LFS) own calculations)

employed in medium-skilled occupations (ISCO groups 4–8). It is also noticeable that EU-3 migrants have a high proportion of unskilled workers at almost 30%, reflecting their lower educational attainments compared to EU-10 migrants. The latter are sometimes overqualified for their jobs, more so than any other group in the Austrian labour market. Although only six percent of all employed EU-10 migrants have a low level of qualification, more than three times as many are found in low-skilled occupations. This is due to difficulties in recognizing and utilizing their qualifications, as well as lack of German language skills (Huber et al., 2017, pp. 52-54). However, the demand side is also relevant. Here it becomes apparent that a segmentation tendency in the Austrian labour market has become more entrenched in the context of recent migration. Especially in sectors such as hospitality, agriculture, and temporary agency work, which have experienced strong inflows from the NMS, there has been a rise in unstable employment (Eppel et al., 2017). This in turn suggests that the supply and demand sides in transnational labour markets can develop in a dynamic relationship, especially when there are significant income differences.

How has the sharp rise in migration affected the labour market? Are there negative effects in terms of employment opportunities and wage development? With regard to the overall labour market, there is little evidence so far, simi-

Table 5.1 Occupational statusby country of birth (main occupational groups ISCO-08) (2018) (in %)[3]. (Source: EU-LFS, own calculations)

	Domestic	EU-15/EFTA	EU-10	EU-3	Third country
1. Managers	5	6.1	2.2	2.6	2.8
2. Professionals	20.8	33.9	17.5	13	11.4
3. Technicians & Associate Professionals	19.1	18.9	8.8	7.8	8
4. Clerical Support Workers	10.6	7.2	6.6	5.2	4.7
5. Services & Sales Workers	17.2	17.8	23.4	20	21.1
6. Skilled Agricultural, Forestry & Fishery Workers	4.4	1.1	0.7	0.9	0.6
7. Crafts & Related Trades Workers	12.7	7.2	16.1	13.9	15.3
8. Plant & Machine Operators & Assemblers	4.9	3.3	5.1	9.6	11.6
9. Elementary Occupations	5.2	4.4	19.7	27	24.4
Total	*100*	*100*	*100*	*100*	*100*

lar to Germany, that an increased labour supply has led to larger displacement effects. Between 2010 and 2018 (when migration was particularly high), the overall employment rate in Austria increased from just under 71% to just under 73% (age group 17 to 62) (EU-LFS). In terms of wage development, there was growing divergence between workers in stable and unstable employment. While the former were able to record a slight increase in real wages between 2009 and 2015, the latter (among them disproportionately migrants) experienced a decline in real wages (Eppel et al., 2017, p. 50–51). Even though the overall economic effects of migration were likely to have been relatively small (Huber & Böhs, 2012), there were different distributional effects in regions, industries, and occupational groups. There are indications that some displacement effects occurred in lower-skilled occupations in border regions in Burgenland, Styria, Lower Austria, and the Vienna outskirts (Schweighofer, 2012, p. 612).

[3] Deviations from the sum of 100% are due to rounding.

Among those groups who were most affected by these effects were long-term migrants in Austria, who were often in direct competition with the newcomers (Eppel et al., 2018, pp. 85–86; Schnauder, 2016). This is also reflected in official labour market data. While the unemployment of domestically born persons decreased from 4.1% to 3.7% between 2010 and 2018, it increased for third-country nationals from 10.5% to 12.3% (age group 17 to 62) (EU-LFS). Especially in labour-intensive occupations in construction, hospitality, and in agriculture, employers seem to have developed a preference for mobile workers from the NMS, who are appreciated for their "flexibility" (Wiesböck, 2016). This illustrates quite well the challenges associated with a transnational mobility space which is characterized by significant income differences. While in Austria the minimum wage for a full-time job is €1700 gross per month (as agreed by collective agreements) (and is usually paid fourteen times a year), a comparable minimum wage in Hungary is €504 and in Slovakia €646 (as of the end of 2022) (Eurostat, 2024).[4] These differences can provide an incentive for mobile workers from the NMS to work for a salary in Austria that is below the minimum wage, but still significantly better than in their home country (Wiesböck, 2016). In other words, a "bad" wage in Austria can be a "good" wage in Hungary or Slovakia. An increase in precarious mobility cannot be considered separately from an expansion of atypical employment.

5.3 Transnational Mobility and "Atypical" Employment

Like most other OECD countries, Austria has experienced an increase in atypical employment over the past two decades. As elsewhere, atypical workers are more affected by low-wage employment than people in a standard employment relationship (Geisberger, 2017). At the same time, a division of the labour market between "insiders" and "outsiders" is less pronounced than in Germany, for example. A major reason for this is the Austrian system of social partnership, which enables worker representative bodies to pursue a policy course that is less "insider"-oriented than elsewhere (Obinger et al., 2012). In 2002, unions succeeded in negotiating a ollective agreement for temporary agency work, the first

[4] The comparison of wages is based on direct exchange rates.

of its kind in Europe, which established the equal pay principle. Further, worker representatives succeeded with their demand for the inclusion of freelancers and other self-employed people into the Austrian social insurance system (Obinger et al., 2012, p. 182).

However, there is some evidence to suggest that a trend towards dualization in the Austrian labour market has become more pronounced in recent years. This is visible in the area of atypical employment and cross-border mobility, particularly concerning posted work. Since the labour market opening, there has been a sharp increase in postings to Austria. While in 2010 just under 60,000 postings (A1 social security forms) were registered, by 2017 this number had risen to over 140,000 postings.[5] This development was almost exclusively due to postings from the NMS, the number of which almost quintupled in the same period from 17,000 to 82,000, the highest increase in the enlarged EU (European Commission, 2012, p. 13; Pacolet & De Wispelaere, 2019, p. 21). The main countries of origin were Slovenia, Hungary, and Slovakia, and the main destinations were border regions in Austria (Schmatz & Wetzel, 2014, p. 28).[6] Thus, geographical proximity is an important predictor of employment-related mobility. In this respect, Austria is in an exceptional position as it is the only old member state to border four of the new member states. In 2018, there was a relatively strong decline in postings from Slovenia, the most important posting country from the NMS. This decline is attributed, among other things, to stricter regulation of posting companies there (De Wispelaere et al., 2021, p. 29).

Service mobility mainly takes place in the construction sector, which accounted for about 60% of all postings in 2017.[7] Within the industry, postings are distributed to roughly the same extent between the main construction trades

[5] A1 social security forms (formerly E101) are issued by the respective insurance carrier in the country of origin for the posted worker and, among other things, indicate the destination of the posting. These certificates do not necessarily provide information about the total number of posted workers, as multiple A1 forms can be issued for one person in the case of repeated postings. At the same time, these certificates may underestimate the extent of cross-border mobility, as companies are occasionally keen to avoid the administrative reporting procedure, especially in the case of short-term postings.

[6] The high number of postings from Slovenia is likely due to the fact that workers from the former Yugoslavia, especially from Bosnia-Herzegovina, are recruited by Slovenian companies for the purpose of posting (Schmatz & Wetzel, 2014, p. 32).

[7] A special case is the transport sector, where the number of posting notifications is many times higher, but this is not considered in more detail here (Riesenfelder & Reichert, 2021).

(building and civil engineering) and ancillary industries (e.g., installation or painting) (Schmatz & Wetzel, 2014, p. 23). When looking at the national composition of posting companies, it is noticeable that it has fundamentally changed. While before the labour market opening the vast majority of all companies came from the old EU, mainly from Germany, in 2018 almost 60% came from the NMS (Riesenfelder & Reichert, 2021, p. 10). This suggests that posted work is increasingly dominated by a logic of "competitive subcontracting" (Harvey, 2003), which aims to reduce labour costs.

Although companies are obliged by the Wage and Social Dumping Prevention Act to pay posted workers the collectively agreed minimum wages, there are problems of enforcement (see below). Even if collective agreements are paid, foreign service providers from the NMS can still offer their services more cheaply, as posted workers are usually covered by the (significantly lower) social security contributions of the country of origin (Krings, 2019). There is also the phenomenon of "bogus postings", where workers are specifically recruited in the country of origin (or occasionally even in the destination country), without any subsequent employment in the country of origin (Riesenfelder et al., 2012, pp. 93–94). Overall, the working environment of transnational posting (as in other countries too (Wagner, 2018) is often characterized by informality and precariousness.

This development has to be seen in the context of Austrian social partnership. Almost 100% of all employees in Austria are still covered by a collective agreement, an internationally outstanding value (OECD/AIAS, 2021). Against this background, Eastern enlargement offered companies in the construction industry and elsewhere an opportunity to use atypical forms of employment such as posted work to circumvent the institutions of collective labour relations and dismissal protection. This seems to confirm the assumption that companies resort to atypical employment in more heavily regulated labour markets, which are characterized by high collective bargaining coverage (Rueda et al., 2015).

At least in part, an increase in atypical employment was also the result of the transitional arrangements. Since regular labour market access was restricted for citizens from the NMS, posted work and (bogus)self-employment offered alternative entry into the labour market (Dølvik et al., 2014; Wagner & Hassel, 2016). Conversely, this does not mean that these employment relationships would disappear once the labour market is open. Rather, the development since 2011 suggests that atypical employment has become more entrenched in the context of cross-border mobility. This is visible not only in relation to the huge rise in postings, but also in terms of temporary agency work, where EU-13 migrants accounted

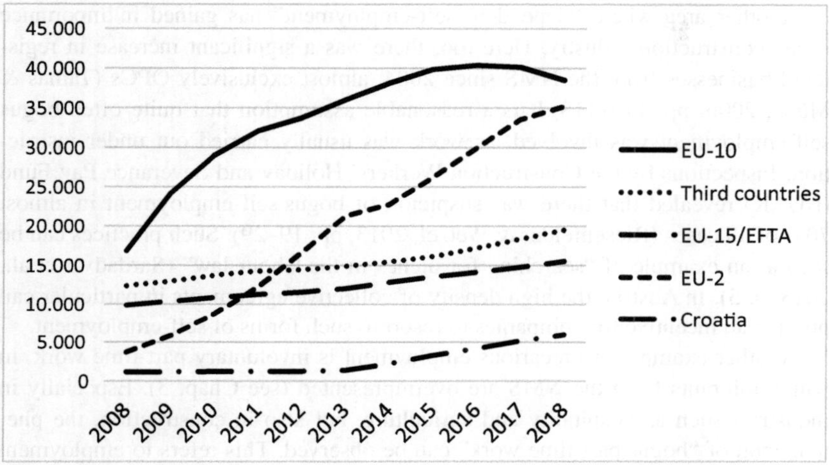

Fig. 5.3 Development of foreign self-employment (2008–2018). (Source: BMA, 2021)

for over six percent of all agency workers in the Austrian labour market in 2018 (EU-LFS).[8] Another example of the growing importance of atypical employment is (bogus)self-employment.

As Fig. 5.3 shows, the number of self-employed workers from the NMS has risen sharply since 2008, even compared to other migrant groups. These are likely to be mostly one-person companies (OPCs) that found employment through contracts for work. No labour law provisions such as collective agreements or working time regulations apply to them. While the increase in self-employed workers from the EU-10 flattened somewhat since the opening (but continued to rise until 2016), there was almost a linear increase among workers from the EU-2. This increase was mainly due to the so-called 24-hour mobile care in Austria, where female workers from Romania and Slovakia are often employed under rather precarious conditions. This so-called self-employed work is in fact a form of "dependent self-employment" (Mühlberger, 2007), as the caregivers are usually integrated into the household of the people they care for and take orders from the latter (see Chap. 6).

[8] Own calculations based on microdata from the European Labour Force Survey.

Another area where "dependent self-employment" has gained in importance is the construction industry. Here too, there was a significant increase in registered businesses from the NMS since 2004, almost exclusively OPCs (Tamas & Münz, 2006, pp. 116–118).It is a reasonable assumption that quite often bogus self-employment was involved, as work was usually carried out under instruction. Inspections by the Construction Workers' Holiday and Severance Pay Fund (BUAK) revealed that there was suspicion of bogus self-employment in almost 70% of all cases (Riesenfelder & Wetzel, 2013, pp. 19–29). Such practices can be seen as an example of "searching for niches in the labour law" (Sardadvar et al., 2013, p. 5). In Austria, the high density of collective agreements in particular can provide an incentive for companies to resort to such forms of self-employment.

Another example of precarious employment is involuntary part-time work, in which migrants from the NMS are overrepresented (see Chap. 3). Especially in industries such as hospitality and agriculture, but also in construction, the phenomenon of "bogus part-time work" can be observed. This refers to employment relationships that are registered on the basis of regular part-time or marginal employment, even though more work is actually being done, often full-time. In this employment, the outstanding part of the wage is sometimes (though by no means always) paid under the table (Riesenfelder & Wetzel, 2014, pp. 50–52). Such informality may be desired by both sides of the employment relationship. However, such employment is characterized by a strong imbalance of power between employer and worker, as the former can change the conditions at any time and almost at will (Krings et al., 2013, pp. 78–79).

As shown in Fig. 5.4, the number of marginally employed workers from abroad has increased significantly since 2008, to a much greater extent than for domestic workers (not shown). The strongest increase can be observed among workers from the EU-10 and EU-2 member states, where marginal employment has quadrupled. As the same figure shows, the number of marginally employed foreigners has been stagnating or even slightly declining since 2017. This may be due to more inspections especially on construction sites, but also to new reporting obligations for part-time employees (Riesenfelder & Reichert, 2021, pp. 24–35, 39). This suggests that changes in the regulatory environment can indeed have an impact on the development of atypical employment.

Overall, there is a certain path dependency of atypical employment, which has outlasted the transition rules. Even in an open labour market, companies resort to posted work, temporary agency work, marginal employment, and self-employment. There is little evidence to suggest that this trend towards outsourcing and subcontracting will be reversed. Hence, the challenge is to "re-embed" these

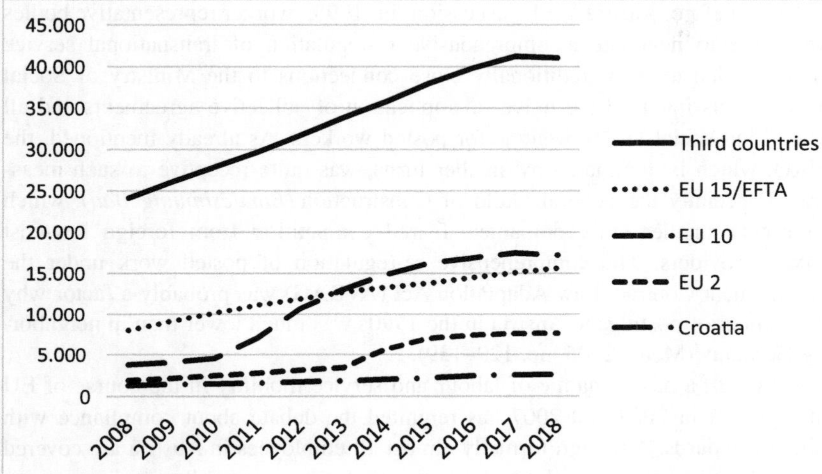

Fig. 5.4 Development of foreign marginal employment (2008–2018). (Source: BMA, 2021)

employment relationships into national labour relations. This is traditionally the domain of the social partners in Austria.

5.4 Re-Regulation, Equal Pay and an Enforcement Problem

Comparative studies on labour relations usually classify Austria as a "paradigmatic case of corporatism" (Traxler, 2001). Despite a significant decline in union density, which almost halved between 1975 and 2015 (OECD/AIAS, 2021), labour representatives still possess considerable institutional resources. Employee and employer associations still conclude collective agreements that cover almost all dependent employees. This high coverage rate is made possible by compulsory membership for companies in the WKO, which binds them to comply with the collective agreements (Pernicka & Hefler, 2015, p. 42). Also in the field of labour market and social policy, the associations are involved in the political process together with the state due to a tripartite tradition of interest representation (Tálos, 2008).

Even before Austria's EU accession in 1995, worker representative bodies were able to negotiate a comprehensive re-regulation of transnational service mobility due to their traditionally close connections to the Ministry of Social Affairs. This involved the universal application of collective agreements and all labour and social law provisions for posted workers. As already mentioned, the WKO, which is dominated by smaller firms, was quite receptive to such measures. Especially the Federal Guild of Construction *(Bundesinnung Bau)*, which represents smaller craft companies, feared competition from foreign low-cost service providers. The comprehensive re-regulation of posted work under the Employment Contract Law Adaptation Act (AVRAG) was probably a factor why the number of postings to Austria in the 1990s was much lower than in neighboring Germany (Menz, 2005, pp. 129–139).

However, a new dynamic of labour and service mobility in the course of EU enlargement in 2004 and 2007 has reignited the debate about compliance with labour standards. Although formally almost all employees in Austria are covered by a collective agreement (KV), until recently there were hardly any instruments to monitor compliance with the KV. This poses a particular challenge in the face of a growing "representation gap at enterprise level" (Hermann & Flecker, 2006) and significant wage differences to the NMS, some of which are in the immediate vicinity of Austria. Especially with smaller domestic and foreign companies, which usually do not have works councils, there is an enforcement problem (Riesenfelder et al., 2012, pp. 405–406). Although there was always the possibility for workers to sue for withheld wages, this is an unlikely scenario in the case of transnational posting, as these workers are usually only deployed abroad for a short period and are often not familiar with the language and labour law conditions in the host country.

Against this background, the ÖGB and the BAK demanded new legal measures to prevent the undermining of collective agreements in view of the opening of the labour market in 2011 and 2014. Their endeavours were helped by the fact that a new grand coalition of ÖVP and SPÖ came into place in 2007, which was once again relying on the involvement of the social partners in the political process (Tálos, 2008). In the summer of 2010, the Minister of Social Affairs, Rudolf Hundstorfer (SPÖ), a former president of the ÖGB, presented a draft of a law against wage and social dumping, which was submitted to the social partners for evaluation. The draft, which among other things provided for more inspections and penalties for underpayment, was welcomed by the employee side.

Initially, there was criticism from the employer's side, as the WKO feared a bureaucratic burden for Austrian companies. However, the WKO eventually gave its consent for the new law. On the one hand, there was still some concern about price-cutting competition by foreign service providers in sectors such as construction, industrial cleaning, or temporary agency work. On the other hand, and this was crucial, the approval of the new law against wage and social dumping was part of a social partnership deal, for which the employer's side received the approval of the employee's side for the introduction of a new system for qualified immigration ("Red-White-Red Card") in return (Krings, 2013). Accordingly, both initiatives were presented to the public as "one" social partner package as part of the Bad Ischler Dialog 2010.

The proposals of the social partners were largely based on the original draft by Minister Hundstorfer and were adopted by the Austrian Parliament as the Wage and Social Dumping Prevention Act (LSD-BG) in the spring of 2011. The LSD-BG states, among other things, that posted workers "are entitled to at least the statutory, through regulation or collectively agreed wage...which is due to comparable workers from comparable employers at the place of work" (§3, Sect. 3, LSD-BG). This also includes special payments and allowances. Thus, the principle of "equal pay for equal work" (as already enshrined in the AVRAG) was reasserted (Krings, 2016). A key component of the new law was the implementation of workplace inspections by the Austrian financial police and in the case of the construction industry by the BUAK, as well as the introduction of administrative penalties for underpayment. This makes Austria an example of a relatively comprehensive re-regulation of transnational labour and service mobility in the EU. Crucial for the enactment of the LSD-BG was the approval of the WKO, which, unlike other employer associations in Europe (Menz, 2005), did not resist stronger regulation of the labour market. This shows that employer associations, which like the WKO have a high degree of centralization and are dominated by smaller firms, support regulatory measures towards establishing a level playing field (Dølvik et al., 2014, p. 182). The Austrian case also shows that corporatist policy patterns are increasingly dependent on the political constellation: Without a renewal of the grand coalition including the SPÖ, a return to a tripartite interest policy would hardly have been possible.

The LSD-BG was originally anchored in the AVRAG, but was established as a law of its own in 2017 as part of an amendment that also served to implement the EU Enforcement Directive 2014/67/EU. In addition to measures for better enforcement of administrative penalties abroad, the new law also includes the creation of a client liability for the wage claims of posted workers in the construction sector.

Despite this relatively comprehensive re-regulation, the enforcement of labour and social standards remains a challenge. For example, in almost 40% of foreign posting companies that were inspected on Austrian construction sites in 2019, there was suspicion of underpayment (Riesenfelder & Reichelt, 2021, pp. 28–29). For such companies, hardly any sanctions can be imposed so far, as penalties are usually unenforcable abroad. Moreover, it remains a dilemma for policy actors in Austria and elsewhere that a wage just below the minimum standards can be attractive for mobile workers from the NMS.

International research suggests that controls and sanctions are a necessary, but not yet sufficient condition for an effective compliance regime (Bosch et al., 2019; Dickens, 2012). Importantly, there is a need for the involvement of non-state actors in the enforcement of labour standards. The social partners play an important role in that regard, as they can contribute to the (self-) regulation of the labour market not least through the negotiation of collective agreements. However, there is some evidence to suggest that the regulatory capacity of collective agreements is limited, especially in the context of cross-border subcontracting. Therefore, it seems advisable to start at higher level by introducing a general contractor liability to rectify the power imbalance within procurement chains "in a strategic manner" (Weil, 2018). Currently, however, there is little indication that such a liability, which already exists for public contracts, will also be extended to the private sector, as the WKO is opposed to it (Szigetvari, 2018).

Therefore, it seems all the more necessary to supplement workplace inspections with initiatives that are aimed at the inclusion of mobile workers. In this context, virtual projects such as the "Posting Platform", which informs in seven different languages about Austrian labour law and collective agreements, are of relevance, as well as the "Fair Work" project, which was set up by the Construction-Wood Union, and includes a bilingual advisory office in the border area between Styria and Slovenia.[9] These initiatives show that to enforce national labour standards more cooperation beyond the nation-state is required.

For the Austrian trade union movement, labour mobility from the NMS poses a challenge. Until the 1990s, trade unions primarily pursued a restrictive approach, which was characterized as "protecting indigenous workers from immigrants" (Gächter, 2000). Nowadays, unions accept that cross-border mobility is part of the transnationalization of labour markets. Already years ago the ÖGB

[9] The corresponding web addresses of the projects are www.entsendeplattform.at and www.faire-arbeit.at.

realized that a strategy based solely on national isolation falls a long way short of what was required in the context of Eastern enlargement (Biffl, 2010). Against this background, unions intensified their cooperation with employee representatives from the NMS in projects such as the "IGR—Future in the Border Region Burgenland—Western Hungary", which emerged from the Interregional Trade Union Council Burgenland—Western Hungary.[10] As part of such projects, since 2004 over 80,000 legal consultations have been provided for mobile workers (Muratovic, 2016). However, for Austrian (and German) unions, it remains a "dilemma" (Penninx & Roosblad, 2000) that such inclusion-oriented initiatives are in tension with their years-long support for transitional arrangements for citizens from the NMS.

5.5 Summary

One of the major developments in the Austrian labour market has recently been a huge rise in migrant inflows from the enlarged EU. Between 2008 and 2018, 90% of all newly created jobs went to foreign workers, who mainly came from the NMS (Szigetvari, 2019). The main dynamic for this was an open labour market in conjunction with a demand for labour, huge income differences, and the geograpical location of Austria in the "middle of Europe". EU migrants are represented in different occupations in the labour market. While migrants from the old EU are overrepresented in highly qualified positions, migrants from the new EU are overrepresented in low-skilled positions. The overall economic impact of this migration is likely to be rather small. However, there were likely to be some displacement effects in some low-wage segments, which mainly affected third country migrants who were already living in Austria for a long time. There is also some evidence that the trend towards segmentation in the Austrian labour market has intensified in connection with cross-border posting of workers, part-time work, and various forms of "dependent self-employment" (Mühlberger, 2007). These atypical employment relationships can be seen as part of a flexibility strategy by companies in some sectors of the labour market to increasingly outsource employment risks to mobile workers from the NMS (and from third countries).

[10] In 2015, the "IGR—Future in the Border Region" project was replaced by the "MIG—Multilingual Advisory Center in the Border Region".

This has raised concerns among employee representatives in particular about price-cutting competition and falling wages.

Against this background, the Austrian social partners together with the federal government agreed on comprehensive re-regulation of cross-border mobility. Essentially, the equal pay principle was enshrined in the Wage and Social Dumping Prevention Act on the basis of collective agreements. With this new act, Austria has one of the most comprehensive regulatory regimes in Europe, which underlines the importance of a tripartite interest policy in shaping labour market policy. However, the influence of the social partners on labour market and social policy increasingly depends on the political constellation, especially on the participation of the SPÖ in a federal government (Krings, 2013). If this condition is no longer in place, a politically induced marginalization of the social partnership process is a realistic scenario, as was last observed with the rightwing ÖVP-FPÖ federal government from 2017 to 2019.

The ideological glue of this government was an anti-immigration agenda. The EU free movement regime was repeatedly questioned by the FPÖ, also with reference to the domestic labour market (Die Presse, 2018). This position did not have a majority within the federal government (and would ultimately only be achievable through an EU exit by Austria). However, the ÖVP-FPÖ government decided on measures such as adjusting the Austrian family allowance (child benefit) to the cost of living in the country of origin, which mainly affected transnational commuters from the NMS (Kurier, 2018). This politically induced stratification of social rights according to origin is quite in line with the central assumptions of the dualization literature, which assumes that divisions are not limited to the labour market, but also include social security systems (Emmenegger et al., 2012).

In general, most observers agree that the Austrian economy has benefited from EU membership like few others (e.g. Die Presse, 2019). This mainly affects the internationalization strategies of Austrian companies and an increase in cross-border trade. The macroeconomic effects of recent intra-European migration may have been smaller, but here too the positive effects are likely to predominate. Migration has not only expanded the pool of available skills and qualifications, but also of labour.[11] From the perspective of the migrants, new employment opportunities have opened up, often associated with the prospect of higher wages.

[11] The fact that migration of qualified workers can also have negative effects in the countries of origin is not discussed in detail here. For an overview of the negative consequences of the emigration of medical professionals in the NMS, see Fóti (2013).

At the same time, the Austrian experience also illustrates the challenges associated with transnational labour mobility in a migration area characterized by significant income differences.

References

Biffl, G. (2010). Gewerkschaften und Zuwanderung in Österreich. In G. Biffl (Ed.), *Migration und Integration: Dialog zwischen Politik, Wissenschaft und Praxis* (Vol. 1, pp. 91–101). Omninum.

Biffl, G. (2019). *Migration and labour integration in Austria: SOPEMI report on labour migration: Austria 2017–2018*. Donau-Universität Krems.

Bosch, G., Hüttenhoff, F., & Weinkopf, C. (2019). *Kontrolle von Mindestlöhnen*. Springer VS.

Bundesministerium Arbeit (BMA). (2021). Arbeitsmarktinformationssystem. https://www.dnet.at/Amis/Datenbank/DB_Index.aspx. Accessed 4 Nov 2021.

Der Standard. (29. February 2016). Burgenland: 56 Prozent mehr Ausländer beschäftigt als vor fünf Jahren.

De Wispelaere, F., & Pacolet, J. (2019). Posting of workers: Report on A1 portable documents issued in 2017. European Kommission, Generaldirektion Beschäftigung, Soziales und Integration.

De Wispelaere, F., De Smedt, L., & Pacolet, J. (2021). Posting of workers: Report on A1 portable documents issued in 2019. European Kommission, Generaldirektion Beschäftigung, Soziales und Integration.

Dickens, L. (Ed.). (2012). *Making employment rights effective: Issues of compliance and enforcement*. Hart Publishing.

Die Presse. (29. Mai 2018). Strache stellt die EU-Personenfreizügigkeit infrage.

Die Presse. (22. Mai 2019). WIFO: Österreich profitierte überdurchschnittlich von EU-Beitritt.

Die Sozialpartner Österreich. (2007). Vorschläge der Sozialpartner für ein Maßnahmenpaket zur Deckung des Fachkräftebedarfs und zur Jugendbeschäftigung. https://docplayer.org. Accessed 4 Nov 2021.

Dølvik, J. E., Eldring, L., & Visser, J. (2014). Setting wage floors in open labour markets: The role of the social partners in Europe's multilevel governance. In S. Evju (Ed.), *Regulating transnational labour in Europe: The quandaries of multilevel governance* (pp. 131–186). University of Oslo, Department of Private Law.

Emmenegger, P., Hausermann, S., Palier, B., & Seeleib-Kaiser, M. (Eds.). (2012). *The age of dualization: The changing face of inequality in deindustrializing societies*. Oxford University Press

Eppel, R., Leoni, T., & Mahringer, H. (2017). *Österreich 2025: Segmentierung des Arbeitsmarktes und schwache Lohnentwicklung in Österreich*. Österreichisches Institut für Wirtschaftsforschung.

Eppel, R., Famira-Mühlberger, U., Horvath, T., Huemer, U., & Mahringer, H. (2018). *Anstieg und Verfestigung der Arbeitslosigkeit seit der Wirtschaftskrise. Entwicklung, Ursachen und Handlungsansätze*. Österreichisches Institut für Wirtschaftsforschung.

Europäische Kommission. (2012). *Posting of workers in the European Union and EFTA Countries: Report on A1 portable documents issued in 2010 and 2011.* Generaldirektion Beschäftigung, Soziales und Integration.

Eurostat. (2024). Monatliche Mindestlöhne—halbjährliche Daten. https://ec.europa.eu/eurostat/databrowser/view/earn_mw_cur/default/table?lang=de. Accessed 20 Mar 2024.

Fóti, K. (2013). Mobility and migration of healthcare workers in central and eastern Europe. European foundation for the improvement of living and working conditions. https://www.eurofound.europa.eu/sites/default/files/ef_publication/field_ef_document/ef1335en.pdf. Accessed 20 June 2020.

Fries-Tersch, E., & Mabilia, V. (2015). Annual Report on Intra-EU Labour Mobility. European Kommission, Generaldirektion Beschäftigung, Soziales und Integration.

Gächter, A. (2000). Austria: Protecting indigenous workers from immigrants. In R. Penninx & J. Roosblad (Eds.), *Trade unions, immigration, and immigrants in Europe, 1960–1993: A comparative study of the attitudes and actions of the trade unions in seven West European countries* (pp. 65–90). Bergham Books.

Geisberger, T. (2017). Niedriglohnbeschäftigung in Österreich 2014: Geringe Dynamik—Verfestigte Strukturen. *Statistische Nachrichten, 11*, 924–939.

Harvey, M. (2003). Privatization, fragmentation and inflexible flexibilization in the UK construction industry. In G. Bosch & P. Philip (Eds.), *Building chaos: An international comparison of deregulation in the construction industry* (pp. 188–209). Routledge.

Hermann, C., & Flecker, J. (2006). Betriebliche Interessenvertretung in Österreich. FORBA Schriftenreihe 1/2006. Forschungs- und Beratungsstelle Arbeitswelt.

Höller, C. (29. September 2015). Arbeitsmarkt: Ansturm der Osteuropäer. *Die Presse.*

Huber, P., & Böhs, G. (2012). Monitoring der Arbeitsmarktauswirkungen der Zuwanderung aus den neuen EU-Mitgliedsländern im Regime der Freizügigkeit—Begleitende Beratung und Analyse. In: Bundesministerium für Arbeit, Soziales und Konsumentenschutz (Eds.), *Arbeitsmarktöffnung 2011. Vol. 12: Sozialpolitische Studienreihe* (pp. 29–234). BMASK

Huber, P., Horvath, T., & Bock-Schappelwein, J. (2017). *Österreich als Zuwanderungsland.* Österreichisches Institut für Wirtschaftsforschung.

Kraler, A. (2012). The case of Austria. In G. Zincone, R. Penninx, & M. Borkert (Eds.), *Migration policymaking in Europe: The dynamics of actors and contexts in past and present* (pp. 21–60). Amsterdam University Press.

Krings, T. (2013). Von der "Ausländerbeschäftigung" zur Rot-Weiß-Rot-Karte: Sozialpartnerschaft und Migrationspolitik in Österreich. *Österreichische Zeitschrift für Politikwissenschaft, 42*(3), 263–278.

Krings, T. (2016). East-West mobility and the (re-)regulation of employment in transnational labour markets. In J. E. Dølvik & L. Eldring (Eds.), *Labour mobility in the enlarged Single European Market* (Comparative Social Research) (Vol. 32, pp. 183–213). Emerald Group Publishing.

Krings, T. (2019). *Posted Workers* in Österreich: Grenzüberschreitende Entsendearbeit im Spannungsfeld von offenen Märkten und (supra)nationaler Regulierung. *Österreichische Zeitschrift für Soziologie, 44*(1), 21–41.

Krings, T. (2022) Sozialpartnerschaft und Migrationspolitik. In S. Pernicka (Ed.), *Sozialpartnerschaftliche Handlungsfelder: Kontinuitäten, Brüche und Perspektiven.* Wiesbaden: Springer VS.

Krings, T., Moriarty, E., Wickham, J., Bobek, A., & Salamońska, J. (2013). *New mobilities in Europe: Polish migration to Ireland post-2004*. Manchester University Press.

Kurier. (4. January 2018). Familienbeihilfe für Kinder im Ausland wird gekürzt.

Menz, G. (2005). *Varieties of capitalism and Europeanization: National response strategies to the single European market*. Oxford University Press.

Mühlberger, U. (2007). *Dependent self-employment: Workers on the border between employment and self-employment*. Palgrave Macmillan.

Muratovic, A. (2016). Grenzenlos dagegenhalten: Schwerpunkt Internationale Solidarität. *Arbeit & Wirtschaft* 5/2016. https://www.arbeit-wirtschaft.at/archiv-posts/grenzenlos-dagegenhalten/. Accessed 27 Oct 2021.

Obinger, H., Starke, P., & Kaasch, A. (2012). Responses to labour market divides in small states since the 1990s. In P. Emmenegger, S. Hausermann, B. Palier, & M. Seeleib-Kaiser (Eds.), *The age of dualization: The changing face of inequality in deindustrializing societies* (pp. 176–200). Oxford University Press.

OECD/AIAS. (2021). Institutional characteristics of trade unions, wage setting, state intervention and social pacts. OECD Publishing. http://www.oecd.org/employment/ictwss-database.htm. Zugegriffen: 5. Juli 2021.

Penninx, R., & Roosblad, J. (Eds.). (2000). *Trade unions, immigration, and immigrants in Europe, 1960–1993: A comparative study of the attitudes and actions of the trade unions in seven West European countries*. Bergham Books.

Perchinig, B. (2010). Ein langsamer Weg nach Europa: Österreichische (Arbeits)migrations- und Integrationspolitik seit 1945. Sozialwissenschaftlicher Fachinformationsdienst soFid. Migration und ethnische Minderheiten 2010/1(11–32).

Pernicka, S., & Hefler, G. (2015). Austrian corporatism—Erosion or resilience? *Österreichische Zeitschrift für Politikwissenschaft, 44*(3), 39–56.

Riesenfelder, A., Schelepa, S., & Wetzel, P. (2012). Monitoring der Arbeitsmarktöffnung—Auswirkungen auf Beschäftigungsformen und auf Lohndumping. In: *Arbeitsmarktöffnung 2011*. In Bundesministerium für Arbeit, Soziales und Konsumentenschutz (Eds.), *Arbeitsmarktöffnung 2011. Vol. 12: Sozialpolitische Studienreihe* (pp. 235–479). BMASK.

Riesenfelder, A., & Wetzel, P. (2013). *Neue Arbeitsformen im Grenzbereich Selbständigkeit/Unselbständigkeit: Analyse der Situation in der Bauwirtschaft—Endbericht*. L & R Sozialforschung.

Riesenfelder, A., & Reichert, H. (2021). *Lohn- und Sozialdumping in Österreich. Eine Studie vor dem Hintergrund der 10- jährigen Arbeitsmarktöffnung*. L & R Sozialforschung.

Rueda, D., Wibbels, E., & Altamirano, M. (2015). The Origins of Dualism. In P. Beramendi, S. Häusermann, H. Kitschelt, & H. Kriesi (Eds.), *The politics of advanced capitalism* (pp. 89–111). Cambridge University Press.

Sardadvar, K., Saunders, E., & Holtgrewe, U. (2013). *"Arbeitnehmer bist du irgendwie trotzdem …"—Grenzbereiche von Selbstständigkeit, Unselbstständigkeit und Scheinselbstständigkeit in der österreichischen Bauwirtschaft. Endbericht.* Forschungs- und Beratungsstelle Arbeitswelt.

Schmatz, S., & Wetzel, P. (2014). *Entwicklungen im Bereich des Lohndumping*. L&R Sozialforschung.

Schnauder, A. (15. Februar 2016) "Mehr Ungarn als Türken am österreichischen Arbeitsmarkt". *Der Standard*.

Schweighofer, J. (2012). Gab es auf regionalsektoraler Ebene Verdrängungseffekte im Gefolge der Arbeitsmarktöffnung vom Mai 2011? *Wirtschaft und Gesellschaft, 38*(3), 601–614.

Szigetvari, A. (20. April 2018). Was gegen Sozialbetrug am Bau helfen könnte. *Der Standard.*

Szigetvari, A. (1. März 2019). Arbeitsmarkt: 90 Prozent der neuen Jobs gingen seit 2008 an Migranten. *Der Standard.*

Tálos, E. (2008). *Sozialpartnerschaft. Ein zentraler politischer Gestaltungsfaktor in der Zweiten Republik.* Studien.

Tamas, K., & Münz, R. (2006). *Labour migrants unbound? EU enlargement, transitional measures and labour market effects.* Institute for Future Studies.

Traxler, F. (2001). Die Metamorphosen des Korporatismus: Vom klassischen zum schlanken Muster. *Politische Vierteljahresschrift, 42*(4), 590–623.

Wagner, I. (2018). *Workers without borders: Posted work and precarity in the EU.* Cornell University Press.

Wagner, B., & Hassel, A. (2016). Move to work, move to stay? Mapping atypical labour migration into Germany. In J. E. Dølvik & L. Eldring (Eds.), *Labour mobility in the enlarged Single European Market* (Comparative Social Research) (Vol. 32, pp. 125–158). Emerald Group Publishing.

Walterskirchen, E. (2009). The dimensions and effects of EU labour migration in Austria. In B. Galgóczi, J. Leschke, & A. Watt (Eds.), *EU labour migration since enlargement: Trends, impact and policies* (pp. 149–170). Ashgate.

Weil, D. (2018). Creating a strategic enforcement approach to address wage theft: One academic's journey in organizational change. *Journal of Industrial Relations, 60*(3), 437–460.

Wiesböck, L. (2016). A preferred workforce? Employment practices of East-West cross border labour commuters in the Central European Region. *Österreichische Zeitschrift für Soziologie, 41*(4), 391–407.

Restricted Mobilities In-Between the COVID-19 Pandemic and Brexit

6

One of the central arguments of this book is that the EU free movement regime structurally promotes mobility. But what are the consequences for employment if this regime is suspended? This will be examined in more detail in this chapter using the example of an exogenous shock (the COVID-19 pandemic) and a politically induced shock (Brexit). Using three case studies on German agriculture, Austrian 24-hour mobile care, and the British transport sector, the "systemic importance" of migrant labour becomes apparent, as well as more broadly the importance of freedom of movement for relatively frictionless transnational labour mobility.

Freedom of movement for individuals has always been considered an unshakeable cornerstone of European integration. Although the free movement of workers can be temporarily restricted, as was most recently the case with Eastern enlargement, the four fundamental freedoms of the internal market have always been considered sacrosanct. Hence, it was all the more surprising how easily these freedoms were effectively suspended in the spring of 2020. When an as yet unknown coronavirus caused infection rates in Europe to rise dramatically, borders were closed almost everywhere. At this point, it will not be further discussed to what extent border closures curtailed the spread of the pandemic.[1] Rather, the focus will be

[1] In multidisciplinary research on the COVID-19 pandemic, the prevailing view seems to be that border closures and travel restrictions did indeed contribute to containing the pandemic. A crucial factor seems to be at what stage of the pandemic the restrictions are imposed. For two recent contributions from different disciplines, see Koopmans (2021) and Nadeau et al. (2021).

T. Krings, *Labour Mobility in the European Union as an Example of the Transnationalization of Employment*,
https://doi.org/10.1007/978-3-658-43977-4_6

on the labour market consequences of mobility restrictions. As will become clear, transnational labour mobility during the pandemic was caught in-between opening and closure. A national logic of pandemic control stood in tension with an economic logic that pushed for a rapid (re-)opening of the borders.

6.1 Europe and the World in Spring 2020

The COVID-19 pandemic was primarily a health crisis, which claimed 900,000 lives in the EU and the United Kingdom alone and brought many national health systems to the brink of collapse.[2] But it was also a global mobility crisis, which temporarily restricted cross-border movement of goods and people in Europe and elsewhere (Manderscheid, 2020). As part of a largely national containment strategy, most countries closed their borders in March 2020, leaving nearly three million migrants, travelers, and students stranded worldwide (the actual number may be much higher) (Benton et al., 2021). The number of international air passengers decreased by over 90% compared to the previous year, and the number of irregular border crossings at the EU's external borders had fallen to a record low (ibid.). Overall, international migration flows were at a "historic low" in the first half of 2020 (OECD, 2020, p. 11).

In Europe, not only were the EU's external borders effectively closed in March 2020, but controls were also reintroduced at the internal borders, thereby partially suspending the Schengen Agreement and freedom of movement. The border measures ranged from travel restrictions (e.g., quarantine regulations) to de facto border closures (Cavazzini, 2020). This was not a "coordinated approach at EU level", as called for by the European Commission (2020a), but a patchwork of national measures. This was due to the fact that health policy is still largely a national matter and the EU has little competence in this area. The following Fig. 6.1 illustrates the improvised ad-hoc nature of these measures, using the example of the German-Polish border (which was closed by the Polish side) and the Portuguese-Spanish border (which was closed by the Portuguese side).

Although the fundamental freedoms are enshrined in EU primary law, which in principle takes precedence over national law, EU law allows member states to restrict freedom of movement for reasons of "public order", "security" or

[2] As of: October 4, 2021, data from the European Centre for Disease Prevention and Control (https://www.ecdc.europa.eu/en/cases-2019-ncov-eueea).

Deutsch-polnische Grenze ©picture-alliance/A. Gora

Portugiesisch-spanische Grenze ©EPA-EFE/NUNO VEIGA

Fig. 6.1 Ad-hoc border closures in Europe in March 2020

"health". In relation to the COVID-19 crisis, the border measures were mainly justified with Regulation (EU) 2016/399 ("Schengen Borders Code") and the EU Free Movement Directive (Directive 2004/38/EC). The latter in particular specifies the conditions under which freedom of movement can be restricted in a pandemic situation. Article 29 (paragraph 1) states:

> The only diseases justifying measures restricting freedom of movement shall be **the diseases with epidemic potential** as defined by the relevant instruments of the World Health Organisation and other infectious diseases or contagious parasitic diseases if they are the subject of protection provisions applying to nationals of the host Member State. (emphasis, T.K.).

It is probably not an exaggeration to claim that before the COVID-19 crisis, Article 29 was not given much attention outside a small circle of politicians and European law experts. In the crisis, the article (and other legal acts) served as the basis for justifying the massive restriction of freedom of movement and cross-border traffic. Arguably, there was a trend towards renationalization in the fight against the pandemic. Supranational institutions such as the European Commission were usually presented with a fait accompli in terms of mobility restrictions and had little choice but to appeal for the restrictions to be "proportional" and "non-discriminatory" (European Commission, 2020a).

6.2 Labour Mobility in -between "Systemic Importance" and Precarity

As a result of the border measures, a significant decline in cross-border labour mobility in the EU was observed (Fries-Tersch et al., 2021, pp. 73–74). This was a "simultaneous supply and demand shock" (Bossavie et al., 2020). On the supply side, transnational commuters, posted workers, and seasonal workers who reside in an EU/EFTA member state and work in another EU/EFTA member state were particularly affected.[3] Many of these workers, who mainly come from Central and Eastern Europe (CEE), lost their jobs during the crisis or did not return to their workplaces for fear of contracting the coronavirus (Rasnača, 2020). Their precarious employment situation was exacerbated by often incomplete integration into social security systems. While the principle of the place of work for-

[3] Third-country nationals were often in an even more precarious situation due to their residence status (Rasnača, 2020).

mally applies to commuters and seasonal workers (and to a limited extent also to posted workers), the inadequacies of *national* security systems for *transnational* workers became apparent in the pandemic. Many cross-border workers had difficulty applying for assistance during the crisis. In addition, they often had inadequate health insurance in either the destination or the home country (ibid., p. 3). This precarious situation was exacerbated by exceptions for short-term workers in countries like Germany, who could work without social insurance for over 100 days during the COVID-19 crisis (see below). Also, closed borders and reduced mobility options sometimes made workers more dependent on their employers (Bejan, 2020). This affected not only the employment situation, but also the housing situation. Especially in connection with posted work, employers are often also providers of housing (not infrequently overcrowded mass accommodation with suboptimal sanitary conditions). Thus, a job loss can be accompanied by a loss of housing and thus lead directly to homelessness (Sepsi, 2021).

Migrants who continued to work during the crisis were overrepresented in sectors of the labour market defined as "systemically important". These sectors mainly included retail, warehouse logistics, home care, the transport sector, and food and agriculture. What many of the jobs in these sectors have in common is not only relatively poor pay but also physical presence on site. In this respect, "divisions of inequality" (Manderscheid, 2020) became apparent in the crisis between domestic workers and EU-15 migrants on the one hand, and migrants from the new member states (NMS) and third countries on the other. While the former were overrepresented in occupations that could be performed from home, the latter were overrepresented in occupations that required direct interaction at the workplace (and thus were exposed to an increased risk of infection) (Bossavie et al., 2020, p. 17).

This has implications for segmentation theory and labour market inequalities. The pandemic showed that there are job clusters that not only covary in terms of pay, employment stability, and promotion opportunities, but also in terms of whether the job can be shifted to home office or not. A "good" job in the primary segment during the crisis not only included job security and income stability, but also the possibility of working from home. In contrast, a "bad" job in the secondary segment often not only included low wages and labour-intensive activities, but also job-related spatial immobility. Arguably, health risks were partly outsourced to people with a migration background, thus confirming a central assumption of dual labour market theory as formulated by Michael Piore over 40 years ago: "[M]igrants provide a way in which workers in the native labour force are able to escape the role to which the system assigns them" (Piore, 1979, p. 42).

The COVID-19 crisis also had a shock-like effect on the demand side of the labour market. As a result of border restrictions, not only were cross-border goods and supply chains in Europe interrupted, but also the transnational mobility of workers. This led to an abrupt shortage of workers in sectors such as the food industry, home care for the elderly, and agriculture, and subsequently to supply bottlenecks of essential goods (Media Service Integration, 2020; Rasnača, 2020). The European Commission (2020a) had already called in its "Guidelines for border management measures to protect health and ensure the availability of goods and essential services" in mid-March 2020 for allowing transnational commuters in the health and food sectors to cross borders. This demand was further specified at the end of March 2020 in the "Guidelines for the exercise of the free movement of workers during the COVID-19 outbreak" and extended to all mobile workers with "systemically important functions" (European Commission (2020b). This group included, in addition to people in health and IT occupations, engineers, security personnel, and workers in the transport sector (ibid.).

National mobility restrictions were partially relaxed as early as April 2020. However, this was less a response to the appeals of the European Commission (2020a, b) to "preserve the integrity of the internal market", but rather due to the fact that a shortage of workers and skilled labour was becoming too acute in some areas of the labour market. In this respect, the COVID-19 pandemic was a "natural experiment" with regard to the availability of labour. What happens when this pool of labour is no longer available? Does this lead to increased mobilization of domestic labour?

As will now be shown using the example of the three case studies, domestic workers played only a minor role as a substitute for foreign workers. Instead, various exceptions and relaxations were issued, which aimed at least at a partial restoration of cross-border mobility. From the perspective of dual labour market theory, these reactions were not that surprising. Since the demand for migrants in the secondary segment is structurally conditioned and originates not only in unattractive working conditions, but also in the social status of the job (Piore, 1979), a short-term mobilization of domestic labour appears as an unlikely scenario.

6.3 Save the German Asparagus with Domestic Harvest Workers?

Every year, the German agricultural sector employs around 300,000 foreign seasonal workers from Central and Eastern Europe, who are mainly used in fruit, vegetable and wine cultivation. The origin of this temporary labour mobility lies in the early 1990s, when a series of new recruitment programs were concluded to channel growing East-West migration into orderly paths (Rudolph, 1996). These programs were continued even after Eastern enlargement, as Germany implemented transitional regulations for workers from the NMS (Wagner & Hassel, 2016, pp. 142–144). Even after the expiry of the transitional period in 2011 and 2014 (and 2015 in the case of Croatia), most migrant harvest workers still came from the NMS, especially from Romania and Poland. This form of temporary labour mobility was always relatively uncontroversial, as there were no displacement issues, unlike, for example, in the construction industry. A major reason for this is likely to be the very labour-intensive work in the fields, which is low-paid, at least by German standards.[4] The often precarious employment conditions in agriculture have been repeatedly criticized by trade unions (e.g., Fair Mobility, 2019).

When cross-border mobility came to a partial standstill in the spring of 2020, the agricultural sector was particularly affected. At the end of March, Interior Minister Horst Seehofer (CSU) announced a ban on entry for citizens from EU member states that are not part of the Schengen Agreement. This included, among others, harvest workers from Romania, who could no longer enter Germany. It was foreseeable that this measure would lead to significant staff shortages in the sector at a time when the asparagus harvest was imminent.

The German Farmers' Association (DBV) launched a massive political campaign for a relaxation of the measures and warned of "supply shortages" and "crop failures" (Deutschlandfunk, 2020). The entry stop was met by incomprehension in the Federal Ministry of Agriculture in view of a tense supply situation. Federal Minister of Agriculture Julia Klöckner (CDU) tried to find replacements for the seasonal workers in the form of unemployed people, refugees and "volunteer harvest helpers" (ZDF, 2020). This was accompanied by temporary measures, including a relaxation of the additional earnings limit for people who were on short-time work. In addition, (despite protests from trade unions) the period for social security-free employment was extended from 70 days to 115 days. By

[4] Since January 1, 2018, the national minimum wage applies as a wage floor for all workers in agriculture, forestry and horticulture. In July 2022, the minimum wage was raised to €10.45 gross per hour. This wage does not seem unattractive for seasonal workers, who mainly come from Romania, where the average gross monthly wage in March 2021 was just under €1200 (in exchange rates (EURES, 2021). Another question, of course, is whether the minimum wage in Germany is always paid, as some farmers seem to have developed creative circumvention strategies (e.g., Fair Mobility, 2019).

the beginning of April, over 40,000 volunteers had registered for field work via the "Das Land hilft" platform. However, the Farmers' Association made it clear that these workers were no proper substitute for the loss of foreign harvest workers. It is instructive how Joachim Ruckwied, the chairman of the DBV, explained the "systemic importance" of the seasonal workers from abroad:

> They are to some extent regular staff in many agricultural plants [...] They are experienced, they know their way around [...] they have been employed in the harvest season in Germany in the plants for years, some for decades [...] (In) the current situation we are happy if we get domestic workers who will then work in the fields, but without the core of Eastern European seasonal workers it will not work (in Deutschlandfunk, 2020).

This quote sheds an interesting light on the role of seasonal workers, who apparently play a more important role in agriculture than the term "auxiliary forces" would suggest. Particularly enlightening is the expression "to some extent regular staff", which points to the long-term and repeated nature of this mobility. The reference to the fact that seasonal workers "are experienced and know their way around" shows that even jobs in the secondary segment require "ethnomethodological competence" (Waldinger & Lichter, 2003, p. 37), which must be learned. Obviously, this is not an unstructured "everyone's labour market" (Sengenberger, 1987).

The massive lobbying of the DBV was not without success. Just ten days after the announcement of the entry ban, it was partially relaxed, not least due to repeated interventions by Agriculture Minister Klöckner with Interior Minister Seehofer. In early April, they finally announced that in April and May, 40,000 seasonal workers each would be allowed to enter Germany by plane under strict hygiene measures and then be placed under a fourteen-day quarantine on the agricultural farms (with work opportunities during that time) (Löhr, 2020). Insightful is a joint concept paper of both ministries, in which the goal of these measures was described as follows: "The core goal of the efforts is, in addition to securing the domestic harvest, to ensure the protection against infection of the population" (BMI/BMEL, 2020, p. 1). The health protection of the foreign seasonal workers obviously did not play a central role in the considerations of the ministerial staff.[5]

[5] In the Federal Ministry of Agriculture, a certain rethinking seems to have taken place, as multilingual digital flyers have been published in cooperation with the EU Equal Treatment Office, which not only inform about hygiene measures and health protection, but also about minimum working conditions in Germany (BMEL, 2021).

Therefore, it was little surprising that by the end of April, the first reports of non-compliance with hygiene measures in overcrowded living cabins on agricultural farms emerged (Edelhoff et al., 2020). The priority was to secure the harvest—and any health risks could be outsourced to mobile migrant workers.

After the border restrictions for EU/EFTA citizens were gradually lifted from June 2020 onwards, seasonal workers could enter again without numerical restrictions. However, even before the pandemic, it had become apparent that especially Polish seasonal workers were less interested in agricultural field work in Germany. This was probably not only due to the challenging working conditions, but also to the rising wages in Poland. Seasonal work also became increasingly unattractive for Romanian migrant workers, not least because there were more attractive job opportunities in Germany since the expiry of the transition period. Against this background, the federal government increasingly tried to attract seasonal workers from third countries for agricultural work. Amid protests from trade unions ("state-ordered social dumping"), an agreement was concluded at the beginning of 2021 between the German government and Georgia for a harvest contingent of 5000 auxiliary workers (Herrmann, 2021). This showed that the "systemic importance" of securing the harvest continues to rely on a low-wage model.

6.4 24-Hour Mobile Care in Austria: Cross-Border Commuters in the Pandemic

In Austria (as in other Western European countries), there has been a growing reliance on female caregivers from Central and Eastern Europe since the 1990s to meet the growing care needs of elderly people in private households. Initially, this was an irregular care model, which was tacitly tolerated by the state authorities. After the irregular status of 24-hour mobile care[6] came under scrutiny due to a care case in the family of the former Chancellor Wolfgang Schüssel (ÖVP) during the 2006 national election campaign, it was finally legalized in 2007 under the Home Care Act (Benazha et al., 2021, p. 29). From then on, it was possible to employ a personal caregiver, as the official designation goes, within the framework of an employment model or a self-employment model. In

[6]The term 24-hour care is not without problems, as it suggests the constant availability of caregivers. However, due to its high recognition value, it is still used here.

practice, almost 100% of caregivers are employed as formally self-employed one-person companies (OPCs). The self-employment model is the more cost-effective option from the perspective of private households, as no labour law such as collective bargaining, minimum wages or working time regulations apply. It is usually a form of "dependent self-employment" (Mühlberger, 2007) as the personal caregivers receive instructions from the private house holds in where they work and the placement agencies. The latter come from both the destination country and the country of origin of the workers and, as "market organizers" (Ahrne et al., 2015), play a central role in the placement of caregivers. The "self-employed" caregivers are often dependent on the agencies, which are the much more powerful actors. The relationship between the caregivers and the cared-for is often characterized by mutual dependencies, which is not least due to the unique working environment in private households (Aulenbacher et al., 2021a, pp. 149–150; Bachinger, 2015).

24-hour mobile care is part of a transnational care market that is stratified by gender, ethnicity, class, and origin (Aulenbacher et al., 2021b). Unlike, for example, agricultural seasonal work, the area of personal care is almost exclusively female-dominated (95%), with a noticeable presence of older women. In Austria, around 62,000 personal caregivers, who mainly come from Romania and Slovakia, offer their service (Famira-Mühlberger, 2018, p. 108). As with other forms of East-West mobility, the central dynamic is a demand for labour against the backdrop of transnational inequalities. With a rotation period of two to four weeks in private households, which usually means round-the-clock deployment in places with often very care-dependent people and a net salary of around €900 per month (Kaltenbrunner et al., 2021), there is little interest from domestic workers. Thus, as with agricultural seasonal work, there is hardly any competition between domestic and foreign workers in this particular transnational labour market. Meanwhile, 24-hour mobile care has become an integral part of the Austrian care system, albeit a precarious one, as the COVID-19 crisis was to show.

The pandemic not only made the "fragility" (Leiblfinger et al., 2021) of this care model visible, but also exposed a network of mutual dependencies, including the dependency of the Austrian care system on personal caregivers. When Hungary, Slovakia, Romania, and finally Austria closed their borders completely or partially in March 2020, the prerequiste for transnational personal care was no longer in place. From then on, personal caregivers could no longer enter or leave Austria unhindered. Faced with an unfolding care emergency, a "collapse of the 24-hour mobile care" (ORF, 2020) was feared. Attempts to recruit waiters or community service workers as replacements for the transnational caregivers were not very successful, nor were ideas to convert vacant hotels into "temporary stationary facilities" (Bachmeier, 2020a). Against this backdrop, the Austrian federal government issued a series of exceptions for commuters to facilitate the entry of personal caregivers. These measures were supplemented by the establishment of a "train corridor" between Austria and Romania and charter flights, which were mostly organized in cooperation with the respective regional representation of the Austrian Economic Chamber (WKO) and the placement agencies. However, these measures remained "insufficient" (Leiblfinger et al., 2021) to enable the transnational commuting of caregivers. On the one hand, Romania did not allow caregivers to leave for a while, on the other hand, entering Austria during pandemic times was linked to considerable hardship, including an initially mandatory two-week unpaid quarantine in hotels (Staudacher, 2020).

The travel exceptions contributed to upholding the 24-hour mobile care. Another (and probably more important) factor was that many of the caregivers already residing in Austria extended their stay, partly under pressure from the agencies, partly out of responsibility towards those being cared for, but also partly due to travel restrictions and the fear of a mandatory state quarantine in Romania (Bachmeier, 2020b; Scherndl, 2020). Financial incentives from the Austrian state in the form of a tax-free "Stay here!-bonus" amounting to 500 € may also have been a factor.

Overall, there can be little doubt that the precarious work and living situation of the caregivers worsened during the pandemic. There was an intensification of care work, not least because relieving visits by care services and family members hardly took place anymore. In addition, the exceptional situation had a negative impact on the mental well-being of the people being cared for. The social isolation of the caregivers at the workplace also intensified during the lockdown, with longer separation from their families being a particular psychological burden (Kaltenbrunner et al., 2021). Although there was a discursive upgrading of "systemically important" occupations during the crisis, this was "largely symbolic" (Leiblfinger et al., 2021). An example of this was the hardship fund set up by the federal government for OPCs and smaller companies. This fund was hardly accessible for personal caregivers who, due to mobility restrictions or concern for their health, could not or did not want to undertake a work trip to Austria (ibid., pp. 97–98). This was no coincidence: The focus of the federal government was on "securing 24-hour mobile care in Austria" (Federal Chancellery, 2020)—and not the specific transnational problems of the caregivers, which played at best a subordinate role in the considerations of policy-makers and most interest groups.

One characteristic of the COVID-19 crisis was that dependencies between caregivers and placement agencies intensified. At the same time, there were some indications that the care workers were increasingly aware of their "systemic importance" (Kaltenbrunner et al., 2021). There were also efforts to organize personal caregivers, starting with union-related initiatives like vidaflex or associations like "IG24", which represent a bottom up-attempt at self-organization.[7] So far, the potential market power of the caregivers has not yet translated into collective organizational power. It remains to be seen to what extent the caregivers will be able to bundle their interests in the future and become a noticeable force in the transnational field of care work.

6.5 Regained Sovereignty? Brexit, COVID-19, and a Shortage of Truck Drivers

In late summer 2021, images of empty supermarket shelves and long queues at petrol stations made the rounds in the UK. In October, the situation escalated to such an extent that the British army was deployed to deliver fuel (BBC, 2021a). A major reason for the supply shortages was a lack of truck drivers from the NMS, who turned their backs on the island in greater numbers since the EU exit and especially since the COVID-19 pandemic. Brexit supporters had campaigned for the "restoration of sovereignty" and the "regaining of control over national borders". The reality in autumn 2021 was different. Instead of celebrating the regained sovereignty, an escalating supply crisis dominated the public debate. At the same time, disillusionment about the withdrawal set in. Even among Brexit supporters, a majority believed, according to a survey, that the crisis was at least partly due to leaving the EU and the internal market (Henley et al., 2021).

When the British population voted by a narrow majority (52% to 48%) on June 23, 2016, for the withdrawal from the EU, this triggered a severe political crisis in the UK and the EU. Prime Minister David Cameron, who had campaigned for remaining in the EU, resigned a few weeks after the result. The European institutions were no less shocked. For the first time in the history of the EU, a member state had decided to leave the union. Among observers, there was

[7] The web addresses of the two initiatives are https://www.vidaflex.at and https://ig24.at.

broad agreement that migration from the enlarged EU was a central, if not the decisive factor for the Brexit vote (e.g. The Economist, 2016).

As already documented in Chap. 3, the UK was by far the most important destination country after Germany since Eastern enlargement. Between 2004 and 2017, the number of EU migrants residing in the UK increased by two million to just under four million (Vargas-Silva & Walsh, 2020, p. 4). This was mainly due to large inflows from the NMS, but also from Spain and Italy, where rising unemployment at the beginning of the second decade was a push factor, especially for younger and better-qualified people.

The search for employment was an important motive for EU migrants to move to the UK (ibid., pp. 8–9). A flexible labour market and the institutional architecture of the free movement regime led to high employment rates for EU migrants (see Chap. 3). Nevertheless, or perhaps for this very reason, the labour market impact was a controversial issue in the run-up to the referendum. In the academic debate, a "consensus" had developed (Portes, 2016) that large inflows from the NMS had no major negative effects on the employment opportunities of British workers (see also MAC, 2018). While there were probably some negative wage effects in low-wage sectors, these were likely to have been relatively small. Overall, the macroeconomic effects of intra-European migration were likely to be slightly positive. This was also reflected in the fiscal balance: NMS migrants paid more in taxes and social contributions than they claimed in corresponding benefits (Dustmann & Frattini, 2014).

The fact that immigration nevertheless took on such a central role in the Brexit referendum was probably due to other factors such as the rapid population change in certain regions and growing competition for scarce public resources such as health services or social housing (MAC, 2014). It is probably no coincidence that the regions in the UK that experienced the highest percentage of inflows voted for Brexit by a clear majority (Buchsteiner, 2020).

Since the Brexit referendum, there has been a sharp decline in new migration from the EU/EFTA. At the same time, net migration remained slightly positive—until 2019, more people moved to the UK from the EU than left (Vargas-Silva & Walsh, 2020, p. 10). This was probably due to the fact that the free movement rules remained in force until the end of 2020 as part of a transition phase. In addition, the vast majority of resident EU migrants seemed to see their future in the UK despite Brexit and a sharpened anti-immigration discourse (Burrell & Schweyher, 2019). Over three million of them applied for settled status by mid-2020, which guarantees them unlimited residence rights and access to the labour market.

For new migrants from the EU (with the exception of Irish citizens), the situation has become more complicated. As a result of the reform of British immigration law since January 1, 2021, they are treated the same as non-EU migrants in terms of residence and labour market. They must apply for a visa for labour market access, with applications for low-skilled jobs (with the exception of a temporary program for seasonal workers in agriculture) generally not permitted (Walsh, 2021). In the words of British Home Secretary Priti Patel, the aim of the reform was to attract "the best and brightest in the world" in the future (GOV.UK, 2020). This stated objective ignores the fact that the UK already attracts an above-average proportion of academics from the enlarged EU and third countries—but due to the high proportion of low-skilled jobs ("low skill equilibrium" (Finegold & Soskice, 1988), many only found employment in the low-wage sector (see Chap. 3).

Ultimately, it was less the end of the free movement regime in the UK, but the COVID-19 pandemic in 2020, which led to a dramatic increase in return migration. As elsewhere, migrants in the UK were more affected by the health and economic consequences of the pandemic, with significant differences between individual migrant groups (Bossavie et al., 2020). Migrants from the EU-15 had the highest probability of all groups in the British labour market to work from home (and thus had a lower probability of losing their job or exposing themselves to an infection). In contrast, other migrant groups, with the exception of Indian-born people, had a lower probability than the domestic population to work from home (Fernández-Reino & McNeil, 2020, p. 6). NMS migrants, who are overrepresented in low-skilled manual jobs, were the least likely to do so (ibid., p. 5). Faced with a rapidly deteriorating economic and pandemic situation, many EU migrants began to re-evaluate their options and eventually turned their backs on the UK and especially London:

> (T)he choice would have been to stay here, with no job, less or no money, and pay for relatively expensive rented accommodation—or return home to family, with lower costs and most likely less risk of catching Covid. Not a hard choice (O'Connor & Portes, 2021).

Estimates suggest that between 2019 and 2020, over a million migrants left the UK (ibid., p. 5). How many of them came from the EU is unclear, but their number is certainly several hundred thousand, especially since this group was overall more affected by job losses than other groups (Sumption, 2021, pp. 6–7).

A crucial question is to what extent migrants will return to the UK in a post-pandemic situation. This is quite possible, especially for people who already have an unlimited right of residence (The Economist, 2021).[8] For other EU migrants, the probability of a return (or new migration) is likely to be significantly lower. One of the central arguments of this book is that the rise in intra-European labour mobility in the 21st century is inconceivable without the EU free movement regime, which structurally promotes mobility. The experience of the UK since the end of freedom of movement seem to confirm this assumption. Since the introduction of a points-based immigration system, the number of new EU migrants has dramatically decreased: Between January and June 2021, only 7000 work-related visas were issued for EU migrants, if cross-border commuters are excluded who only come to the UK for a short time. For comparison: In 2019, when the free movement rules were still in force, longer-term migration of EU citizens was still almost 80,000 people (Sumption, 2021, pp. 8–9). Without a doubt, the COVID-19 pandemic in-between and uncertainties linked to it were a factor why migration did not pick up strongly again in 2021. However, there can be no doubt that the end of freedom of movement was a decisive factor for the drastic decline in new migration. One consequence was a worsening labour shortage in the summer of 2021. Employers at home and abroad pointed out that the UK had lost the advantages of relatively frictionless mobility due to leaving the EU. This was evident not least in the lack of truck drivers from the NMS, who, according to Mario Digioia, Secretary General of the European Freight Transport Association UETR, will not return to the island as easily:

> Tempting European drivers back to the UK when they also have to face the reality of customs and border checks, all the uncertainties of Brexit ... We have to be realistic... a lot of money is being thrown at this whole problem in Europe right now. There's a level playing field (in the EU), and none of the Brexit-related hassle (in Henly et al., 2021).

To remedy the acute shortage of truck drivers, the British government announced a temporary quota of 5000 visas for foreign drivers at the end of September 2021, as well as for workers in the meat industry. These "exceptions" implicitly acknowledge that the strategy announced in 2020 to attract only "the best minds" in the future has failed. It should be mentioned that interest in the temporary quotas has so far been limited. By mid-October, only 300 drivers had applied for a

[8] The prerequisite for an unlimited right of residence is a stay of at least five years in the UK.

corresponding visa (BBC, 2021b). Obviously, the UK has lost its attractiveness for new migrants from the EU due to the termination of the free movement regime in conjunction with the COVID-19 crisis. At the same time, the high number of EU citizens who have applied for settlement in the UK suggests that the island and especially London has not completely lost its reputation as a "prime destination for European free movement" (Favell, 2008, p. 30).

6.6 Summary

All three case studies on transnational mobility in times of COVID-19 and Brexit point to a certain "path dependency" (Anderson & Ruhs, 2010) in the employment of migrants. Neither in German agriculture, nor in Austrian 24-hour mobile care, nor in the British transport sector, could a short-term absence of migrant workers be sufficiently compensated by locals. Instead, various "exceptions" for migrants were sought everywhere to rectify the labour shortages, however incomplete. This suggests that there is still a structural demand for foreign workers in secondary segments of the labour market in the 21st century (Piore, 1979). This demand may be socially constructed, but it is nonetheless real. In particular, the experience of the UK suggests that a disruptive change in the immigration system produces suboptimal outcomes if it is not embedded in a longer-term transformation of the main labour market and welfare state institutions (Anderson & Ruhs, 2010). These "systemic effects" were also visible in Germany and Austria, where the shortcomings of a low-wage model became apparent in relation to the role of migrant labour. What all three case studies have in common is not only the "systemic importance" of migrant work, but also the importance of the EU free movement regime for transnational labour mobility. As it turned out, the importance of this regime becomes particularly visible when it no longer applies or is suspended, be it due to an exogenous shock (COVID-19) or a political shock (Brexit).

References

Ahrne, G., Aspers, P., & Brunsson, N. (2015). The organization of markets. *Organization Studies, 36*(1), 7–27.

Anderson, B., & Ruhs, M. (2010). Migrant workers: Who needs them? A framework for the analysis of staff shortages, immigration, and public policy. In M. Ruhs & B. Anderson (Eds.), *Who needs migrant workers? Labour shortages, immigration, and public policy* (pp. 15–52). Oxford University Press.

Aulenbacher, B., Leiblfinger, M., & Prieler, V. (2021a). Anforderungen und Ansprüche in der Live-in-Betreuung. Oder: "… ab und zu denken wir uns, das ist eigentlich unmöglich". In B. Aulenbacher, H. Lutz, & K. Schwiter (Eds.), *Gute Sorge ohne gute Arbeit? Live-in-Care in Deutschland, Österreich und der Schweiz* (pp. 147–171). Beltz.

Aulenbacher, B., Lutz, H., & Schwiter, K. (Eds.). (2021b). *Gute Sorge ohne gute Arbeit? Live-in-Care in Deutschland, Österreich und der Schweiz*. Beltz.

Bachinger, A. (2015). 24-Stunden-Betreuung als Praxis. Identitätskonstruktionen, Arbeitsteilungen und Ungleichheiten—Eine Intersektionalitätsanalyse. *SWS-Rundschau, 55*(4), 476–495.

Bachmeier, A. (2020a). Grenzen dicht: Pflege in Österreich droht zusammen zu brechen. https://www.moment.at/story/grenzen-dicht-pflege-oesterreich-droht-zusammen-zu-brechen. Accessed 24 Sept 2021.

Bachmeier, A. (2020b). 24-Stunden-Betreuung: 90 Tage arbeiten und kein Mal frei. https://www.moment.at/story/24-stunden-betreuung-90-tage-arbeiten-und-kein-mal-frei. Accessed 24 Sept 2021.

BBC. (2021a). Fuel supply: Military to deliver petrol to UK garages from Monday. https://www.bbc.com/news/uk-58766648. Accessed 15 Oct 2021.

BBC. (2021b). Why is there a shortage of lorry drivers? https://www.bbc.com/news/uk-england-northamptonshire-57587253. Accessed 17 Oct 2021.

Bejan, R. (2020). Covid-19 and disposable migrant workers. https://verfassungsblog.de/covid-19-and-disposable-migrant-workers. Accessed 13 Aug 2021.

Benazha, A. V., Leiblfinger, M., Prieler, V., & Steiner, S. (2021). Live-in-Care im Ländervergleich. In B. Aulenbacher, H. Lutz, & K. Schwiter (Eds.), *Gute Sorge ohne gute Arbeit? Live-in-Care in Deutschland, Österreich und der Schweiz* (pp. 20–45). Beltz.

Benton, M., Batalova, J., Davidoff-Gore, S., & Schmidt, T. (2021). COVID-19 and the state of global mobility in 2020. Migration Policy Institute and International Organization for Migration.

Bossavie, L., Garrote Sanchez, Daniel, Makovec, M., & Ozden, C. (2020). Do immigrants shield the locals? Exposure to COVID-related risks in the European Union. Policy Research Working Paper No. 9500. World Bank.

Buchsteiner, J. (19. February 2020). Einwanderung nach Britannien: Nur noch die "richtigen Talente". *Frankfurter Allgemeine Zeitung*.

Bundesministerium für Ernährung und Landwirtschaft (BMEL). (2021) Saisonarbeit in der Pandemie. https://www.bmel.de/SharedDocs/Downloads/DE/Broschueren/Flyer-Poster/flyer-saisonarbeitskraefte-corona-deutsch.pdf?__blob=publicationFile&v=5. Accessed 20 Sept 2021.

BMEL/BMI. (2020). Konzeptpapier Saisonarbeiter im Hinblick auf den Gesundheitsschutz [Coronavirus (SARS-CoV-2). 2. April 2020. https://www.bmel.de/SharedDocs/Downloads/DE/Presse/pm062-corona-saisonarbeitskraefte.pdf?__blob=publicationFile&v=1. Accessed 20 Sept 2021.

Bundeskanzleramt. (2020). Edtstadler/Bode: "Grünes Licht für Korridorzug aus Rumänien". https://www.bundeskanzleramt.gv.at/bundeskanzleramt/nachrichten-der-bundesregierung/2020.html?page=9. Accessed 22 Sept 2021.

Burrell, K., & Schweyher, M. (2019). Conditional citizens and hostile environments: Polish migrants in pre-Brexit Britain. *Geoforum, 106*, 193–201.

Cavazzini, A. (2020). Corona im Schengen-Raum: Grenzen dicht zur Virusbekämpfung? https://www.annacavazzini.eu/corona-im-schengen-raum-grenzen-dicht-zur-virusbekaempfung. Accessed 17 Aug 2021.

Deutschlandfunk. (2020). Bauernverband in der Coronakrise: "Wir befürchten Engpässe bei Obst und Gemüse." https://www.deutschlandfunk.de/bauernverband-in-der-coronakrise-wir-befuerchten-engpaesse.694.de.html?dram:Article_id=473423. Accessed 20 Sept 2021.

Dustmann, C., & Frattini, T. (2014). The fiscal effects of immigration to the UK. *Economic Journal, 124*(580), 593–643.

Edelhoff, J. Ghassim, A., & Hurst, F. (2020). Die Ernte ist sicher—Nur die Erntehelfer nicht. https://daserste.ndr.de/panorama/archiv/2020/Die-Ernte-ist-sicher-nur-die-Erntehelfer-nicht,erntehelfer198.html. Accessed 20 Sept 2021.

EURES. (2021). Lebens- und Arbeitsbedingungen: Rumänien. https://ec.europa.eu/eures/main.jsp?catId=8681&acro=living&lang=de&parentId=7826&countryId=RO&living. Accessed 20 Sept 2021.

Europäische Kommission. (16. March 2020a) Mitteilung der Kommission: Leitlinien für Grenzmanagementmaßnahmen zum Schutz der Gesundheit und zur Sicherstellung der Verfügbarkeit von Waren und wesentlichen Dienstleistungen (2020a/C 86 I/01).

Europäische Kommission. (30. March 2020b) Mitteilung der Kommission: Leitlinien zur Ausübung der Freizügigkeit der Arbeitskräfte während des COVID-19-Ausbruchs (2020/C 102 I/03).

Faire Mobilität. (2019). Betrogen bei der Ernte. https://www.faire-mobilitaet.de/faelle/++co++242a1146-ce00-11e9-8d8b-52540088cada. Accessed 20 Sept 2021.

Famira-Mühlberger, U. (2018). Die Rolle der 24-Stunden-Betreuung im österreichischen System der Altenpflege. *Wirtschaftspolitische Blätter, 65*(1), 101–115.

Favell, A. (2008). *Eurostars and Eurocities: Free movement and mobility in an integrating Europe*. Blackwell.

Fernández-Reino, M., & McNeil, R. (2020). Migrants' labour market profile and the health and economic impacts of the COVID-19 pandemic. Migration Observatory report, COMPAS, University of Oxford.

Finegold, D., & Soskice, D. (1988). The failure of training in Britain: Analysis and prescription. *Oxford Review of Economic Policy, 4*(3), 21–53.

Fries-Tersch, E., Jones, M., & Siölan, L. (2021). Annual Report on Intra-EU Labour Mobility 2020. Amt für Veröffentlichungen der Europäischen Union.

GOV UK. (2020). Home Secretary announces new UK points-based immigration system. https://www.gov.uk/government/news/home-secretary-announces-new-uk-points-based-immigration-system. Accessed 22 Sept 2021.

Henley, J., Savage, M., & Tapper, J. (25. September 2021). Emergency visas won't tempt European lorry drivers to UK, say haulage chiefs. *The Guardian*.

Herrmann, S. (2021). Von Georgien auf deutsche Spargel- und Erdbeerfelder. *Frankfurter Rundschau, 26*(4), 2021.

Kaltenbrunner, V., Kern, J., Nisavic, J., & Oeggl, M. (2021). Grenzgänger*innen in der Pandemie: 24-Stunden-Betreuer*innen und die (In)Visibilisierung transnationaler Ungleichheiten. Forschungsbericht Institut für Soziologie, Johannes Kepler Universität Linz. https://www.jku.at/fileadmin/gruppen/119/WOS/Forschung/Forschungsberichte/Endbericht_24-Stunden-Betreuung.pdf. Accessed 10 Nov 2021.

Koopmans, R. (2021). A virus that knows no borders? Exposure to and restrictions of international travel and the global diffusion of COVID-19. Discussion paper SP VI 2020-103. Wissenschaftszentrum Berlin für Sozialforschung.

Leiblfinger, M., Prieler, V., Schwiter, K., Steiner, J., Benazha, A. V., & Lutz, H. (2021). Auswirkung der Maßnahmen zur Eindämmung der COVID-19-Pandemie auf Live-in-Betreuer*innen. In B. Aulenbacher, H. Lutz, & K. Schwiter (Eds.), *Gute Sorge ohne gute Arbeit? Live-in-Care in Deutschland, Österreich und der Schweiz* (pp. 92–103). Beltz.

Löhr, J. (2. April 2020). 80.000 Arbeiter bis Ende Mai. Erntehelfer sollen doch einreisen dürfen. *Frankfurter Allgemeine Zeitung*.

Manderscheid. K. (2020). Über die unerwünschte Mobilität von Viren und unterbrochene Mobilitäten von Menschen und Gütern. In M. Volkmer & K. Werner (Eds.), *Die Corona-Gesellschaft: Analysen zur Lage und Perspektiven* (pp. 101–110). Transcript.

Mediendienst Integration (2020) Geschlossene Grenzen: Wo die meisten Arbeitskräfte fehlen. https://mediendienst-integration.de/artikel/wo-die-meisten-arbeitskraefte-fehlen.html. Accessed 20 Sept 2021.

Migration Advisory Committee (MAC). (2014). Migrants in low-skilled work: The growth of EU and non-EU labour in low-skilled jobs and its impact on the UK. Full Report July 2014. https://www.gov.uk/government/publications/migrants-in-low-skilled-work. Accessed 13 Oct 2020.

Migration Advisory Committee (MAC). (2018). EEA migration in the UK: Final report. September 2018. https://www.gov.uk/government/publications/migration-advisory-committee-mac-report-eea-migration. Accessed 17 Oct 2020.

Mühlberger, U. (2007). *Dependent self-employment: Workers on the border between employment and self-employment*. Palgrave Macmillan.

Nadeau, S. A., Vaughan, T. G., Scire, J. S., Huisman, J. S., & Stadler, T. (2021). The origin and early spread of SARS-CoV-2 in Europe. *PNAS, 118*(9). https://doi.org/10.1073/pnas.2012008118.

O'Connor, M., & Portes, J. (2021) Estimating the UK population during the pandemic. Economic Statistics Centre of Excellence. https://www.escoe.ac.uk/estimating-the-uk-population-during-the-pandemic/. Accessed 17 Oct 2020.

OECD. (2020). *International migration outlook 2020*. OECD Publishing.

ORF. (2020). 24-Stunden-Betreuung vor Zusammenbruch. https://orf.at/stories/3162384. Accessed 24 Sept 2021.

Piore, M. J. (1979). *Birds of passage: Migrant labour and industrial societies*. Cambridge University Press.

Portes, J. (2016). Immigration, free movement and the EU referendum. *National Institute Economic Review, 236*(1), 14–22.

Rasnača, Z. (2020). Essential but unprotected: Highly mobile workers in the EU during the Covid-19 pandemic. ETUI Policy Brief No. 9/2020. Brüssel.

Rudolph, H. (1996). The new Gastarbeiter system in Germany. *New Community, 22*(2), 287–300.

Scherndl, G. (11. Mai 2020). Pflege: 24-Stunden-Betreuung im Sonderzug: "Jetzt wird allen klar, wie wichtig unser Job ist". *Der Standard.*

Sengenberger, W. (1987). *Struktur und Funktionsweise von Arbeitsmärkten: Die Bundesrepublik Deutschland im Vergleich*. Campus.

Sepsi, S. (3. April 2021). Sie sind zuhause nicht willkommen. *Der Freitag*.

Staudacher, A. (9. April 2020). 24-Stunden-Betreuung steht vor dem Kollaps. *Der Kurier*.

Sumption, M. (2021). *Work visas and migrant workers in the UK*. COMPAS, University of Oxford.

The Economist. (27. August 2016). Brexit and immigration: Raising the drawbridge.

The Economist. (12. June 2021). The Job market: Something to celebrate.

Vargas-Silva, C., & Walsh, P. W. (2020). *EU Migration to and from the UK*. COMPAS, University of Oxford.

Wagner, B., & Hassel, A. (2016). Move to work, move to stay? Mapping atypical labour migration into Germany. In J. E. Dølvik & L. Eldring (Eds.), *Labour mobility in the enlarged Single European Market* (Comparative Social Research) (Vol. 32, pp. 125–158). Emerald Group Publishing.

Waldinger, R., & Lichter, M. I. (2003). *How the other half works: Immigration and the social organization of labour*. University of California Press.

Walsh, P. W. (2021). *The UK's 2021 points-based immigration system*. COMPAS, University of Oxford.

ZDF. (2020). Einreiseverbot für Saisonkräfte. Klöckner: Bürger als freiwillige Erntehelfer. https://www.zdf.de/nachrichten/panorama/corona-erntehelfer-kloeckner-das-land-hilft-100.html. Accessed 20 Sept 2021.

European Labour Mobility in Uncertain Times

7

In this concluding chapter, the central findings of the book are summarized and a look is taken at the future of cross-border labour mobility in Europe. As shown in the book, intra-European migration in the 21st century has become the dominant form of employment-related migration in Western Europe. The central dynamics for the increase in this mobility were relatively strong economic push and pull factors (especially a demand for labour), which interacted with the EU free movement regime, social networks, and new transport and communication opportunities. EU enlargement, which enabled the largest voluntary migration movement in recent European history, was of particular importance.

In the first decade of the 21st century, new (and partly also old) EU citizens were mainly drawn to the UK, Ireland, Spain, and Italy. While the two islands opened their labour market as early as 2004 and became the favored destination for Polish migrants, the two southern European countries implemented transitional rules. Nevertheless, hundreds of thousands of mainly Romanian migrants were drawn to Spain and Italy. This suggests that intra-European labour mobility is largely demand- and network-driven and that linguistic factors (Romance language tradition) also have an influence on migration movements.

In the course of the economic and financial crisis of 2008/2009, a diversion effect of intra-European migration towards Northern Europe was observed, especially to Germany, where a booming and open labour market offered new employment opportunities. There was also an increase in South-North migration, although this mobility was nowhere near as high as an "optimal currency area" (Mundell, 1961) would suggest, given sharply rising unemployment in Southern Europe. Also in the second decade, intra-European mobility mainly took place

T. Krings, *Labour Mobility in the European Union as an Example of the Transnationalization of Employment*,
https://doi.org/10.1007/978-3-658-43977-4_7

from countries *outside* the Eurozone to countries *inside* the Eurozone. The main exception here was the UK, which remained a central destination for mobile individuals from the new and old EU, at least until the free movement rules were terminated in 2020. The latter represented a disruptive change in the British migration system, which, in conjunction with the COVID-19 pandemic, exacerbated a supply crisis in some economic sectors.

Thus, in theoretical terms, Piore's (1979) assumption of a structural demand for migrant labour in the highly developed societies of the West was confirmed. However, segmentation theory with its focus on secondary labour markets cannot explain why an increasing differentiation of intra-European (and extra-European) labour mobility is observable. As shown, the employment of mobile EU/EFTA citizens is by no means limited to the secondary labour market, but also includes higher qualified positions. This mobility is also more supply-side driven, as EU/EFTA workers and services providers no longer depend on an employment permit, but can offer themselves on their own in a cross-border context. This illustrates the dynamic interaction between supply and demand in a transnational mobility space, which is characterized by significant inequalities.

What were the labour market effects of this mobility? Most studies conclude that an increase in intra-European migration has not led to major disruptions in European labour markets (e.g. Bonin et al., 2020). Although it can be assumed that some sectors and occupational groups were more affected by this migration than others, overall there is no compelling evidence to suggest it had a negative impact on the employment opportunities of domestic workers. However, there have been incidents of wage and social dumping, especially in connection with atypical employment such as posted work and self-employment, which have intensified concerns about a "race to the bottom". Such incidents mainly occured in labour-intensive sectors such as construction, the transport sector, shipbuilding, and the meat industry (Dølvik et al., 2014).

Against this background and also because labour market regulation at EU level was largely blocked in the first decade of the 21st century, a trend towards national re-regulation was observable, as illustrated by the country case studies on Germany and Austria. While the former responded to the challenges of transnational mobility, among other things, with the introduction of a statutory national minimum wage, the latter pursued a regulatory strategy based on collective agreements and equal pay as negotiated by the social partners. This suggests that even within the same "variant of capitalism" there are significant differences in terms of labour market regulation, which are not only due to different institutional power resources of worker representative bodies, but also to different organizational structures of employer associations. While larger companies play an important role in the BDA in Germany, which are less receptive to labour market regulations, the WKO in Austria is dominated by smaller companies, which are more open to regulatory measures due to concerns about price-cutting competition by foreign service providers (Menz 2005).

After a market-liberal agenda came under pressure in the wake of the 2008/09 economic crisis and also after the Brexit referendum in 2016, the space for supranational regulatory measures opened up again at the European level. Since 2017, the "social dimension of Europe" has been emphasized again under the Juncker Commission. This program included, among other things, a reform of the Posting of Workers Directive, which largely enshrined the principle of equal treatment. This undoubtedly marked a success for the advocates of a "social Europe". However, this has not eliminated the challenges in terms of enforcing labour standards in a transnational mobility space. In this context, it remains to be seen to what extent the newly established European Labour Authority, which formally started its work in 2019, can contribute to improved cross-border cooperation between national authorities (Cremers, 2020). Evidently, compliance with and enforcement of *national* labour standards increasingly require *transnational* coordination.

The importance of intra-European labour mobility was most recently demonstrated in the COVID-19 pandemic in the spring of 2020, when the temporary suspension of freedom of movement led to supply shortages of essential goods and services in some countries. Migrants from the NMS as well as from third countries were overrepresented in sectors of the labour market that were defined as "systemically important" and required physical presence at the workplace. At the same time, the crisis intensified a trend towards home office in higher qualified positions. In this context, the question arises to what extent such mobile work will also be an option for intra- and extra-European migrants in a cross-border context in the future (The Economist, 2021).

On the whole, the EU free movement regime is a success story, especially from the perspective of most EU migrants, but also from the perspective of the destination countries. The balance sheet is less clear for the countries of origin, where especially in Southeast Europe the emigration of young and often qualified workers has exacerbated a demographic crisis (Atoyan et al., 2016). Even though it is too simplistic to reduce this mobility to a "brain drain" (Koser & Salt 1997), there can be little doubt that the emigration of medical professionals in particular poses a significant challenge for these countries.

In this book, intra-European labour mobility was analyzed as an example of the transnationalization of work. The latter poses a challenge not only for labour market actors, but also for social science research on work and employment, which has so far mostly analyzed the structure and functioning of labour markets within national societies. In order to bring transnational linkages in the world of work more into focus, it is necessary to adopt a multi-scalar perspective with regard to concepts such as supply and demand, qualifications and skills, social inequality and regulation of employment, which emphasizes the simultaneity of different spatial dimensions (Fibich et al., 2016).

This book argued that the EU free movement regime structurally promotes mobility and facilitates entry into the labour market. It is to be expected that intra-European labour mobility will continue to play an important role in the future in order to meet a demand for workers with different qualifications. This is all the more true as the borders of the European mobility space are variable and could change in the coming years with regard to EU accession of the Western Balkans or even concerning the recent application for membership of war-torn Ukraine. In this context, an extension of mobility rights without full EU membership is being discussed as one option (The Economist, 2022).

Some studies (e.g. Fuchs et al., 2015, 2019) postulate on the basis of statistical modelling that intra-European labour mobility will decline significantly in the coming decades. These studies rightly point to demographic change, which is increasingly affecting the new member states. Apart from the fact that these studies are based on rather optimistic assumptions about economic convergence (which have so far hardly shown up in the area of wages), they suffer from some methodological deficits. Labour migration is operationalized as net migration, which significantly underestimates transnational forms of labour mobility such as posted work or commuter migration, which, as shown in this book, play an important role. The crucial point, however, is that such studies generally neglect the importance of the EU free movement regime for cross-border mobility. This regime enables various mobility patterns such as return migration, circular migration, repeated border crossings and onward movements, which do not always appear in official statistics and modelling. To paraphrase an article from the Economist (2019), "the world is messier than conventional models assume".

Of course, immigration from third countries will also play a significant role in Europe in the future. However, there are indications that even in a post-pandemic era, intra-European mobility will remain the dominant form of employment-related migration. A major reason for this is that this migration take place as part of a free movement regime, which allows for relatively frictionless cross-border mobility.

References

Atoyan, R., Christiansen, L., Dizioli, A., Ebeke, C. Ilahi, N., & Ilyina, A. (2016). Emigration and its economic impact on Eastern Europe. IMF Staff Discussion Note 16/07. https://www.imf.org/external/pubs/ft/sdn/2016/sdn1607.pdf. Accessed 25 May 2020.

Bonin, H., Krause-Pilatus, A., Rinne, U., & Brücker, H. (2020). Wirtschaftliche Effekte der EU-Arbeitskräftemobilität in den Ziel- und Herkunftsländern. Ein kurzer Literaturüberblick. IZA Research Report Nr. 102. Forschungsinstitut zur Zukunft der Arbeit.

Cremers, J. (2020). The European Labour Authority and rights-based labour mobility. *ERA-Forum, 21*, 21–34.

Dølvik, J. E., Eldring, L., & Visser, J. (2014). Setting wage floors in open labour markets: The social partners in Europe's multilevel governance. In S. Evju (Eds.), *Regulating transnational labour in Europe: The quandaries of multilevel governance* (pp. 131–186). University of Oslo, Department of Private Law.

Fibich, T., Flecker, J., & Pernicka, S. (2016). Transnationalisierung: Multiskalare Analysen in der Arbeitsforschung. In H. Staubmann (Eds.), *Soziologie in Österreich—Internationale Verflechtungen* (pp. 69–88). Österreichische Gesellschaft für Soziologie.

Fuchs, J., Kubis, A., & Schneider, L. (2015). Wie viel Zuwanderung aus Drittstaaten benötigt Deutschland für ein konstantes Erwerbspersonenpotenzial? Szenarien des Zuwanderungsbedarfs bis 2050 unter Berücksichtigung der Entwicklung der inländischen Erwerbsbeteiligung und der Migration aus EU-Staaten. Bertelsmann-Stiftung.

Fuchs, J., Kubis, A., & Schneider, L. (2019). Zuwanderung und Digitalisierung: Wie viel Migration aus Drittstaaten benötigt der deutsche Arbeitsmarkt zukünftig? Bertelsmann-Stiftung.

Koser, K., & Salt, J. (1997). The geography of highly skilled international migration. *International Journal of Population Geography, 3*(4), 285–303.

Menz, G. (2005). Varieties of capitalism and Europeanization: National response strategies to the single European market. Oxford: Oxford University Press.

Mundell, R. A. (1961). A theory of optimal currency areas. *The American Economic Review, 51*(4), 657–665.

Piore, M. J. (1979). *Birds of passage: Migrant labour and industrial societies*. Cambridge University Press.

The Economist. (23. November 2019). Boots on the ground: When the method is in the meaning.

The Economist. (30. January 2021). How the pandemic reversed old migration patterns in Europe.

The Economist. (26. May 2022). Reheated plans for a multi-tiered Europe revive familiar suspicions.

FSC
www.fsc.org
MIX
Papier aus verantwortungsvollen Quellen
Paper from responsible sources
FSC® C105338

If you have any concerns about our products,
you can contact us on
ProductSafety@springernature.com

In case Publisher is established outside the EU,
the EU authorized representative is:
Springer Nature Customer Service Center GmbH
Europaplatz 3, 69115 Heidelberg, Germany

Printed by Libri Plureos GmbH
in Hamburg, Germany